Related Books

Titles	*Authors*
• 21st Century Biology and Agriculture: Textbook Series *(Vol. 1-5)*	A. Raman
• A Handbook of Environment Impact Assessment	V.S. Kulkarni
• A Manual for Dryland Afforestation and Management	G. Singh
• A Text Book of Environmental Science	Vidya Thakur
• Afforestation of Arid Lands	A.P. Dwivedi
• Agrochemicals in Plant Disease Management	N.G.Ravichandran
• Agroforestry for Climate Resilience and Rural Livelihood	Inder Dev
• Agroforestry Theory and Practices	A.J. Raj
• Essentials of Plant Nursery Management *2nd Ed.*	*P.K. Ray*
• Forest Ecology	S.S..Sagwal
• Forest Fires and their Control	E.S. Artsybashev
• Forest Nursery Manual: Production of Bareroot Seedlings	Mary L.Duryea
• Forest Nursery: How to Raise and Manage	S.S. Sagwal
• Forest Tree Seeds: Handbook	S.S. Sagwal
• Forest Tribology and Anthropology	V.M. Mhaiske
• Forestry Principles and Applications	A.J. Raj
• Forestry: A Subjective Guide for IFS Aspirants	K.T. Parthiban
• Fundamentals of Agriculture	R.L. Arya
• Industrial Agroforestry Perspectives and Prospectives	K.T. Parthiban
• Introduction to Forestry & Agroforestry	K.T. Parthiban
• Nursery and Plantation Practices in Forestry	Vinod Kumar
• Objective Forestry: For All Competitive Examination *3rd Ed***	K.T. Parthiban
• Plantation and Agroforestry Pulpwood Value Chain Approach	K.T. Parthiban
• Practical Manual on Plantation Forestry	Pankaj Panwar
• Research Methodology in Plant Science	P.S. Narayana
• Sacred Groves of Rajasthan	G. Singh
• Tree Breeding and Improvement: Theory and Techniques	K.T. Parthiban
• Universal Objective Forestry *2nd Ed.* for UPSC, PSCS, SRF/JRF/AFO, STATE PG & PH.D. Entrance Examinations and Interviews of All Forest Services	Mohit Husain
Universal Forestry for *2nd Ed.* (UPSC, PCS, ARS/SRF /JRF/AFO, STATE PG & PH.D. Entrance Examinations and Interviews of All Forest Services)	Mohit Husain
• Wildlife Ecology and Conservation (21st Century Biology and Agriculture: Textbook Series)	M.Balakrishnan
• राजस्थान के पवित्र उपवन: संभावित खतरे तथा प्रबंधन नीतियाँ	G. Singh
• राजस्थान के प्रमुख वन्यजीव	Rakesh Vyas

Forest Pathology

Principles and Practice in Forestry

By

Bimal Kumar Bakshi

M.Sc. (Gold Medalist) (Cal.), Ph.D. (Edin)

Forest Pathologist and Director, Biological Research,

Forest Research Institute and Colleges, Dehradun, INDIA

United Book Prints

UNITED BOOK PRINTS
(Imprint of Scientific Publishers, INDIA)
5-B, Bhagat ki Kothhi, Near Police Chowki
New Pali Road, Jodhpur 342 001 (INDIA)
E-mail: info@scientificpub.com

Print, 2026

ISBN: 978-93-90199-98-3

Printed in India

Affectionately dedicated to

my wife Mamata

for her inspiration in writing the book

PREFACE

When the book was under preparation, the late Dr. John Shaw Boyce of the United States saw the general outline and some chapters then written and remarked : "this will be a valuable addition to the literature of forest pathology, since there is no other book dealing with tropical tree diseases or dealing with forest pathology in the Eastern Hemisphere". The lack of a consolidated account of diseases of trees of the Indian subcontinent is the main reason that prompted the author to write this book with his experience as an investigator and teacher for over 3 decades.

In natural forests, a biological balance is normally maintained between the vegetation and its diseases and insects which generally remain at an endemic level. Therefore, wherever feasible, natural forests should be maintained within the limits of economics and utility of the crop. The natural stocking may, however, be too slow-growing to meet the requirements of intensive management seeking maximum financial gain. Nearly 40 per cent of world's forests are unproductive according to the FAO forest inventory published in 1963. Since the latter half of this century, there has been a rapid expansion of forest-based industries, which is particularly high for pulp and paper. To meet this changing pattern of demand and also to meet the deficit of wood throughout the world, plantations of quick-growing species are being raised, particularly in developing countries. Diseases and insects assume considerable significance in plantations. The prime objective of forestry is not only to grow forest stands through to maturity in a healthy state but also to protect wood from biological degradation after conversion and during use. In this integrated approach, the role of a pathologist in forest management and timber utilisation can be appreciated in its proper perspective.

With the above in view, the book is organised in three parts. General principles of forest pathology are discussed in Part I. Following an account of the causes of the diseases, the plant parts affected are dealt with separately because the causal organisms attacking seedlings, and also different parts of the tree belong to ecologically distinct groups. Part II deals with diseases of trees of the Indian subcontinent. An account is presented for all the diseases of forest trees in India occurring in other countries to interest international readers. For exotic conifers, which are mostly in trial stage in India, diseases recorded in India only are described, since diseases on them are fully described in other books. A feature that will be useful to readers who may not be familiar with Indian forests is the inclusion of silvicultural and management practices of major forest tree species as these may have a direct relation to the development of the disease. The causes, effects and control of microbial degradation of wood form the subject matter of Part III. Mycological details of pathogens have been published elsewhere by the author and are purposely avoided in this book to facilitate easy reading. It is hoped that the book will be useful not only to students of plant pathology in general and forest pathology in particular but also to scientists and foresters concerned with the problems of increasing wood production and its protection.

Cooperation and help received by the author from his colleagues, Messers Y.N. Puri and Sujan Singh, at different stages in writing the book is thankfully acknowledged. The illustrations are mostly taken by the author except as otherwise indicated in explanations to the plates. The author also takes pleasure to thank Messers P.C. Pandey, Balwant Singh, M.D. Dhoundiyal and Ujagar Singh for assistance in various ways. The cooperation of the press in preparation of the blocks and printing is thankfully acknowledged.

BIMAL KUMAR BAKSHI

CONTENTS

PART I

FOREST DISEASES—GENERAL CONSIDERATIONS

PART II

DISEASES OF TREES

PART III

PART I

FOREST DISEASES - GENERAL CONSIDERATIONS

1. INTRODUCTION

Forests provide multi-purpose benefits as sources for timber, fuel, fodder and minor forest produce, for conserving soil and water, moderating climate, offering food and shelter for wild life and adding to the aesthetic value and recreational needs of man. To develop and protect forests which are one of the world's greatest renewable natural resources for maximum productive and protective functions are the prime objectives of forestry. Protection of forests from diseases constitutes one such objective.

In the natural forest, the productivity is generally low due to inherent slow growth of the species and mixed composition where all species may not be valuable. Considerable effort is therefore made to increase the productivity of existing forests and to raise plantations to meet the increasing demand of industrial wood and high grade timber. Diseases and insect pests constitute major biological determinants of forest productivity, particularly in plantations where due to drastic change in the ecosystem, catastrophic losses may occur in the event of an outbreak of a disease or an insect pest. The forester is, therefore, interested to know the causes of the diseases and the remedial measures that have to be adopted. Because of the insidious nature of pathogenic and non-infectious diseases, the losses due to them are not fully realised, unlike those from fire and insects, where the damage becomes at once evident. However, timber resource survey in the U.S.A. showed that losses from forest diseases exceeded the combined losses from fire and insects (Hepting and Jemison 1958). While no such country-wide surveillance has been made for India, a critical appraisal of losses would show much the same trend (Boyce 1959).

Forest pathology embraces study of diseased condition of forests due mainly to pathogens, for which a knowledge of mycology, bacteriology and virology is necessary. It is also concerned with the study of non-parasitic diseases. A knowledge of related sister disciplines is necessary towards a proper understanding of forest diseases. These disciplines include plant taxonomy to identify the host for the pathogen, ecology to study environmental factors, physiology to study the disturbance in the normal physiological functions of the host, silviculture to understand the development of the disease in relation to the composition and growth of the crop, biochemistry to understand pathogenesis and defensive mechanism of the host, chemistry for evolving chemicals for disease control and genetics for selection and breeding disease resistant plants. A study of soil characteristics and microbiological aspects is important in investigating root diseases. A collaborative study with the entomologist becomes necessary in case of diseases which are disseminated by insects. A knowledge of statistics is necessary for analysis of data. Finally, advanced technology, like phase and electron microscopy, use of radio-isotopes and remote sensing are some of the useful aids which have opened up new vistas in forest disease research.

A plant should be known in health in order to study it in a diseased condition. Silviculture is defined as the theory and practice of controlling forest establishment, composition and growth (Spurr 1945). In dealing with forest diseases in this book, a short account of the silviculture and management of species is, therefore, included in so far as it has a relation to disease development. This information will be particularly useful to foresters and to pathologists in other countries not familiar with Indian forests. This is a treatise on forest diseases rather than of pathogens and therefore a host-wise and not a pathogen-wise account appears to be the logical approach and this is followed in the book.

FORESTS IN INDIA

In the FAO forest inventory (Anon. 1963), the area under forest on the earth is 3779 million ha of which 2120 million ha occur in the developing countries. About 40 per cent of world's forests are at present unproductive which in less industrialised regions comprise about 50 per cent of the total forested land.

The area under forests in India is about 75 million ha and constitutes about 23 per cent of the total land area. The National Forest Policy (Anon. 1952) lays down that India as a whole should aim at maintaining one-third of its land under forests. The forests of India may be put into three classes ; (i) reserved forests, (ii) protected forests and (iii) unclassed forests, forming 47.1, 30.3 and 22.6 percentages respectively of the total forest area. Reserved and protected forests constitute prominent forest estates maintained for the purpose of producing timber and other produce and for protective reasons. As regards functions, forests in India may be classified into (i) production forests, to meet the needs of defence, communication, railways and industry, (ii) protection forests, to protect the land and ameliorate climate, (iii) tree-lands, to grow more trees in areas outside forests like avenues, parks, canal banks and farm lands, and (iv) village forests, to meet the need of the population for fuel and fodder.

The number of timber species available in India is about 320, of which 65 are of major importance, 106 have a fair demand while the demand for 149 is little (Pearson and Brown 1932). Among indigenous species, conifers comprise important species like deodar (*Cedrus deodara* (D. Don) G. Don f.), chir (*Pinus roxburghii* Sargent), blue pine (*Pinus wallichiana* A.B. Jackson), khasi pine (*Pinus kesiya* Royle ex Gord. syn. *P. insularis* Endl.), spruce (*Picea smithiana* (Wall.) Boiss.) and fir (*Abies pindrow* Royle and *A. spectabilis* (D. Don) Spach.). Conifers occur mostly in the Himalayas and occupy an area of 4.2 million ha. The broadleaved species are predominant and occur both in the hills as well as in the plains and occupy about 70 million ha. Some important species are teak (*Tectona grandis* Linn. f.), natural in southern India and also planted elsewhere, sal (*Shorea robusta* Gaertn. f.) occurring extensively in northern and central India, sissoo (*Dalbergia sissoo* Roxb.) in northern India; khair (*Acacia catechu* Willd.) and babul (*Acacia nilotica* (L.) Willd. ex Del. ssp. *indica* (Benth.) Brenan) in northern and central India, rosewood (*Dalbergia latifolia* Roxb.) in central India, gurjan (*Dipterocarpus* spp.) in Assam and padauk (*Pterocarpus dalbergioides* Roxb.) in Andamans. The natural occurrence of the different species is influenced by topographical, climatic and biological factors which have a direct bearing on the types of forests.

In India, the growing stock of the forests is estimated at 2400 million m^3. The annual increment is only 0.53 m^3/ha as against world average of 2.1 m^3/ha (Anon. 1961 b). The unclassed forests comprising 22.6 per cent of the total forested area in the country are highly degraded and unprofitable, rarely surveyed or subjected to any organised protection or management. While it may be difficult to increase the area under forests, the desired objective of increasing wood production may be achieved by clearfelling degraded forests and raising plantations of economic species for conventional timbers like teak as well as for industrial wood towards which the pattern of demand is changing rapidly. Different species of *Eucalyptus* are planted extensively all over the country exclusively for industrial wood. Trial plantations of hard pines are also raised to find suitable species that may be widely planted to meet the demand of long-fibred industrial wood particularly pulpwood. By 1973, an area of 1.98 million hectare comprising 2.64 per cent of total forest area was planted.

The forests in India are largely owned or managed by the states. State forests which constituted 77.2 per cent of the total forest area in 1949-50 rose to 92.3 per cent during 1957-58. Those owned by corporate bodies and private individuals formed 0.3 and 22.5 per cent of the total forest area in 1949-50 and these figures changed to 5.4 and 2.3 percentages respectively during 1957-58. The increase in State ownership and simultaneous reduction in private forests is due largely to abolition of the proprietary rights over forests and their acquisition by states (Anon. 1961 b).

FOREST TYPES

The distribution of the natural forests in India is governed principally by climate in which the most important single factor is rainfall, which may vary from 125-12,500 mm annually in different parts of the country resulting in a wide variety of vegetation, from xerophytic scrub forests in the arid zones as in Rajasthan to evergreen rain forests in the tropics as in the Western Ghats, Assam, North Bengal, Andaman and Nicobar Islands and conifers in the temperate region. In between the extremes, there are many intermediate types and sub-types. Latitude, altitude and aspect among others influence the climate and these factors together with soil characteristics also influence the character of the vegetation. Biotic factors, like the influences of man and cattle and also fire affect the distribution of forests. The major forest types are (a) moist tropical forests, (b) dry tropical forests, (c) montane sub-tropical forests, (d) montane temperate forests, (e) sub-alpine forests and (f) alpine scrub forests (Champion and Seth 1968).

a. Moist tropical forests

These forests constitute some of the most valuable forests in India. The wettest group of this type, tropical wet evergreen forests, occurs along Western Ghats, Andaman Islands and Upper Assam and consists of many evergreen species forming a dense many storeyed growth.

The tallest of the most important amongst them are the dipterocarps which often attain a height of 45 m and a girth at breast height of over 6.1 m. The second group of this major type, tropical semi-evergreen forests, occurs in a strip in the western coast and Andamans and on a fairly wide area in eastern India comprising Assam-Bengal-Orissa states with an annual rainfall of over 1,500 mm. On the West Coast, vellapine (*Vateria indica* Linn.), hopea (*Hopea parviflora* Bedd.) and irul (*Xylia xylocarpa* Taub.) are the important species while in the north eastern region, sal is predominant. This species occurs throughout North India from Himachal Pradesh in the west to Assam in the east and rises to an altitude of about 1,000 m in the Himalayas. In places where sal does not occur, the important species are champ (*Michelia champaca* Linn.), chilauni (*Schima wallichii* (DC.) Choisy), toon (*Toona ciliata* Roem.), gamari (*Gmelina arborea* Roxb.) and different species of bamboo. Tropical moist deciduous forests constitute the third group. The dominant species are mainly deciduous though some evergreen dominants occur in the top canopy. The lower storey consists of evergreen species. In the southern and central regions, teak forms the important species, while other valuable species include bijasal (*Pterocarpus marsupium* Roxb.), rosewood and extensive formations of *Bambusa arundinacea* Retz. In the Andamans, important timber species are padauk, white chuglam (*Terminalia bialata* (Roxb.) Steudel ex Kurz), black chuglam (*Terminalia manii* King) and white bombwe (*Terminalia procera* Roxb.) and important soft hardwoods prized for matchwood are white dhup (*Canarium euphyllum* Kurz), papita (*Pterocymbium tinctorium* (Blanco) Merr.) and *Bombax insigne* Wall.

The moist tropical forests contain the tidal mangrove forests in river deltas, of which those at the Ganges delta are important. The important species belong to *Heritiera*, *Bruguiera*, *Rhizophora*, *Ceriops*, *Excoecaria*, *Carapa*, *Sonneratia* and *Avicennia*. Due to high salinity in the substrate, the soil is physiologically dry and trees have special adaptations like leathery and glossy leaves to prevent excessive transpiration. Stilt roots for anchorage, pneumatophores for aeration and viviparous germination to ensure reproduction are other special adaptations that may develop.

Khair and sissoo constitute the important pioneer species in recently deposited river gravels. Khair-sissoo forests are, however, more closely related to the dry tropical forest types.

b. Dry tropical forests

These forests occur in areas with rainfall 750 to 1250 mm containing tropical deciduous forests and from 125 to 750 mm containing thorn forests. Dry tropical forests are distributed all over the country along the major river courses. They contain pure formations of poor quality sal and teak, red sanders (*Pterocarpus santalinus* Linn. f.), anjan (*Hardwickia binata* Roxb.), salai (*Boswellia serrata* Roxb.), *Butea monosperma* (Lam.) Taub., kardahi (*Anogeissus pendula* Edgew.), axlewood (*A. latifolia* (Roxb.) Bedd.), *Shorea talura* Roxb., laurel (*Terminalia alata* Heyne ex Roth, *T. coriacea* (Roxb.) W. & A. and *T. crenulata* Heyne ex Roth) in addition to khair-sissoo mixtures. Due to proximity to human habitation, such forests have suffered from fire and severe biotic factors such as over-grazing and cutting. Annual fires result in a dense growth of grass in the rainy season. Bamboo (*Dendrocalamus strictus* (Roxb.) Nees) occurs extensively and constitutes the main source of pulpwood in the south (Karnataka and Tamil Nadu states) and sandal (*Santalum album* Linn.) is the economically important species yielding sandalwood oil.

Tropical thorn forests are found in the desert and arid zones of Rajasthan, Punjab, western U.P., Gujarat and in the Central Peninsula. Species of *Acacia* like *A. nilotica* (L.) Willd. ex Del. ssp. *indica* (Benth.) Brenan, *A. catechu* Willd., *A. chundra* Willd., *A. leucophloea* Willd. and others constitute the major species forming open low forests.

c. Montane sub-tropical forests

These forests occur at elevations of 750 to 1800 m in the Himalayas and also at similar elevations in the Nilgiris and Palni Hills in the South, Pachmarhi (M.P. state), Parasnath (Bihar) and Cherapunji (Assam). The rainfall is above 1,500 mm reaching as high as 11,500 mm in Cherapunji. The forest flora is represented by sub-tropical pines, chir in the western and central Himalayas and khasi pine in Assam and by *Engelhardtia spicata* Blume, *Castanopsis tribuloides* A. DC., chilauni, *Betula cylindrostachys* Gamble, *Phoebe attenuata* Nees and others. However, their composition has been altered in most places due to human interference, by shifting cultivation or by plantations like tea, coffee and rubber, particularly in the eastern and southern India.

d. Montane temperate forests

This type occurs at elevation of 1,800 to 2,700 m and with a rainfall of about 1,250 to 2,000 mm. Deodar, blue pine, spruce and fir are the important conifers in the western Himalayas mixed with different hardwoods comprising of oaks, maples, walnut, *Corylus*, *Celtis* and *Ulmus* among which three species of *Quercus* are silviculturally important as nurse crop to conifers. In the eastern Himalayas, *Quercus* spp., laurels, maples and other hardwoods are dominant. In the South, in the Nilgiris, Anamalai and Palni Hills, the evergreen species are confined in small areas (sholas). The species comprising sholas are *Ternstroemia japonica* Thunb., *Syzygium calophyllifolium* Walp., *S. montanum* Gamble, *Ilex wightiana* Wall. and *Rhododendron arboreum* Sm.

e. Sub-alpine forests

The altitudinal range of this type is between 2,900 to 3,500 m. The vegetation consists of a dense growth of small crooked trees or large shrubs with patches of coniferous overwood comprising *Abies* and blue pine. Birch and *Rhododendron* constitute the commonest broadleaved trees.

f. Alpine scrub forests

The alpine scrub forests occur above 3,500 m along the entire length of the Himalayas and in locations with ample snow on the ground. The vegetation consists of low evergreen forests mostly of *Rhododendron* but with some birch and other deciduous trees forming a dense growth with short branchy stems.

2. DISEASE

DEFINITION

Disease refers to the abnormal state of the individual. Disease in plants may be defined as "sustained physiological and resulting structural disturbances of living tissues and organs, ending sometimes in death" (Ehrlich 1941). A characteristic of the disease is that it is a continuous process and not transient and temporary one, such as caused by injury during lopping of branches, thinning and so on. Furthermore, it is necessary to distinguish a disease from the symptom or condition of a disease. For example, a leaf spot is not a disease but a condition or a symptom of a disease just as fever is not a disease but a condition (Horsfall 1956). According to Horsfall and Dimond (1959), condition is a symptom complex or syndrome and results from the disease but not synonymous with it. The disease is more than the symptom. Plant pathology is that branch of science which deals with diseased condition in plants (Pathos (Gr.) = suffering). Forest Pathology is a specialised branch of plant pathology dealing with tree diseases. In a broader sense, it also includes microbial degradation of timber, the science of which is termed timber or products pathology.

CAUSES

A living plant is called the host when it harbours a parasite. A parasite is termed pathogen when the latter is able to cause disease in the host (Chapter 4). Diseases in trees may occur due to many causes. Parasitic or infectious diseases are caused due to infection by pathogens like fungi, bacteria and viruses and infestation by parasitic seed plants. Non-infectious malfunctioning of an individual caused by one or more adverse environmental factors is referred to as a disorder and is also broadly known as a non-pathogenic or physiogenic disease. Such disorders are not caused by any pathogen. Diseases caused by insects generally come within the domain of entomology. In fact any disorder in the tree and its products, except those caused by insects, comes within the scope of Forest Pathology.

Koch (1882) formulated certain postulates to establish experimentally the connection between a bacterial parasite and the disease which it causes in a man or an animal. These rules are : (i) constant association of the organism with the disease; (ii) isolation of the organism in pure culture; (iii) reproducing the disease by inoculation with pure culture of the organism; and (iv) re-isolation of the organism from the inoculated diseased host. Plant pathologists have generally accepted Koch's postulates in establishing the parasitic cause of a disease though with obligate parasites which cannot be cultured on artificial media, the procedure cannot be followed fully.

DISEASE DEVELOPMENT

Under suitable conditions forests, both natural and man-made, are expected to grow healthy. However, under adverse conditions of growth which include edaphic, climatic and biotic factors, and in a locality where a disease is present, forests may become diseased. Non-parasitic causes may lead to unhealthy growth of forests. They may also predispose or tend to weaken the plant or in some way make it more susceptible to attack by a parasite. For example, injury to the tree from various causes results in open wounds through which decay fungi or canker fungi may establish in the tree. However, if the primary parasitic cause is absent, no combination of predisposing factors is able to cause pathogenic disease in the plant. Except for obligate parasites which may attack trees in full vigour, most pathogenic diseases in forests are the result of the trees being exposed to certain predisposing factors, unfavourable to tree growth and favourable to development of the disease. An analysis of such factors is, therefore, important to understand the epidemiology of the disease and also for taking suitable measures to control or minimise losses from the disease.

EPIDEMICS

Diseases of isolated plants are relatively unimportant except in urban and recreational forestry (Chapter 8). Plant pathology is essentially a community aspect or plant public health science (Stakman and Harrar 1957). When a disease is regularly found in an area and remains more or less at a constant level, it is said to be endemic. Sporadic dying of trees may occur in any forest. Such dying may be considered normal in the development of a forest and no action may appear necessary. However, it becomes epidemic when abnormal large-scale dying of trees in the forests occurs either in time or area due to disease or insect attacks. Immediate efforts are called for to diagnose the cause, appraise damage, determine predisposing factors towards epidemic outbreak of the disease and finally evolve effective and feasible measures to control the disease or reduce losses to an accepted level. The study of the development and/or decline of an epidemic, generally a mathematical analysis of the population of both the host and the pathogen and the various factors affecting them, comprises the science of epidemiology (Van der Plank 1963).

SYMPTOMS

Symptoms are the apparent expressions of disease processes in the plant. When a plant becomes infected, it takes time, known as incubation period, for symptoms to develop. Every disease manifests itself with one or more particular symptoms which are described more fully while describing specific diseases (Part II). Here, general symptoms of forest diseases are described.

It may be mentioned that two or more different causes may produce identically same set of symptoms. Therefore, identity of the cause from such symptoms may not be assured. However, if a parasite is present on the host plant in the form of fruiting structures, then the cause could be immediately identified. The cause may also be isolated in pure culture in which it produces certain diagnostic characteristics from which it may be identified and its role in disease development established following Koch's postulates. A herbarium and a culture collection of authentic microorganisms in conducting disease investigations is, therefore, important in identifying parasites.

Two distinct types of symptoms may be recognised: (i) systemic, in which the entire plant exhibits symptoms of the disease, and (ii) localised, where the symptoms are restricted to particular plant parts affected. In the former case, the cause may not necessarily be present over the entire individual but usually restricted to certain plant parts, while in the latter, the cause is generally associated with the affected plant part showing symptoms of the disease. Localised diseases affect particular plant parts like leaf, shoot, stem and the root which exhibit symptoms of attack on the affected parts. Disease symptoms manifested by any plant part may be classified into three types namely, necrotic, atrophic and hypertrophic (Boyce 1961).

NECROTIC SYMPTOMS

Necrosis means death of affected tissues. This may occur in the green leaf where usually during early stages of attack, there is a change in colour to yellow or brown or red followed finally by the death of affected tissues. The entire organ may be killed as in leaf blight and bud blight or only a portion of the leaf may be affected as in leaf spots or shot holes which represent places where the affected tissues had fallen off. Die-back or top-dying or top-drying refer to dying in trees from top dowuwards (Pl. **10**, 1 ; **11**, 2 ; **15**, 1).

Wilt

Pathological wilt should be distinguished from physiological wilt which is generally a normal phenomenon. In the latter case, wilting occurs particularly when moisture is deficient in the plant as during a hot summer day when the transpiration is more than what the plant absorbs from the roots. When transpiration is cut down as during the evening, the plants recover. Pathological wilting is due to attack by a root or a stem fungus causing permanent wilt in plants and such wilting is irreversible. In seedlings, damping-off is a common disease in the nursery. In this case, plants are attacked at their bases near the ground level by fungi belonging to *Phytophthora*, *Pythium*, *Rhizoctonia*, and *Fusarium* which are normal soil inhabitants. Due to fungus attack, the bases of seedlings become rotted and seedlings suddenly fall on the ground (Chapter 5). In trees, wilting may take place due to attack by pathogenic soil fungi as in case of wilt of sissoo (*Dalbergia sissoo* Roxb.). As a result of the attack, the leaves gradually turn yellow and droop down (Pl. **11**, 1). Eventually the trees become dead.

Canker

Localised lesions on woody stems formed as a result of killing of bark tissues or cambium generally resulting in an open wound exposing the wood from beneath are termed cankers (Pl. **6**, 2; **15**, 3, 4). If the cambium is repeatedly killed, cankers with stratified margin develop as a result of repeated callusing. In woody plants, cankers may be limited in size because of the slow growth of the parasite and also the reaction of the host which restricts spread of the parasite by producing callus tissues around edges of the canker. When, however, cankers are able to girdle the twigs, branches or stems, death of parts above the cankers occurs. Besides being pathogenic in origin, cankers may also be formed due to physiogenic causes, like frost, sun scorch, fire and drought (Chapter 9). The resultant injury may form infection courts for pathogenic fungi to establish and cause disease in trees.

Decay

Decay refers to breaking down of tissues. In trees, decay fungi may occur in the dead heartwood causing heart rots (Chapter 7). In the living sapwood, such fungi kill the tissues and may result in death of the tree. Usually two types of decay in the wood are distinguished, namely, white rot and brown rot in addition to soft rot. In white rot, all components of cell wall material including lignin, which imparts the normal brown colour to the wood, are generally broken down by decay fungi and the resultant rot is termed white rot. Such decay can be fibrous or spongy (Pl. **18**, 2, 3) or develops in pockets (Pl. **18**, 1). In brown rot, fungi attack mainly the cellulose while lignin remains mostly unaffected. Therefore, the decayed wood gives a brown appearance and the term brown rot is applied to the decay. The wood showing brown rot shrinks and tends split in radial, tangential and cross-sectional ends into cuboidal blocks (Pl. **18**, 4). Details of microbial degradation of wood are given in Part III.

ATROPHIC OR HYPOTROPHIC SYMPTOMS

These symptoms appear when there is a slowing down in the development of the plant or plant parts, resulting from subnormal cell division (hypoplasia) or from cell degeneration. The common symptom in such cases is known as dwarfing. Chlorosis is another symptom where there is subnormal development of chlorophyll apparatus resulting in yellowing of green tissues. Chlorosis may be distinguished from etiolation where yellowing is due to lack of sunlight.

HYPERTROPHIC SYMPTOMS

Hypertrophy indicates overgrowth due to abnormal increase in the number of cells i.e., excessive cell division or abnormal increase in cell size. The symptoms are distinguished by their size, form and appearance into galls and burls or tumors when plant parts become abnormally swollen; erinosis when there is a profuse development of hairs on leaves forming felt-like patches and witches'-brooms, when dormant buds are induced to develop many branches in close clusters giving appearance of brooms (Pl. **17**, 1).

SIGNS

In addition to symptoms described above, other evidences of diseased condition in the tree include vegetative or fruiting structures of the disease-causing organisms which usually develop at or near the points of infection and also various types of exudations. Trees attacked by heart rot fungi generally do not mainfest any external symptoms of the disease but only signs in the form of fruiting bodies.

FRUITING BODIES

The disease may manifest either as vegetative (Pl. **15**, 5) or reproductive structures (Pl. **8**, 5; **15**, 7) of the organism causing the disease. It is important to know the different stages in the life-history of disease-causing organisms, particularly fungi, in order to recognise and evaluate the different signs. The fruiting structures or sporophores or fructifications of fungi commonly develop on diseased plants and are important signs of the disease. In heart rots in living trees, the sporophores are the only reliable external evidences of decay in the wood.

EPICORMIC BRANCHES

When trees grow under adverse conditions, branches usually of limited growth may develop in clusters on the main stem due to the activity of dormant or adventitious buds. Such clusters are termed epicormic branches which may develop from infectious causes (Pl. **10**, 1; **15**, 2, 8) or due to attack by insects and also from non-infectious causes.

EXUDATION

Exudations may be the result of normal physiological process in the plant, e.g. guttation, where water is forced out upon uninjured free surface of leaves. However, other exudations may be caused by infection due to bacteria, or may also arise from non-infectious causes as in water blister in teak (Pl. **13**, 6, 7). In *Eucalyptus*, clear amber coloured exudate may ooze out from the main bole. This phenomenon, termed gummosis, can result from injury due to parasitic or non-parasitic causes. In conifers, small quantity of resin may normally exude on the bark. However, an excessive out-flow of resin, known as resinosis, is a sign of disease and may be caused due to wounding or due to attack by pathogenic fungi.

3. THE PLANT KINGDOM

The plant kingdom is divided into four classes : Thallophyta, Bryophyta, Pteridophyta and Spermatophyta, arranged in an increasing order of evolution. Spermatophyta (spermos (Gr.) = seed, phyton (Gr.) = plant) are seed plants whose vegetative body is usually differentiated into roots, stems and leaves, meant for carrying on diverse functions. The remaining classes are characterised by the absence of seeds. Among them, the Pteridophyta (pteris (Gr.) = feather) include ferns and their allies where the vegetative plant body is similarly differentiated as in the Spermatophyta. The Bryophyta (bryon (Gr.)=a moss) which include moss and its allies are usually differentiated in the vegetative body into stems and leaves (except the liverworts) but roots are absent and their function is performed by rhizoids. In the Thallophyta (thallus (Gr.) = a young shoot), the plant body is undifferentiated into roots, stems or leaves. They are subdivided according to presence or absence of chlorophyll into two groups, namely, algae and fungi (sing. fungus (L.) = a mushroom) respectively. Mycology (mykes (Gr.) = a mushroom) is the science of fungus. Fungi may attack plants and cause diseases in them, the study of which falls under the realm of plant pathology (pathos (Gr.) = suffering).

FUNGI

Most tree diseases are caused by fungi, the study of which falls under a distinct science—mycology. For a comprehensive knowledge in mycology, the reader is referred to various standard works like 'The structure and development of the fungi' by Gwynne-Vaughan and Barnes (1930), 'A text book of mycology' by Bessey (1939), 'Physiology of fungi' by Hawker (1950), 'Comparative morphology of the fungi' by Gaumann and Dodge (1928), 'Introductory mycology' by Alexopoulos (1962) and 'Ainsworth & Bisby's dictionary of the fungi' by Ainsworth (1961) among others. Only an introductory knowledge of fungi is given here.

The fungi constitute one of the lowest groups of plants devoid of chlorophyll in the absence of which they are unable to synthesise carbohydrates unlike green plants. Fungi, therefore, derive nutrition from extraneous organic sources either as saprophytes from dead substrata or as parasites, from the living. Cottony growth on moist stale bread, cheese, rotten vegetable and fruit and mushrooms or toadstools developing on humus during the rains are common examples of saprophytic fungi. The parasitic fungi may cause diseases in plants Fungi are common everywhere as spores which are microscopic, float in air and water or occur in soil.

VEGETATIVE BODY

When a spore gets a suitable substrate, it germinates under favourable conditions of moisture and temperature and finally develops into the mycelium (pl. mycelia) which constitutes the vegetative body of the fungus. The mycelium is a much branched filamentous structure, the individual filament or thread of which is termed hypha (pl. hyphae) which contains protoplasm when living. The hyphae are walled, thin or thick, and may not possess any cross partition wall or septum so that the plant body is continuous as in Oomycetes and Zygomycetes or may develop septa as in Ascomycetes, Basidiomycetes and Fungi Imperfecti. The mycelium ramifies within the tissues of the host or other substrates and derives nutrition for the fungus.

The hyphae secrete enzymes, particularly at the growing regions. The enzymes convert the complex organic material of the substrate into easily assimilable forms for the fungus. Also, they possess the power to dissolve the hard lignified cell walls and thereby spread from cell to cell. Within the host, the hyphae may grow within the cells (intra-cellular) or between the cells (inter-cellular). In the former case, the hyphae come into direct contact with host protoplasm. In the latter case, the hyphae develop special branches, known as haustoria, which penetrate into the cell cavity and absorb food.

The mycelium may undergo modification characteristic of the species. It may develop into a mycelial felt which spreads on the substrate as occurs in decay fungi like *Fomes annosus* (Fr.) Cke., *F. caryophylli* (Racib.) Bres., *Armillaria mellea* (Vahl.) Quel. and others. The mycelium may develop into a sclerotium in which the hyphae cluster together into a hard woody structure e.g., *Claviceps purpurea* (Fr.) Tul. Sclerotia develop further after a period of rest. In some fungi, the hyphae aggregate to produce long, firm and branching strands with a hard rind (rhizomorph) e.g., *Armillaria mellea.* Rhizomorphs spread to long distances by means of apical growth and may infect trees.

REPRODUCTIVE STAGES

After the vegetative body has undergone full development, reproductive stages develop for the purpose of perpetuation or multiplication of the species. The reproductive structures usually develop on the surface of the substrate to facilitate dispersal of reproductive units, termed spores. Spores are microscopic in size and may be formed by vegetative, asexual or sexual methods. Vegetative propagation occurs during budding of yeast. In the asexual method, no sexual stage is involved. In the lower fungi, spores may be produced by the division of the protoplasm of the vesicle (sporangium). Such spores may be naked masses of protoplasm provided usually with two flagella or cilia to facilitate movement in water (zoospores) as in Oomycetes (Pl. 1, 1) or they may be walled and non-motile as in Zygomycetes (Pl. 1, 2). The cells in the hyphae may enlarge and become thick-walled to form chlamydospores (Pl. 1, 16) ; or remain thin-walled to form endoconidia which develop within specialised hyphae (endoconidiophores) (Pl. 1, 18). The hyphae may break up into segments to form oidia which fall apart (Pl. 1, 17). In higher fungi, conidia are generally borne on vegetative hyphae or on conidiophores or grouped in different types of asexual fruiting bodies (pycnidium) (Pl. 1, 10). Pycnidia are globose or flask-shaped containing a hard pseudoparenchymatous wall and may be closed or may have a small opening (ostiole) and vary greatly as to mode of formation, size, shape, colour and the type of spores developed in them. This great variation in pycnidial structure serves to delimit the different genera among pycnidial Deuteromycetes (Fungi Imperfecti). The genera under Deuteromycetes are regarded as form-genera in absence of sexual reproductive stages. When, however, the sexual stages become known, the genera are referred to the perfect groups, either Ascomycetes or Basidiomycetes.

Sexual reproduction is well defined in lower fungi, where sexual organs develop and fertilisation results in formation of zygote, either oospore or zygospore, depending on whether or not the male and female reproductive organs are differentiated. Sexual reproduction is well pronounced in the lower Ascomycetes, while in the higher, even though sex organs may develop, the union may not take place. This degeneration of sex becomes more pronounced in Basidiomycetes where sex organs do not develop. In this group, sexual strains exist in a species. If the strains occur in the same mycelium, the species is homothallic and if they are separated in different mycelia, the species is heterothallic in which case the mycelia are to be brought together in suitable combinations to initiate pairing of nuclei of opposite strains (diploidization). This act may also be brought about by oidia. In the homothallic mycelium and also in the diploidized or dikaryotic mycelium of heterothallic species, the cells usually contain two nuclei (conjugate nuclei) of opposite strains, comparable to sexual units. The conjugate nuclei remain and move together, divide simultaneously and the union does not occur till in the basidium (in Basidiomycetes) or in the ascus (in Ascomycetes), which are spore mother cells. The union of the two nuclei (karyogamy) is regarded as equivalent to a sexual act and leads to doubling of the number of chromosomes, in the same manner as in flowering plants. However, unlike the latter, the diploid nucleus does not develop into a diploid individual but undergoes reduction division, in the ascus or basidium, into haploid nuclei. Segregation of sex in different nuclei occurs at this stage in heterothallic species while in the homothallic ones, the strains are contained in the same nucleus. As a result of the reduction division, spores are formed, within the ascus (endogenously), usually 8 in number in Ascomycetes and outside the basidium (exogenously) on projections called sterigmata, usually 4 in number, in Basidiomycetes. In higher fungi, therefore, the diplophase or dikaryotic stage is predominant and this stage is organised into fruiting bodies (sporophores or fructifications) which eventually develop the fertile region (hymenium) in which reproductive structures, either asci or basidia develop. These fruit bodies are analogous to fruits of flowering plants in function as both are eventually concerned in reproduction. However, unlike fungi, the fruit in flowering plants is formed after fertilisation. The dikaryotic mycelium of many Basidiomycetes is characterised by the presence of clamp connections (Pl. 1, 19) which ensure that the sister nuclei arising from conjugate division of the dikaryon become separated into two daughter cells.

In Ascomycetes, asci are produced in fruiting bodies (ascocarps). The asci may lie scattered within a closed ascocarp or cleistothecium (Pl. 1, 3) as in Plectomycetes or may be arranged in the hymenium in a closed ascocarp or perithecium (Pl. 1, 5) as in Pyrenomycetes or in an open ascocarp or apothecium (Pl. 1, 4 a, b) as in Discomycetes. The Basidiomycetes comprise a diverse variety of fungi with generally highly organised fruiting bodies (basidiocarps) of various sizes, forms and structure.

SPORE FRUITS, SPORES AND CLAMPS

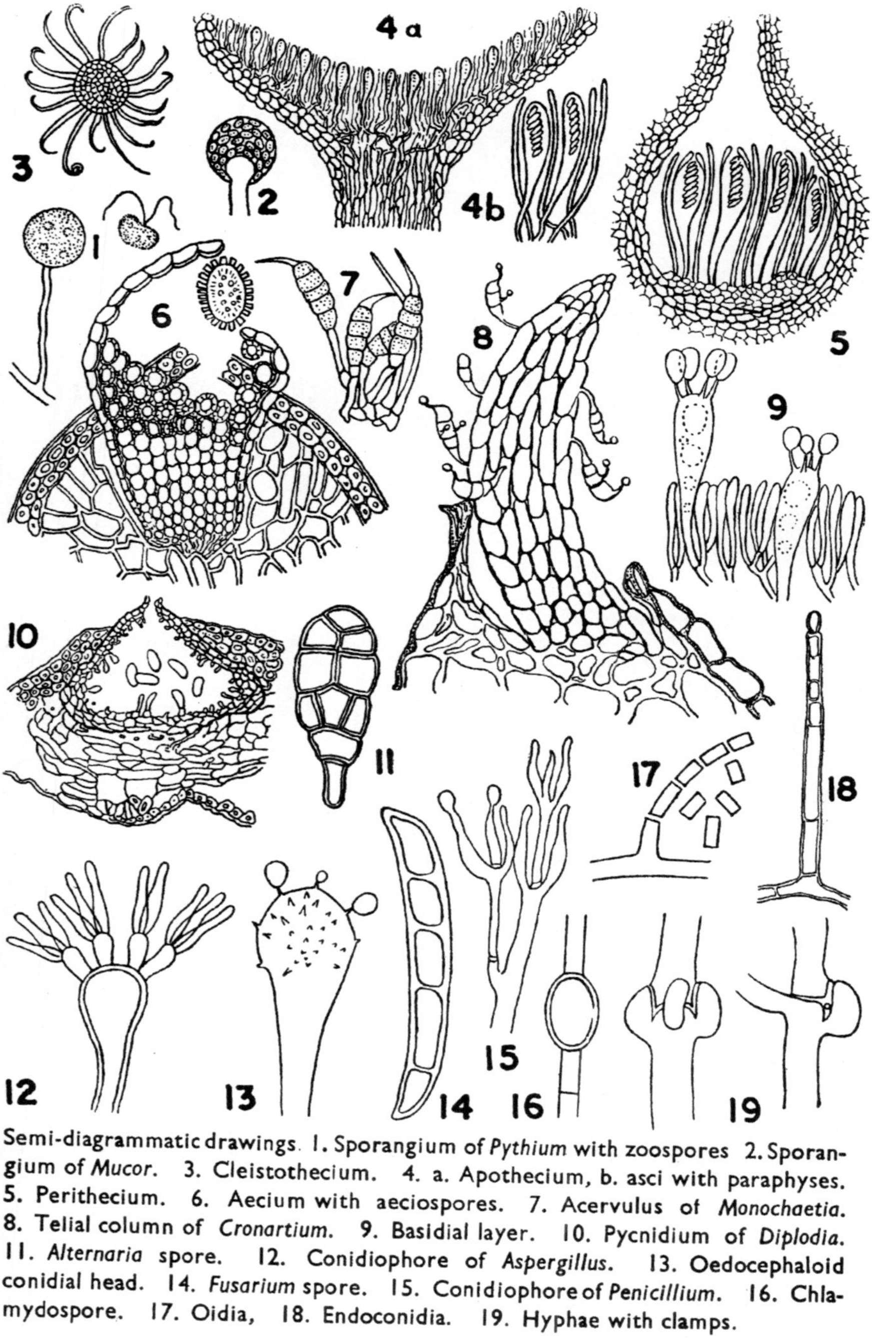

Semi-diagrammatic drawings. 1. Sporangium of *Pythium* with zoospores 2. Sporangium of *Mucor*. 3. Cleistothecium. 4. a. Apothecium, b. asci with paraphyses. 5. Perithecium. 6. Aecium with aeciospores. 7. Acervulus of *Monochaetia*. 8. Telial column of *Cronartium*. 9. Basidial layer. 10. Pycnidium of *Diplodia*. 11. *Alternaria* spore. 12. Conidiophore of *Aspergillus*. 13. Oedocephaloid conidial head. 14. *Fusarium* spore. 15. Conidiophore of *Penicillium*. 16. Chlamydospore. 17. Oidia, 18. Endoconidia. 19. Hyphae with clamps.

PLATE 2

CHIR PINE BLISTER RUST—LIFE CYCLE

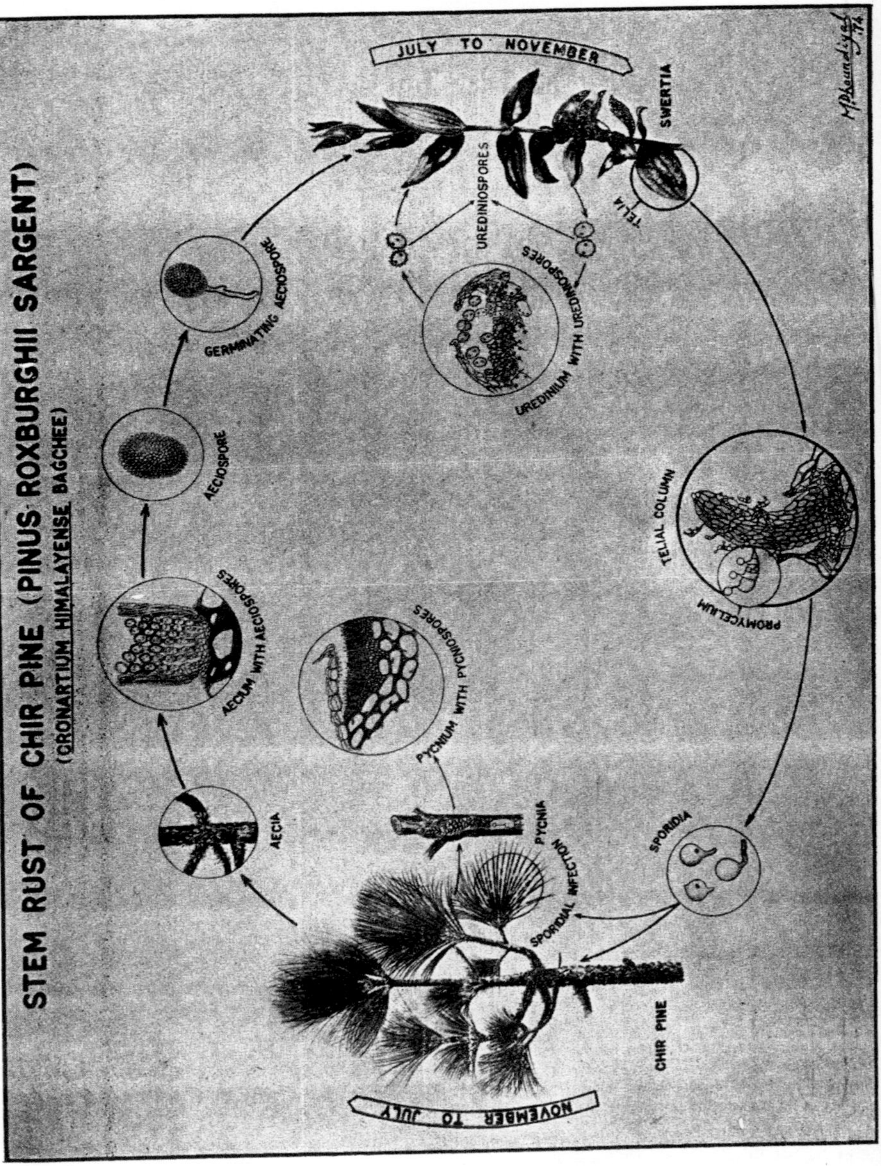

The Basidiomycetes are distinguished into two sub-classes depending on whether the basidium is septate (Heterobasidiomycetidae) or non-septate (Homobasidiomycetidae). The former includes important plant parasites, the rusts (Uredinales) and smuts (Ustilaginales) and in which the characteristic teleutospores develop into septate basidium (promycelium). Smuts cause diseases in agricultural crops, while the rusts in both agricultural and tree crops.

FRUCTIFICATIONS IN HOMOBASIDIOMYCETIDAE

Homobasidiomycetidae, in most cases, possesses highly organised fruit bodies where the hymenium may be exposed from the beginning on the free surface of the fructification (gymnocarpus) or the hymenium may be enclosed at the beginning but becomes free before spores are mature (hemiangiocarpus). Both forms occur in the hymenomycetes which include fungi causing decays in trees and timber. In the third form, the hymenial layer remains enclosed within the peridium until the basidia mature (angiocarpus) as in Gasteromycetes, which are mostly saprophytes though some genera are known to be mycorrhizal.

In its simplest form, the fructification represents a thick hyphal mat which develops closely appressed to the substrate, grows radially in an unlimited manner and bears hymenium behind the growing region. Such fructifications are called 'resupinate' or 'effused' (Pl. **15**, 7). The growing margin of such fructifications may tend to grow away from the substrate and form characteristic brackets which bear hymenium only on the lower side (Pl. **16**, 1). Such effuso-reflexed forms may later develop typical reflexed or bracket-shaped fructifications (Pl. **10**, 2; **11**, 4; **12**, 1; **17**, 4) by elimination of the resupinate portion. From reflexed forms, there is a transition to stipitate forms, either laterally or centrally, where the expanded cap or pileus bears hymenium on the lower surface. The disposition of the hymenium on the lower surface in all forms of fructifications is guided by the fact that it is protected from desiccation due to direct insolation and also from rains. Furthermore, when the spores are discharged from the hymenium, they are carried by the wind.

THE UREDINALES

The Uredinales or the rusts belong to the Basidiomycetes in which the basidium (promycelium) is normally 4-celled by transverse septa and produces a definite number (usually 4) of basidiopores (sporidia). All rusts are obligate parasites. The hyphae ramify in the intercellular space and send in haustoria into the cell, by means of which the nutrition for the rusts is secured. Some species of rusts may complete their life-cycles on one host, when they are termed autoecious; while others require two hosts to complete their ontogeny, in which case, the rusts are called heteroecious. In a heteroecious rust, the pycnia and aecia occur on one host and uredia and telia on the other.

Spore forms

For reproduction, four sori or fruiting structures may be recognised in the life-cycle of a rust having full complement of spore forms. The sori, succeed in the order pycnia, aecia, uredia and telia, in the life-cycle of a rust and are designated for convenience as O, I, II and III. These sori, respectively produce the pycniospores, aeciospores, urediospores and teliospores, the last invariably giving rise to the basidial state (IV) with basidia and basidiospores (sporidia). Thus there are 5 spore forms in the rusts. One or more sori may be present in the life-cycle of a rust, of which the telia must be one and is the indispensible stage. The spore fruits may be subcuticular or subepidermal and may be non-erumpent or erumpent.

O. Pycnia. The pycnia or spermogonia are borne on the haploid thallus which is produced as a result of infection by the basidiospore. They may be globular, lenticular or conical and open to the surface as minute pore. Within the cavity, spermatial hyphae cut off from their ends pycniospores or spermatia which are liberated through the opening in a drop of honey. The paraphyses, if produced, arise from the pycnial cavity and extend through the opening. The pycniospores cannot cause any fresh infection but a sexual function, in the way of diploidization of the aecial initial, has been ascribed to them (Buller 1931). Pycnia do not appear alone but are accompanied by aecia in macrocyclic rusts or telia in microcyclic forms.

I. Aecia. Aecia are distinguished by their usual association with pycnia and by their occurrence in groups of definite pattern. Aecial stage closely follows the pycnial stage and like the latter is produced on the haploid thallus. However, the full development of aecia and formation of aeciospores occur only after diploidization of the hyphae constituting the aecial sorus. In a heterothallic species, this diploidization occurs due to mating of the spermatia from the pycnia with the flexuous hypha of the aecial initial resulting in the migration of nucleus from the former to the latter to initiate paired or dikaryotic condition. The paired nuclei then undergo a series of conjugate divisions. In the homothallic species, pycnia are not necessary and may not be present. Here, the dikaryotic condition is initiated by pairing of two nuclei at some stage in the life-cycle. Aeciospores are unicellular, dikaryotic and germinate to produce a dikaryotic thallus, on which uredia followed by telia are produced.

Aecia may be recognised into different types like (1) aecidioid or cluster cup on angiospermous host; (2) peridermioid, tongue-shaped (Pl. 3, 11) or blister-like (Pl. 3, 5, 7) on gymnospermous host and (3) roestelioid or horn-like (Pl. 3, 2) applied to the aecial stage of *Gymnosporangium*. In all these, there is a peridium and aeciospores are catenulate and verrucose. Other types are recognised as (4) caeomoid where aecium lacks peridium but has catenulate spores and (5) uredinoid where aecia appear like uredia but unlike the latter, are closely grouped, accompanied by pycnia and occupy the position of aecia in the life-cycle. Uredinoid aecia are without peridium and bear pedicellate, usually echinulate aeciospores singly, morphologically similar to urediospores. Each of the aecial types forms the basis of a form genus, when the perfect or the telial stage in the rust is not known.

II. Uredia. The uredium is produced from the dikaryotic mycelium developing from infection by aeciospores. Uredia may have sessile, catenulate or pedicellate spores, which are usually echinulate and always 1-celled. The sorus may be with or without peridium and paraphyses. The uredial sori with reddish yellow spores present a rusty appearance, from which the common name rust is derived. Urediospores are wind disseminated and infect the same species of the host on which they are borne. These fresh infections produce a fresh crop of urediospores. The spores are, therefore, called repeating spores or also summer spores as they develop during summer. They are, therefore, asexual in nature.

III. Telia. The telia develop on the same host which produces the uredia either from the uredial sorus at later stage of its development or independent of uredia. Teliospores may be variously arranged, scattered in the mesophyll (*Uredinopsis*) or in epidermal cells (*Milesia*) or raised on stalks above the epidermis by rupture of the latter (*Puccinia*). They may be grouped in one-spore deep layer (*Malampsora*) or strongly laterally adherent forming spore heads raised on pedicel (*Ravenelia*) or in hair-like columns (*Cronartium*) (Pl. 1, 8 ; Pl. 3, 6). Teliospores may germinate immediately on maturity or may require a period of rest, usually during winter in cold areas, or during summer in hot regions, when it serves to tide the fungus over through the unfavourable period. The two nuclei (dikaryon) in the teliospore then fuse to form the only diploid nuclear stage in the life-cycle of the rust. The teliospore germinates to form an external promycelium (basidium) into which the diploid nucleus migrates and undergoes reduction division to form 4 haploid nuclei. The promycelium becomes four-celled by transverse septa, each cell containing one haploid nucleus. In *Coleosporium*, the promycelium is internal where the one-celled teliospore is converted into a four-celled promycelium. From each cell of the promycelium, a stalk or sterigma arises, at the tip of which a sporidium (basidiospore) develops. The sporidia are wind disseminated and infect the host which produces pycnia and aecia to initiate the haplophase.

Life-cycles

The different life-cycle stages in rusts are given below (Ainsworth 1961).

LIFE-CYCLES OF RUSTS

States (IV comes after III in every case)	Name of life-cycle	
	Schroeter	Arthur
O, I, –II, III, heteroecious	hetereu- } Eu-form	Macrocyclic
O, I, II, III, autoecious	auteu- } Eu-form	Macrocyclic
O, I,–III, heteroecious	} Opsis-form	Demicyclic
O, I, III, autoecious	} Opsis-form	Macrocyclic
O, II, III, autoecious	Brachy-form	
II, III, cycle uncertain	Hemi-form	
(O), III, germinating without rest .	Lepto-form	} Microcyclic
(O), III, germinating after rest . .	Micro-form	} Microcyclic

Except for the telia, there may be omission of one or more spore forms in the different rusts and life-cycles are named according to the complements of the sori that a species possesses. The life-cycle of a heteroecious macrocyclic rust, *Cronartium himalayense* Bagchee, with full complements of spore-forms is given in Pl. 2.

Host relationship

All rusts are obligate parasites. Some genera of rust fungi, namely, *Puccinia* and *Uromyces*, have a wide range of host plants belonging to many orders and families. There are, however, certain genera which are restricted as to their hosts and sometimes to the plant part attacked, like *Cronartium* and *Coleosporium*, both of which produce aecial stages on pines, the former on stems and never on the needles, while the latter on needles and not the stem. A rust may be further restricted to the host by the species or its groups like *Cronartium ribicola* J.C. Fisher restricted to certain 5-needle pines only and *C. himalayense* on 3-needle pines. Other instances of host specificity are the *Chrysomyxa*, the heteroecious species of which produce aecia on needles or cones of spruce and telia on *Rhododendron*. For the heteroecious rusts on conifers, aecia occur on conifers. The only exception to this is *Gymnosporangium* for which the coniferous plants belonging to Cupressaceae serve as the telial hosts and aecia occur mostly on the Rosaceae. Conifers may serve as the telial host when the rust is microcyclic, like *Chrysomyxa deformans* (Diet.) Jacz. (Pl. 3, 10) and *Ceropsora piceae* (Barclay) Bakshi & Sujan Singh (Pl. 3, 9) on spruce. *Ravenelia* are mostly known to parasitise Leguminoseae. A rust genus may, therefore, be identified fairly closely on the basis of the host.

Alexopoulos (1962) has given detailed keys for classification of fungi. Based mainly on these, a simple key is presented to indicate the taxonomic position of the genera of fungi commonly referred to in the book.

EUMYCOTINA
(Filamentous Fungi)

Flagellated cells (zoospores) characteristically produced, flagella two, nearly equal, one whiplash (directed backward) and other tinsel (directed forward). ...OOMYCETES
(*Phytophthora*, *Pythium*)

Flagellated cells lacking.

Hyphae aseptate or if septate, sexual reproduction resulting in the formation of resting spores...ZYGOMYCETES
(*Mucor*)

Hyphae septate. Sexual reproduction not resulting in the formation of resting spores.

Sexual reproduction resulting in the formation of ascospores or basidiospores.

Spores resulting from karyogamy and meiosis, borne endogenously in ascus. ...ASCOMYCETES

Ascocarps or ascogenous hyphae not produced, asci naked. ...HEMIASCOMYCETIDAE (*Saccharomyces*, *Taphrina*)

Asci produced in ascocarps, asci typically unitunicate; if bitunicate, then borne in an apothecium; ascocarps of various types. ...EUASCOMYCETIDAE

Asci scattered within the ascocarp or united in spore balls within a spore cyst; evanescent. ...PLECTOMYCETES (*Ceratocystis*)

Asci in a basal or peripheral layer forming a hymenium or in basal tufts; typically persistent.

Ascocarp a closed perithecium mostly ostiolate. ...PYRENOMYCETES (*Chaetomium*, *Claviceps*, *Diaporthe*, *Endothia*, *Erysiphe*, *Glomerella*, *Hypoxylon*, *Meliola*, *Nectria*, *Phyllachora*, *Phyllactinia*, *Uncinula*, *Ustulina*, *Xylaria*)

Ascocarp an open apothecium. ...DISCOMYCETES (*Bifusella*, *Hypoderma*, *Hypodermella*, *Lophodermium*, *Monilinia*, *Morchella*, *Rhytisma*)

Asci produced in ascocarps, asci bitunicate. ...LOCULOASCOMYCETIDAE (*Capnodium, Dothidella, Elsinoe*)

Spores resulting from karyogamy and meiosis, borne exogenously on basidia. ...BASIDIOMYCETES

Basidium septate, or deeply divided or consisting of a teleutospore germinating into promycelium. ...HETEROBASIDIOMYCETIDAE

Basidiocarp well developed, mostly saprophytic. ...TREMELLALES

Basidiocarp lacking, mostly parasitic. Teleutospores present.

Basidiospores sessile, passively discharged. ...USTILAGINALES (*Pericladium*)

Basidiospores borne on sterigmata, ejected by force. ...UREDINALES

Teleutospores germinate into external promycelium.

Teleutospores free or variously united but never in layers or crusts. ...PUCCINIACEAE (*Gymnosporangium, Hapalophragmium, Maravalia, Puccinia, Ravenelia*)

Teleutospores laterally united into layers, columns or crusts. ...MELAMPSORACEAE (*Ceropsora, Chrysomyxa, Cronartium, Dasturella, Melampsora, Olivea, Pucciniastrum, Tunicopsora*)

Teleutospores becoming transversely septate after karyogamy and meiosis. ...COLEOSPORIACEAE (*Coleosporium*)

Basidium continuous, basidiospores producing germ tubes on germination. ...HOMOBASIDIOMYCETIDAE

Basidiocarp not developed. ...EXOBASIDIALES (*Exobasidium*)

Basidiocarp present.

Hymenium exposed before maturity of spores. ...HYMENOMYCETES

Hymenium variously borne; if lining pores or gills, sporocarp not soft or putrescent, gills frequently branched or anastomosing. ...POLYPORALES

Hymenium smooth or roughened.

Sporocarp typically hard or leathery. ...THELEPHORACEAE (*Corticium*, *Hymenochaete*, *Peniophora*, *Stereum*)

Sporocarp club- or coral shaped. ...CLAVARIACEAE

Sporocarp funnel shaped. ...CANTHARELLACEAE (*Cantharellus*)

Hymenium lining spines or teeth. ...HYDNACEAE (*Echinodontium*, *Hydnum*)

Hymenium lining pores.

Pores shallow, fertile on ridges. ...MERULIACEAE (*Merulius*)

Pores deep, sterile on ridges. ...POLYPORACEAE (*Fomes*, *Ganoderma*, *Hexagonia*, *Polyporus*, *Poria*, *Trametes*)

Hymenium lining gills, if lining pores, sporocarps soft and putrescent. ...AGARICALES

Sporocarp soft, putrescent; hymenophore separating readily from context, hymenium lining pores. ...BOLETACEAE (*Boletinus*, *Boletus*)

Sporocarp fleshy or tough; hymenophore not separating from context.

Sphaerocysts present in context. ...RUSSULACEAE (*Lactarius*, *Russula*)

Sphaerocysts absent, gills not waxy. ...AGARICACEAE (*Agaricus*, *Amanita*, *Armillaria*, *Collybia*, *Pholiota*, *Pleurotus*)

Hymenium enclosed until basidia mature. ...GASTEROMYCETES (*Geastrum*, *Hysterangium*, *Lycoperdon*, *Rhizopogon*, *Scleroderma*)

Sexual reproduction not observed, reproduction by conidia, oidia or budding. ...FUNGI IMPERFECTI

Conidia borne in pycnidia. .. SPHAEROPSIDALES (*Botryodiplodia, Diplodia, Dothistroma, Phoma, Phomopsis, Phyllosticta*)

Conidia in acervuli .. MELANCONIALES (*Colletotrichum, Marssonina, Monochaetia, Pestalotia*)

Conidia borne otherwise ...MONILIALES (*Alternaria, Cercospora, Corynespora, Cylindrocladium, Fusarium, Stigmina, Trichosporium*)

Reproductive structures not known MYCELIA STERILIA (*Rhizoctonia*)

DISSEMINATION

The purpose of reproduction is not complete till the reproductive units are dispersed to a suitable substrate where new individuals of the species continue to grow under suitable environmental conditions, like temperature and humidity. The spores are mainly disseminated by wind and insects. Wind-borne spores are light, powdery, dry and produced in large numbers to account for many of them failing to germinate as they do not find a suitable substrate. Spores dispersed by insects are produced in much smaller quantity and may be embedded in mucilage which sticks to the body of the insect and carried to different places. The pycniospores of rusts are liberated from the pycnia in a drop of honey which attracts insects which disperse spores. In addition, the dispersal of fungi may occur when the infected seeds, fruits, vegetative stock etc. are transported to other places in various ways.

BACTERIA

Bacteria are minute 1-celled plants with a definite cell wall, devoid of chlorophyll and a typical nucleus but chromatin material is present. The cells may be rod-like, spherical, curved or they may be joined together temporarily in filaments. Many bacteria are motile by means of one or more flagella. Bacteria are able to multiply rapidly by means of simple fission, due to which they are named fission fungi or Schizomycetes. In some bacteria, reproduction may also occur by spores, usually endospores.

Bacteria are classified into three types based on the form of the vegetative cell: (1) coccus type, with globular or spherical forms, (2) bacillus type, with short or long cylindric rod-shaped forms and (3) spirillum type, with short or long spiral cylindrical forms. Only the bacteria of the bacillus type are known to cause plant diseases. Bacteria cause important diseases in man and animals. While many bacterial diseases are recorded in agricultural crops on which damage may become significant, they are not generally important in forest trees. In teak, bacteria may cause wilt in seedlings.

Bacteria sometimes cause stain in wood in standing trees as in watermark disease of willow and poplar in the U.S.A. and Europe. In red oak, the heartwood is infected by certain anaerobic bacteria which produce enzymes capable of degrading and weakening the compound middle lamella and thus predispose the wood to shake formation in the tree and cause honeycomb and ring failure in lumber during kiln drying (Ward, Hahn, Baltes and Bulgrin 1972). Bacteria also cause degradation in timber (Part III).

VIRUSES

A virus is an infective principle. Plant viruses are particular nucleoproteins which possess the ability to multiply within the appropriate hosts when infected and frequently derange the normal metabolism. Viruses do not respire or undergo metabolism. They are considered intermediate between living and non-living. The infective principle will pass through Berkefeld and Chamberland filters without loss of virulence. Viruses are particulate, and for each virus, the size and shape of the particles are uniform. Virus particles cannot be seen under light microscope but may be observed under electron microscope.

Virus infections may result in a variety of host reactions like chlorosis, mosaic, mosaic-mottling, necrosis, vein-cleaning, yellowing, atrophy, leaf curl, dwarfing and others. Internally, the living cells show peculiar inclusions which may be of two kinds, amorphous or X-bodies and crystalline material. Inclusions are formed as a result of combination of the virus particles with certain cell contents.

Virus diseases are readily transmissible by budding or grafting. Some viruses are sap transmissible through mechanical means while most viruses are spread naturally by insects, especially aphids, leaf-hoppers (jassids), white-flies, thrips and plant bugs. The insect vector takes the virus during feeding on a diseased plant and transmits to a healthy plant during course of feeding. Instances of virus transmission through true seed or contaminated soil are rare.

Viruses respond to serological tests. They produce antibodies when injected into blood stream of animals. An animal antiserum prepared against a particular plant virus will produce a precipitate when mixed with another. Serological tests are widely used for identification of plant viruses.

Measures to control virus diseases include planting of disease-free stock, sanitation cutting of infected trees, search for resistant or tolerant individuals and propagating through them and breeding of resistant varieties.

MYCOPLASMA

Mycoplasma like bodies (MLBs) have been found associated with a particular group of diseases known as the 'yellows'-type. A list of about 40 such diseases of plants including sandal spike is given by Hull (1971) who also reviewed the work on mycoplasma-like organisms in plants. The number of diseases with MLBs is increasing rapidly as more of the 'yellows' group are examined. The symptoms include excessive and premature development of buds leading to witches'-broom, reduction in leaf size with leaves often becoming brittle and chlorotic, shortening of internodes, phyllody followed by decline or death of affected branches or the plant. The disease agents are filterable and are transmitted by insect vectors and by grafting—all characteristic of virus diseases, for which reason such diseases were formerly believed to be caused by viruses.

However, on purification diseased agents yield particles which do not resemble virus. The basic structure of plant mycoplasma agents is similar to that of the mycoplasmas infecting animals or with L-form bacteria. Mycoplasmas are pleomorphic. The smaller infective form consists of particles (elementary bodies) about 100 nm diameter. After entering a susceptible host cell, the elementary bodies enlarge to initial bodies, which are ovoid or spherical with a diameter of 1 μ or more (Hull 1971). Mycoplasmas are bound by unit membranes, contain cytoplasm and ribosomes but no nuclear membrane or nucleolus.

The MLBs are found in the phloem of diseased plants located mostly in sieve tubes and occasionally in phloem parenchyma. Most of the 'yellows'–type diseases are transmitted by leaf hoppers and the disease agent multiplies in the vector which in thin sections shows MLBs under the electron microscope. Most diseases are also transmitted by grafting. Except in rare instances, MLBs from plants have not been isolated in pure culture to satisfy Koch's postulates and thereby prove them as the cause of the disease though strong circumstantial evidences like the constant association of the MLBs with the diseases and transmission by vectors suggest that MLBs are the etiological agents.

Unlike most viruses, mycoplasmas are sensitive to certain antibiotics of the tetracycline group which cause remission of symptoms and of vector transmissibility of some 'yellows'–type diseases. Tetracyclines are often mycoplasmastatic. No permanent cure has yet been obtained. 'Yellows' infected plants and infected leaf hoppers are amenable to heat cure treatment more readily than plants infected with viruses (Maramorosh, Shikata and Granados 1968).

PHANEROGAMIC PARASITES

Members of the Loranthaceae are important phanerogamic partial stem parasites on forest trees including commercially important species. The family consists of two sub-families, namely Loranthoideae and Viscoideae, though some advocate family rank for each. A review of the Loranthaceae has been done by Gill and Hawksworth (1961), while the dwarf mistletoe (*Arceuthobium*, Viscoideae), many species of which are parasitic on conifers, has been the subject of taxonomic treatment by Kuijt (1955), and Hawksworth and Wiens (1972) among others. Diseases of trees caused by important parasites are dealt with under respective tree species.

The members of Loranthaceae are perennial evergreen parasites on stems and branches of trees and shrubs and stand out distinct on the host, particularly if the host species becomes leafless during normal leaf-fall. While there is a great variation in floral structure, in gross features the flowers are hermaphrodite or unisexual (monoecious or dioecious), with or without a corolla or its counterpart. Pollination is mostly secured through birds which also disseminate seeds. The fruit is characterised by a leathery or fleshy coat enclosing a layer of mucilaginous viscin which serves as an adhesive to attach the seed to the host and also presumably aids germination in some species by providing better moisture conditions. Seeds of certain Viscoideae, particularly *Arceuthobium*, are equipped with explosive mechanisms which shoot the seed with considerable force to a distance of 5 to 10 m and thus aid in local spread. True ovules are absent and the seed is a naked embryo. On finding a suitable substrate, the seed germinates to develop a hypocotyl which grows along the surface of the host stem and eventually develops a holdfast from which the haustorium penetrates the host during which process both chemical and mechanical forces play a part. The typical haustorial system consists of cortical strands which invade the cortical region of the host and sinkers which form a close union with the host xylem and extract water and minerals from the host. Both cortical strands and sinkers develop xylem but no phloem elements. Haustoria may live for long period without the benefit of aerial parts.

In loranths, shoots commonly arise from the holdfast which develops as an outgrowth of the hypocotyl or as a secondary structure at the tip of stolon-like runners or suckers. Shoots commonly arise from adventitious roots taking their origin opposite the sucker. The morphology and life history of *Dendrophthoe falcata* (L. f.) Ettingsh., the mistletoe on a wide range of 319 hosts including teak, sal, mango, semul, salai, casuarina, *Anogeissus* spp., tendu, *Bischofia javanica* Bl., *Bridelia retusa* Spreng., *Acacia* spp. including babul, khair, *Albizia* spp., palas, *Cassia* spp., *Dalbergia* spp. including sissoo and others have been described and listed by Singh (1962). In *Arceuthobium*, the stems arise from buds in the endophytic system. The attack due to mistletoe is generally not serious in dry locations but may become damaging in warm, damp and sunny locations. The attack may result in loss of vigour and growth increment, malformation on the stem at the region of infection, death of affected branches and sometimes the tree.

Mistletoe control may be achieved either through sanitation or through chemical treatments. Both these methods are directed to remove or kill the endophytic system which may be extensive in the host. The haustoria may live for considerable time without the shoot and, therefore, the endophytic system is vital for propagating the parasite.

A systematic removal of infected parts of the host has been practised generally in horticulture and forestry. Once the parasite is eradicated or materially reduced, it may be kept under check through continuous vigilance. However, pruning of infected aerial parts may not ensure the removal of the haustoria and, therefore, sanitation cutting may not be always effective in keeping the mistletoe under control.

Chemical control of mistletoe may be achieved through use of various chemicals either as sprays or as injections. Formulations of 2, 4-D are particularly successful in killing the mistletoe though in some cases the chemicals may have variable effects on different hosts. Injections of copper sulphate and fernoxone (a 2, 4-D preparation) in the host control mistletoe of sissoo (Kadambi 1954). Silviculture and management procedures to control dwarf mistletoe are discussed under diseases of blue pine (Chapter 29).

4. FUNGUS NUTRITION

On the basis of nutrition, the plant kingdom may be divided into two broad classes, namely autophytes and heterophytes. The autophytes which include green plants are able to utilise simple materials, particularly carbon in inorganic form (carbon-dioxide) and also water and mineral salts and synthesise them into more complex substances for use as food. In this, they significantly differ from non-green plants, the heterophytes such as fungi and bacteria, which are unable to synthesise food. Therefore, such plants depend on extraneous food base for their subsistence. For fungi, carbon must be provided in an organic form. In addition, fungi also need an extraneous supply of nitrogen, phosphorus, potassium and other elements. Fungi also have definite requirements of micro-nutrients and vitamins. Fungi like all other living organisms produce extracellular enzymes which act on different food substances in the substrate and render them into simpler forms. The action of any particular enzyme is specific and is identified by the class of food substances on which they act namely, amylases which attack starches and sugars, cellulase on cellulose, protease on protein. The enzymes responsible for decomposition of wood by wood destroying fungi are discussed in Part III.

Fungi establish on a food substrate through the process of infection which is defined as the condition of being infected by a foreign micro-organism. When the foreign organism is large e.g., mistletoe, the term infestation is generally applied. On the living plant, fungi secure penetration in different phases, the physiology of which has been described by Brown (1915, 1916, 1917, 1922). For a detailed account on the nutrition in fungi, the books by Hawker (1950) and Cochrane (1958) may be referred.

PARASITES AND SAPROPHYTES

The heterophytes are distinguished into parasites, when they grow on living individuals, termed the host, and saprophytes when they grow on dead organic substrate. The term parasite indicates a one-sided food relationship, from the host to the parasite, that is never beneficial and is often harmful to the host (Boyce 1961). Parasites have been defined variously because of their different modes of food habit. According to Stevens and Young (1927) "a parasite is an organism that lives in or is attached to some other species of living organism from the living matter of which it secures part or all of its food materials". This definition excludes those fungi which grow on dead parts of a living plant from being considered as a parasite e.g., heart rot fungi on living trees. Nevertheless, such fungi have the potential pathogenic implications for the tree as a whole. For this reason the wider concept of parasitism as given by Link (1933) is more useful in forest pathology. Link defines parasitism as "the dependence of one individual upon another individual for food, irrespective of whether material used by the dependent individual or parasite is actually derived from the living or dead parts of the host".

Parasitism connotes an inter-specific interaction in which the parasite grows at the expense of the host which is debilited. This, therefore, excludes intra-specific reaction, as in some Bryophytes, where the sporophytic generation while dependent on the gametophyte for nutrition is a useful and sometimes necessary means of disseminating and perpetuating the host species, unlike a parasite which contributes nothing to the welfare of the host (Neish 1972).

Parasites and saprophytes are further distinguished into 'obligate' which strictly possess one mode of existence and 'facultative' which possess two phases of life. An obligate parasite must live parasitically and draw nutrition from living parts of the host. The host of the obligate parasite may live for sometime in a symbiotic state but later this relationship switches in favour of the parasite which then continues to live at the expense of the host. The mycelium of the obligate parasite is characterised by the presence of well-marked haustoria which enter the living cells and draw nutrition or in case of lower fungi, the whole thallus is intracellular. An obligate parasite is not known to exist without the host and cannot be cultivated in artificial media. An obligate saprophyte has always a saprophytic existence and never has any parasitic mode of living. These are easily cultivated in artificial media.

A facultative parasite lives saprophytically but under certain conditions when the host is debilited and the host resistance is lowered or when external conditions predispose the host or favour the development of the parasite, it becomes parasitic. Similarly a facultative saprophyte usually lives as a parasite but if the host dies, it continues to live saprophytically for sometime. Both the facultative parasite and the saprophyte may be cultivated on a variety of artificial media. The hyphae grow indiscriminately through the host tissue for nutrition and are, therefore, intracellular as well as intercellular.

A parasite may not necessarily cause any serious disturbance in the host. However, when it does, diseased condition is produced. The parasite is then termed the pathogen and the affected plant is called the suscept. A pathogen, therefore, is an organism capable of causing disease in a plant liable to infection by a pathogen. Therefore, while a pathogen is always parasitic, a parasite is not necessarily pathogenic (Boyce 1961). Consequently, there are pathogenic and non-pathogenic parasites. A pathogen may exhibit variable degrees in its pathogenicity when it is said to contain different races or strains and the terms like pathogenic strains or biological or physiologic races are used.

Soil microorganisms play a useful role in Nature's indispensable and beneficial cyclic processes that decompose plant litter, debris and other organic matter and thereby replenish carbon dioxide, ammonia and other substances essential for plant growth which in turn supports animal life. Ammonia is then changed by oxidation, first into nitrites by Nitrosomonas and then into nitrates by Nitrobactor. The action takes place when there is sufficient air (as the bacteria are aerobic) and soil is alkaline.

Some of the nitrates may be reduced to gaseous nitrogen by denitrifying bacteria leading to a loss of combined nitrogen in the soil. These bacteria are more active at higher temperatures of the soil as compared to nitrifying bacteria and this explains why tropical soils are deficient in humus.

The vast quantity of free nitrogen in the air is ordinarily inert. However, this free gaseous nitrogen becomes fixed in the soil by nitrogen fixing bacteria which are able to manufacture proteins from carbohydrates and free nitrogen. Some bacteria live free in the soil e.g., *Azatobactor* obtaining carbohydrate from the humus, while *Rhizobium* lives in symbiosis in small nodules or swellings on the roots of plants belonging to the family Leguminosae. They obtain carbohydrate from plants and in return, the plant obtains nitrogenous compounds from the bacteria. The nitrogen

RUSTS ON FOREST TREES

1-2. *Gymnosporangium cunninghamianum* : 1. Telia on *Cupressus torulosa*. 2. Roestelioid aecia on *Pyrus pashia*. 3. Witches'-brooms in deodar caused by *Peridermium cedri*. 4. *Peridermium thomsoni* on spruce, aecia on deformed needles. 5-6. *Cronartium himalayense* : 5. Aecia on chir pine. 6. Telia on swertia. 7. *Cronartium ribicola*, aecia on blue pine. 8. Rosette formation in *Rhododendron arboreum* due to *Chrysomyxa himalensis*. 9. Telia of *Ceropsora piceae* on spruce. 10. *Chrysomyxa deformans* on spruce. 11. *Coleosporium campanulae*, aecia on chir pine. (3, 5, 6, 7 Photo-FRI).

PLATE 4

MYCORRHIZA IN FOREST TREES

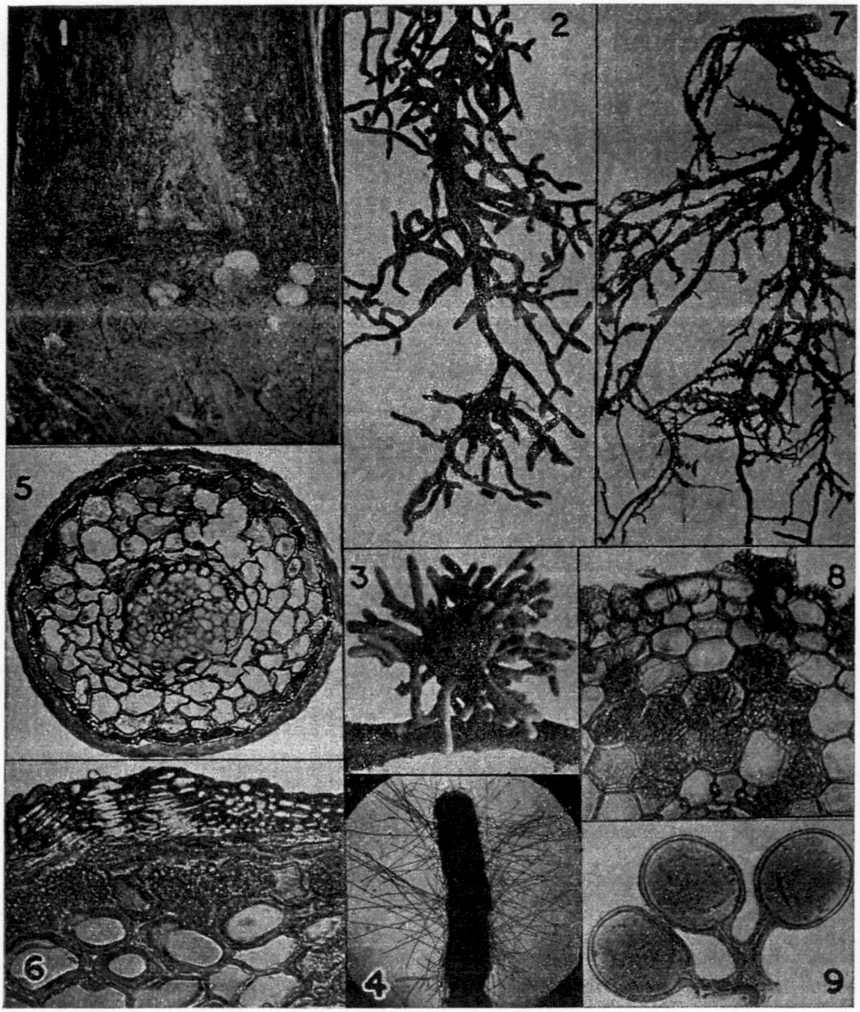

I. *Scleroderma geaster*, sporophores around *Pinus elliottii*. 2. Racemose mycorrhizae in silver fir. 3. Coralloid mycorrhiza of monterey pine. 4. Mycorrhiza showing hyphae of *Cenococcum gran iforme*. 5. Mycorrhiza of chir pine with mantle and Hartig net. 6. Palisade mantle due to *Cenococcum graniforme*. 7. Nodular endomycorrhiza in *Podocarpus gracilior*. 8. Endomycorrhiza in *Taxus baccata*. 9. *Glomus fasciculatus*, spores.

containing compounds are ultimately released into the soil thus enriching it. It may be mentioned that in wet hot soils such as found in the tropics, certain blue-green algae can fix nitrogen. Soil contains numerous micro-organisms many of which are beneficial and play an important role in the nitrogen cycle of the Nature. Among soil fungi, some enter into symbiosis with roots of plants forming a distinct morphological structure called mycorrhiza.

MYCORRHIZA

In forest trees, most absorbing roots are invaded by specific non-pathogenic soil fungi forming a distinct morphological structure called mycorrhiza. Physiologically, the mycorrhiza represents a case of symbiosis where the higher and lower symbionts live to mutual benefit or reciprocal parasitism but generally not of pathogenesis. A comprehensive account of mycorrhiza in Indian forestry is given by Bakshi (1974).

TYPES OF MYCORRHIZA

All mycorrhizae may by grouped into 3 types, ectotrophic, endotrophic and ectendotrophic. In ectotrophic type, the root system is heterozoic and distinguished into long root and short root. The latter may become infected and turn mycorrhizal. Mycorrhizal roots are limited in growth and short-lived, swollen when fresh and active and variously coloured due to the colour of the fungal symbionts. Short roots may be simple or monopodial as in spruce, fir (Pl. 4, 2) and sal or forked and often exhibit repeated dichotomy as in pines (Pl. 4, 3). In short roots, the root hairs are absent and become replaced by a fungal mantle enveloping the short roots (Pl. 4, 5) and hyphae usually radiate from the mantle into the soil (Pl. 4, 4). The morphology of the mantle can be varied and depends on the species of the fungal symbionts (Pl. 4, 6). Fungi enter the cortex by secretion of pectolytic enzymes, thus permitting the hyphae to grow in the intercellular space to form a net work of hyphae called the Hartig net (Pl. 4, 5). Ectotrophic mycorrhiza occurs in the Fagaceae, Betulaceae, Salicaceae among hardwoods and Abietineae among conifers.

In endotrophic mycorrhiza, the morphology of infected roots may remain unchanged or may become beaded (Pl. 4, 7). The fungal mantle is absent and root hairs normally develop. The fungi are present in the form of individual hyphae on the root surface and penetrate the cells of the cortex by secreting cellulolytic enzymes. The hyphae are intracellular (Pl. 4, 8) and produce coils and swellings that eventually disappear as a result of digestion by the invaded cells while the hyphae follow the advancing meristematic root tip. Endotrophic mycorrhiza is present in majority of plants including *Taxus*, *Podocarpus*, *Cupressus* and *Araucaria* among others in forest tree genera.

Ectendotrophic mycorrhiza represents a condition where the typical ectotrophic infection is accompanied by intracellular penetration of hyphae. They are represented on beech, lodgepole pine and ponderosa pine.

MECHANISM OF SYMBIOSIS

The mechanism of infection of roots by fungi to form mycorrhiza is not fully understood yet. Bjorkman (1942, 1949) postulated the carbohydrate theory in which it is explained that mycorrhiza is formed in soil which is deficient in available nitrogen and phosphorus and under conditions of higher light intensity. In strong light, there is an accumulation in the roots of soluble carbohydrates which favour infection. While this carbohydrate theory is a valuable contribution towards determining the establishment of mycorrhizal symbiosis, the mechanism of the initiation and establishment of the symbiotic relationship is not fully clear. Slankis (1961) found that at low light intensities at which mycorrhiza is not formed because of the lack of surplus soluble carbohydrates in the roots, indoleacetic acid added to the nutrient solution at relatively high concentrations did not induce morphological differentiation in the roots in the pines. Addition of glucose or sucrose to the nutrient medium does not bring about this change. Slankis concluded that at low light intensities, neither auxin nor auxin together with soluble sugars induces the profound morphological differentiation resembling ectotrophic mycorrhizae. Possibly excess of carbohydrates in tree roots together with exuded metabolites from the roots induce the mycorrhizal fungi in the soil to infect roots. These exudates are possibly sugars, a large number of different amino acids, vitamins (Slankis 1957) or other specific substances the nature of which is still unknown–M-factor of Melin (1954). Seedlings exposed to an atmosphere containing C^{14} labelled CO_2 transported photosynthate to the fungal associate of pine mycorrhiza (Melin and Nilsson 1957). The photosynthate and its metabolic conversion products and growth promoting metabolites from roots influence the growth and development of the mycorrhizal mycelia.

The fungal symbionts exude substances which are responsible for forking of roots as in *Pinus* or initiation and proliferation of unbranched rootlets in other genera. These substances are probably related to auxins since synthetic auxins at certain concentrations induce similar morphogenesis (Slankis 1957). The infection results in an increase in the absorbing surface of roots.

FUNGAL SYMBIONTS

Fungi forming ectomycorrhiza belong mostly to Basidiomycetes and some to Ascomycetes and Fungi Imperfecti. The mycorrhiza forming fungi are known to belong to *Amanita, Boletus, Cantharellus, Clitopilus, Clitocybe, Cortinarius, Lactarius, Leccinum, Russula, Suillus, Tricholoma, Xerocomus, Corticium* and *Thelephora* among the Hymenomycetes; *Calvatia, Geastrum, Lycoperdon, Scleroderma*, and *Rhizopogon* among the Gasteromycetes *Elaphomyces*, among Ascomycetes and *Cenococcum graniforme* (Sow.) Ferd. & Winge which is the imperfect stage of an Ascomycetes (Melin 1963). These fungi occur in the soil in the form of spores, free living mycelia, hyphal strands (Levisohn 1956) and sclerotia (Mikola 1948). These fungi can be identified from sporophores which usually develop in association with mycorrhizal roots (pl. 4, 1). Such fungi are difficult to isolate and maintain in culture in which they are slow growing. A fungus is proved to be mycorrhizal under pure culture synthesis.

The fungi forming endomycorrhiza belong mostly to Phycomycetes except in Orchidaceae, Ericales and Gentianaceae where the lower symbionts may belong to species of *Rhizoctonia*. The most common endophytes of vesicular-arbuscular mycorrhizae are attributed to *Endogone* which has now been divided into 4 genera, namely, *Endogone*. *Glomus* (Pl. 4, 9), *Gigaspora* and *Modicella* (Gerdemann and Trappe 1974). The name *Endogone* is retained only for zygosporic species. Because of the difficulty of cultivating the endophyte in pure culture, it is studied in the soil where the endophyte occurs as free spores (chlamydospores, zygospores, azygospores), or sporocarps or it is studied in the extramatrical and intramatrical phases of the host. *Endogone* spores are isolated from the soil by the wet sieving and decanting technique (Gerdemann and Nicolson, 1963) or by floatation-adhesion technique (Sutton and Barron 1972). The fungi are maintained in pure pot culture on the roots of host plants in sterilised soil.

The fungi of ectendomycorrhiza are designated as E. strains and are non-sporing (Mikola 1965). These fungi are predominant in old forest nurseries raised on agricultural sites, and have weak competitive ability in forest soils. The ectendomycorrhizal symbionts are regarded as a group distinct from other ectomycorrhizal symbionts.

While some fungi may be specific as to host in developing ectomycorrhizal associations, majority of plants possesses a wide spectrum of fungi with which mycorrhiza is formed. There is also little evidence of host specificity among fungi forming endotrophic mycorrhiza (Mosse 1963).

FUNCTIONS

Mycorrhizal fungi are regarded as the most highly specialised of all root infecting fungi (Garrett 1944). Such fungi use carbohydrates provided by the tree and are, therefore, parasitic but not pathogenic and are beneficial to the higher symbiont. The benefits include increase in the absorbing area of the roots as stated before. Mycorrhiza plays an important function in tree nutrition as demonstrated mostly for ectotrophic forms, in which the mantle and hyphae radiating from the mantle act in a manner similar to root hairs. Using radioisotope technique, Melin and his co-workers demonstrated that the fungi translocate various nutrients from the soil to the tree. Thus, *Boletus variegatus* (Sw.) Fr. as fungal symbiont was able to transfer to pine roots and other parts of seedlings phosphorus (P^{32}) from a phosphate solution (Na_2HPO_4) (Melin and Nilsson 1950); nitrogen (N^{15}) from a solution of ammonium salt (NH_4NO_3), (Melin and Nilsson 1952) and from amino-acids as well as glutamic acid (Melin and Nilsson 1953) and calcium (Ca^{45}) from a solution of $CaCl_2$ (Melin and Nilsson 1955). The same authors (1953a) demonstrated transfer of radioactive phosphorus as H_3PO_4 using *Cortinarius glaucopus* (Schaeff. ex Fr.) Fr., as the fungal symbiont. Melin, Nilsson and Hacskaylo (1958) demonstrated the transfer of sodium (Na^{22}) from NaCl into pine seedlings through mycorrhizae produced by *Rhizopogon roseolus* (Corda) Th. Fr. Inorganic phosphate ($KH_2P^{32}O_4$) was shown to pass into the host tissue from the fungus (Harley and Loughman 1963). The passage of ion from the fungus to the host tissues was increased with increased transpiration of the higher symbiont (Melin and Nilsson 1958). Increased phosphorus absorption by endotrophic fungi was demonstrated by Daft and Nicholson (1966).

The mycorrhizal fungi afford protection to plants to ward off attack by root diseases either by acting as a physical barrier or by producing antibiotics. *Phytophthora cinnamomi* Rands, the cause of little leaf disease of short leaf pine, is unable to secure infection on mycorrhizal structures by means of zoospores (Marx 1971). The infection due to *P. cinnamomi* was absent in plants with mycorrhizae as compared to 90–100 per cent infection on non-mycorrhizal short roots. In laboratory studies, antibiotics isolated from culture filtrates of mycorrhizal fungi were shown to inhibit growth of many root pathogens (Marx 1971). Furthermore, mycorrhizal plants are said to resist drought and frost (Cromer 1935, Harley 1940).

NURSERY MANAGEMENT PRACTICES IN RELATION TO MYCORRHIZA

Age and previous land uses are important for selection of the nursery site. Growth tends to be poor in nurseries established on agricultural land and which had been cropped for many years. Nurseries may be established after clear-felling the forest cover and this will ensure presence of mycorrhiza which occurs naturally in the site. The nursery soil should be either acid or neutral. Conifers are known to grow best at soil pH between 4.5 to 5.5. (Eldin 1964) when the development of mycorrhiza is encouraged. In arable site with high soil pH, conifer seedlings are known to have a poor ability for the uptake of inorganic fertilizers from the soil (Benzian 1965). It is, therefore, necessary that for conifers, the pH is lowered which may be possible with chemicals like sulphuric acid and ammonium sulphate (Chapter 5). However, no lasting effect from acidification can be expected in the soil containing high free calcuim carbonate (Benzian 1965).

The local nursery soil is generally enriched with nutrients and for this, fertilizers may prove more useful than the much more expensive and variable composts which are largely in use.

The seedlings may be raised directly on the nursery bed on a soil which may comprise of the local soil to which fertilizers or composts are added sparingly to permit adequate development of mycorrhiza. The seedlings may remain in the bed for the full period before they are transplanted in the field or the seedlings may be pricked early within 2–3 months after germination to another nursery bed giving spacing or in polythene tubes.

Introduction of mycorrhiza

In the natural forest, mycorrhiza occurs naturally in the site. If a nursery is established after deforesting an area and indigenous species are raised, it may not be necessary to introduce mycorrhiza. However, when nurseries are established on an afforested site or where exotic species are raised in the nursery, it is important that mycorrhiza is introduced. Such introduction is preferably done in the nursery due to high concentration of plants and ease in the mode of application. Once such a nursery is established where seedlings develop mycorrhiza and 2 to 3 crops are harvested, it may be assumed that the mycorrhizal fungi have established in the site and further introduction may not be obligatory unless proved otherwise.

Mycorrhiza may be introduced through (i) soil inoculum, (ii) interplanting 'mother' or 'wild' seedlings in the nursery bed, and (iii) inoculation with mycorrhizal fungi as fruiting bodies, spores or as pure cultures. Soil is collected from a successful stand of the same species or in its absence, closely allied species from the top 10-25 cm of soil where mycorrhizae are more frequent than at lower depths. The soil inoculum should contain, besides soil, active mycorrhizal short roots. The inoculum should be used within 10-15 days after collection to ensure viability of mycorrhizal fungi. It should not be exposed to direct sun which may be harmful for the viability of mycorrhiza. The inoculum is spread on the nursery bed in the proportion of 10 to 20 parts of inoculum and 80 to 90 parts of local soil and the soil is raked up in the top 15 cm. The seeds are broadcast in the bed and the germinated seedlings will eventually develop mycorrhiza. Soil inoculation in the nursery bed may appear unnecessary if the seedlings are pricked within 4-8 weeks after germination into another nursery bed or in polythene tubes where the mycorrhiza may be introduced. In case of tube seedlings, 2 tablespoonful of inoculum may be added to every tube prior to pricking. In some cases, the local soil and the mycorrhizal soil inoculum are mixed in a heap and raked and filled in polythene tubes. This may not ensure even distribution of the inoculum into every tube resulting in erratic development of mycorrhiza in seedlings.

In the interplating method, seedlings collected from a natural stand and containing appropriate mycorrhiza are transplanted in seed beds at regular spacing (about 0.3 m$\times$0.3 m) and the beds are subsequently sown with seeds. The germinated seedlings eventually get mycorrhizal infection from the mother seedlings. This is said to be the only successful method of introducing mycorrhiza in *Pinus merkusii* Jungh. and de Vriese where the usual method of soil inoculation is unsuccessful (Mikola 1969).

In the method of introduction of mycorrhiza either by soil inoculum or using mother seedlings, there is an inherent danger in the transference in soil of injurious microorganisms and this may not be permitted under quarantine regulations if soil is imported from another country. The use of pure cultures of mycorrhizal fungi is, therefore, recommended (Anon. 1964). The use of pure cultures does not pose any bulk handling and transportation problems due to comparatively less quantity of inoculum needed for introduction as compared to soil inoculum. It also ensures introduction of appropriate fungi. Since appropriate fungal symbionts for different tree species are not generally known and also the site conditions under which the fungi will thrive need to be determined, Mikola (1973) recommended mixing pure cultures of several fungi. However, due to difficulties in germinating spores or tissues of mycorrhiza forming fungi and in cultivating them in pure culture for field inoculation where such fungi have to establish against microbial competition of the natural soil microflora, the use of fungi as fruiting bodies, spores and pure cultures has not found wide field application.

Effect of biocides

In the nursery, there is an obvious need to control diseases like damping-off and also insect attack on seedlings by treatment of the nursery soil with toxic chemicals. The depressing or retarding effect of such chemicals on development of mycorrhiza is reported by Wilde and Persidsky (1954) and Hacskaylo and Palmer (1957), though Sobotka (1968) observed that biocides caused little or no reduction in the formation of mycorrhiza. Bakshi and Dobriyal (1970) showed that fungicides applied to the soil to control damping-off in pines had an initial retarding effect on formation of mycorrhiza, particularly when the roots were restricted within the treated zone. Mycorrhiza eventually developed in seedlings when their roots penetrated beyond the treated zone or when the toxic levels of chemicals were lowered in the treated zone in course of time.

Effect of soil nutrients

Within limits, formation of ectotrophic mycorrhiza varies inversely with soil fertility. They are profuse in relatively infertile soils and in soils rich in raw humus where nutrients are not present in a readily available form. Development of mycorrhiza may become suppressed with addition of high dosages of inorganic fertilizers. Dinesh Kumar,

Sujan Singh, Thapar and Misra (1968) showed that mycorrhizal development was inversely proportional to fertilizer dosages. In Sweden, the recommended practice is to apply fertilizers sparingly to tree seedlings to permit mycorrhizal development if fertilizers are not applied to transplants in the field. Fertilizers applied at high rates produce large root stocks but poor mycorrhizae and later when fertilizers are not applied in the field, plants may become stunted and die (Hacskaylo 1958).

Physical soil conditions

Mycorrhizae are abundant when soil moisture is adequate but not too high and also in loose well-aerated and acid soils. Their development is retarded when conditions are otherwise. Some fungi e.g., *Cenococcum graniforme*, are present on predominantly dry site (Worley and Hacskaylo 1959).

Effect of light

Development of mycorrhiza is directly proportional to the amount of sunlight (Bjorkman 1942, Harley and Waid 1955, Hacskaylo and Snow 1959). Development of mycorrhiza in seedlings exposed to full sunlight and 3 hours shade was significantly better than those under 6 hours shade and full shade (Bakshi 1974). If shading of the nursery becomes necessary to protect seedlings against sun scorch, it should be done for a limited period during the hottest part of the day (Bakshi 1974).

SIGNIFICANCE IN FORESTRY

As efficient nutrient absorbing organ, mycorrhiza exerts a beneficial effect on tree growth, particularly in nutrient-deficient soils. However, it may not be correct to assume that the presence of mycorrhiza would enable healthy tree growth on all poor sites. Failures in new nurseries have resulted due to lack of mycorrhiza (Rayner 1945, Pryor 1956, Gilmour 1958, Briscoe 1959, Lundquist 1964 and others) and success is achieved when mycorrhiza is introduced. Plants may grow successfully without mycorrhiza under certain limited conditions such as soilless cultures (Olson 1944) or highly fertile soils (Harley 1972) as may occur usually in the nursery but when such seedlings are transplanted in an afforested site or on a poor site where further nutrient is not added, failure may occur. Even fertilizer applications and cultural practices may not lead to success in plantings as happened in Puerto Rico while introducing pines and where no native pines exist. Success was achieved only when mycorrhiza was introduced which also resulted in better plant growth in comparison to poor growth where mycorrhiza was not introduced (Briscoe 1959). Failure of certain tree species to establish on *Calluna* heathland in England is due to a factor in *Calluna* raw humus which inhibits the growth of mycorrhizal fungi (Handley 1963).

Notwithstanding the failures met with afforestation projects possibly due to absence of mycorrhiza, there are instances where success is achieved without mycorrhizal introductions, even for exotics like eucalypts which are raised successfully in South India on poor afforested sites. However, in all such cases mycorrhiza develops in association with the fungi occurring naturally in the site. In reforestation projects, with the same or allied species, mycorrhiza is likely to develop in plants with fungi normally occurring on the site. In such cases, therefore, introduction of mycorrhiza may not appear necessary.

5. NURSERY DISEASES

In India, most forests have been usually naturally regenerated, though economically important species like teak and deodar have also been raised in plantations for which the seedling stock is raised in the nursery. During the last decade, large scale afforestation projects, and rehabilitation of degraded forests have been taken up to meet the industrial requirement of wood which is in short supply. Large scale plantation programmes have necessitated a demand for planting stock for which nurseries are established. The propagation of most forest trees from seed is not by itself difficult but with increasing scale of the operation, nursery management and skilled supervision become important. An adequate supply of seedling stock of plantable height should be available at as low a price as possible. The nursery should be easily approachable by all-weather motor roads to facilitate supervision and transport of the seedling stock to planting site. In the nursery, disease problems may arise and which may seriously affect the planting stock and disturb the planting programme.

Important requirements for a nursery site are good fertile soil with light texture and moderately acid reaction. Damping-off diseases are kept at a low level in light-textured well-drained soils and in acid soils. If the soil is near neutral or alkaline, it may be acidified to a desired level with any one of the following : concentrated sulphuric acid (100 to 400 ml in 4.5 l water), sulphur (25 to 300 g flowers of sulphur), aluminium sulphate (powdered, iron free, 17–18 per cent Al_2O_3 at 340 to 680 g) and ammonium sulphate (fertilizer grade) (100 to 450 g) may be applied per sq yd of seed bed in every case (Benzian 1965). The exact dosage of the chemical will depend on the buffering capacity of the soil. If, however, the soil contains high content of calcium carbonate, acidification will not have any lasting effect as observed in the spruce and fir nurseries at Shamshi (Kulu, Himachal Pradesh) (Bakshi, Reddy, Thapar and Khan 1972). Nursery management practices in relation to development of mycorrhiza are discussed earlier (Chapter 4).

DAMPING-OFF DISEASE

Damping-off is probably the most serious of all the nursery diseases and is prevalent all over the world. It is primarily a disease of conifer seedlings, although many broadleaved species may be attacked. The loss due to damping-off cannot be predicted. The disease may destroy the entire nursery stock in one season and may be unimportant in other years. In India, the disease is prevalent in all conifer nurseries of indigenous and also exotic pines which are under trial in tropical and subtropical locations where temperature conditions favour the development of the disease.

SYMPTOMS

Mortality in seedlings may occur at three stages during their growth: due to pre-emergence blight, post-emergence mortality soon after germination and root rot in older seedlings resulting in poor growth and in extreme cases even death. In pre-emergence blight, the developing seedlings are attacked and killed before the seedlings emerge above the ground. The emergent crop is sparse. This may be wrongly attributed to poor viability of seeds. The losses due to pre-emergence blight can be estimated by locating the killed seedlings in the soil while germination is still in progress. In post-emergence mortality, the seedlings are attacked on the hypocotyl or on roots or both while still in the cotyledon stage. The fungi spread rapidly in the tissues and the seedlings either wilt completely or fall over on the ground due to rotting of tissues at the base. Fungal attack on succulent root tips in older seedlings and transplants is not usually fatal. If soil moisture conditions are favourable, new roots develop and the seedlings continue to live though with less vigour and poor growth. If the damage of this type is followed by drought or by excessive soil moisture, both of which discourage formation of new roots, mortality may eventually occur in seedlings due to root rot.

CAUSE

The disease is caused by a number of fungi which are normally soil saprophytes but under conditions favourable for the development of fungi and unfavourable to growth of seedlings, the fungi may become pathogenic. The most important fungi causing damping-off in conifer seedlings belong to *Pythium*, *Rhizoctonia*, *Phytophthora*, and *Fusarium*. Damping-off due to *Pythium* and *Rhizoctonia solani* Kuhn. was studied by Roth and Riker

(1943, 1943 a, 1943 b) and due to *Fusarium* by Tint (1945, 1945 a). *Phytophthora* is relatively less important in forest nurseries. In India, *R. solani* is the important cause of damping-off in nurseries. *Pythium* and *Fusarium* are recorded less commonly (Reddy 1969).

PREDISPOSING FACTORS

Damping-off is favoured in clayey and wet soils, because of anaerobic conditions in the rooting zone (Vamos 1954) and relatively high temperatures (Hartley 1921, Roth and Riker 1943 a, Akai and Takeuchi 1954, Wright 1957). Because of the effect of warmth, damping-off tends to develop more during summer on late sowings than in the seeds sown early in spring or in autumn. Soil pH influences development of the disease, which is more serious in alkaline or neutral soils than in acid soils (Jackson 1940, Roth and Riker 1943 a, Tint 1945, Griffin 1958). Low light intensity acting both by reducing the metabolism and increasing the soil moisture favours development of damping-off (Hansen, Kenety, Wiggin and Stakman 1923, Tint 1945 a, Vaartaja 1952). In addition, dense sowings, application of nitrogen in high dosages during early stages of growth, and addition of raw undecomposed humus favour disease development.

CONTROL

Control can be effected by modification of cultural practices in the nursery and by fungicidal treatment of the seed and also soil before and after sowing.

Incidence of damping-off can be minimised through proper selection of site. Conifers make their best growth when nurseries are established on sites with light soil texture to encourage good root development and with soil pH 4.5–5.5 (Benzian 1965, Griffin 1958). In acid soil, the risk of damping-off disease is less. If however, nurseries have to be raised on sites with heavy soil, sand may be mixed to increase porosity and drainage. Soil pH may be lowered by acidification. Damping-off is generally heavy in wet soils and, therefore, during sprouting, watering should be kept to the minimum. Nurseries should be located on well drained site without overhead shed.

Partial sterilisation of the soil by steam or through use of fumigants has the advantage of lowering or eliminating microorganisms harmful to plant growth. Steaming is costly and laborious. The soil for seed boxes may be put under running steam in an autoclave or a closed container. In seed beds, pipes with steaming holes may be led into beds which should be covered with gunny cloth. The steam is let into beds for about one hour. Partial chemical sterilants, like formalin and methyl bromide, are highly phytotoxic and need to be applied to beds 10-15 days in advance of sowing by which time the toxic effects of chemicals are removed through volatilization. Commercial grades of formalin (38 per cent formaldehyde), 250 ml diluted in 4 l water, may be applied as soil drench per sq m of bed. The standard procedure for the use of methyl bromide is to release it beneath a thick polyethylene sheet spread over the nursery bed with edges of the sheet buried in the soil. Ampules containing methyl bromide are broken underneath the sheet. The vapours thus released gradually diffuse into the soil. The sheet is removed after 48 hours. Methyl bromide is poisonous. The chemicals have a useful side effect in killing the weed seeds, nematodes and insects. In the absence of weed competition, the seedling vigour is improved.

Damping-off and other soil-borne diseases may be controlled by seed and soil treatment with chemicals, a review of the work on which is given by Vaartaja (1964). There are many proprietary chemicals with low phytotoxicity that may be applied to the soil at the time of sowing to control damping-off. Such chemicals are broadcast and forked into top 15 cm of soil one day prior to sowing. To control damping-off in nurseries, Reddy and Misra (1970) tested some proprietary chemicals. The names of these chemicals, their active ingredients and dosages per sq yd of seed bed at which they were used are listed below :-

Captan : 20.5 g of a 83.0% dust of N-trichloromethyl-mercapto-4-cyclohexene-1, 2 diocarboximide.

Thiride : 22.5 g of a 75.0% dust of tetramethyl thiuram disulphide.

Zineb : 26.0 g of a 65.0% dust of zinc ethylene-bisdithiocarbamate.

Cuman : 22.5 g of a 80.0% dust of zinc dimethyl-dithiocarbamate.

Brassicol : 22.5 g of a 75.0% dust of pentachloronitrobenzene.

Blitox : 19.3 g of a 50.0% dust of copper oxychloride.

Zinc oxide : 17.0 g of a 100.0% dust of zinc oxide.

In their fungicidal soil treatment to control damping-off in pines, the authors found that all chemicals except brassicol and zinc oxide controlled damping-off effectively.

Fungicidal seed treatments give effective control of damping-off and also seed-borne diseases in many broadleaved species. However, for conifers, soil treatment is generally advocated, particularly against pre-emergence blight and because of the prolonged susceptibility of conifers to the disease. Seed treatment is effective for species which have hypogeous germination, as the treated seed coat remains *in situ* to release the toxic chemical slowly and thus afford protection to the seedlings. In epigeous germination, e.g., in conifers, the seed coat is taken above the ground and, therefore, any seed treatment may not become effective.

6. ROOT DISEASES

The underground part of the plant, the root system, may suffer from diseases, which generally remain unnoticed until manifested by symptoms on the above-ground parts. Diseases may be due to parasitic causes like fungi, bacteria, nematodes, insects etc. or may be of physiogenic origin. Fungi parasitic on plant roots, the root-infecting fungi, constitute an important group of organisms causing root diseases of forest trees.

ROOT-INFECTING FUNGI

The root-infecting fungi constitute an ecological group distinct from other parasitic fungi in that in the former, it involves a triple biological relationship between the host plant, the parasite and the general soil microflora under varying conditions of soil texture, soil moisture, soil reaction, soil nutrients etc. The soil is a reservoir of numerous micro-organisms like fungi, bacteria, actinomycetes, nematodes, protozoa, and insects which are in a state of continuous dynamic biological equilibrium. The root-infecting fungi living in the soil may attack plant roots to cause diseases. Waksman's conception (1917) of a basic cosmopolitan fungus flora of the soil consisting of two groups, namely, the cosmopolitan 'soil inhabitants' and the exotic 'soil invaders' was recognised by Garrett (1960) to distinguish between two contrasting types of behaviour of root-infecting fungi, namely, the 'soil-inhabiting fungi' and 'root-inhabiting fungi'. Garrett characterised the root-inhabiting fungi as possessing "an expanding parasitic phase on the living host, and a declining saprophytic phase after its death", due to their inability to compete with other soil microorganisms, unlike the soil-inhabiting fungi which are characterised by their ability to survive indefinitely as soil saprophytes, but under certain conditions may attack plant roots.

SAPROPHYTIC PHASE

The saprophytic life of parasites in nature in absence of their hosts depends not only on their ability to colonise different substrates but more directly on their capacity for competitive colonisation of substrates in the presence of other microorganisms. The soil-inhabiting fungi can compete successfully for colonisation of dead organic substrates in the soil and are relatively primitive parasites than the more specialised parasites, the root-inhabiting fungi, whose parasitic specialisation makes them less adaptable for saprophytic colonisation of dead substrates.

During active saprophytic colonisation by soil-inhabiting pathogens, the fungi may grow as mycelia obvious to the naked eye as in several species of *Rosellinia* or as individual hypha as in *Rhizoctonia solani* Kuhn, as observed by direct microscopic examination or by the use of Rossi-Cholodny slides. In such cases, a certain amount of food potential of the inoculum may be necessary to initiate saprophytic growth through the soil (Blair 1943). Species of *Fusarium,* however, do not make any outward mycelial growth through the soil from a food base, though they actively colonise any suitable substrate as demonstrated in *Fusarium solani* (Mart.) Appel & Wr. f. *dalbergiae* Gordon, causing wilt in sissoo (Bakshi 1955) where the fungus was shown to colonise dead sissoo roots buried in naturally infested soils, modifying the Cambridge method (Butler 1953, Lucas 1955). In such heavily sporing fungi such as Fusaria, it is suggested that the food reserves of the substrate mycelium are utilised in sporulation upon the substrate rather than in supplying outward mycelial growth away from it (Chesters 1948). Environmental conditions and competition from the population of microorganisms in the soil are likely to prevent outward mycelial growth of such soil-inhabiting fungi. The success in competitive saprophytic colonisation is innate or genetically determined characteristic of the fungus, like the pathogenicity of a parasite (Garrett 1960).

In root-inhabiting fungi, two kinds of saprophytic behaviour in nature may be distinguished, the saprophytic survival and saprophytic colonisation. All root-infecting fungi possess considerable power of saprophytic survival, in which case, the fungi persist in tree stumps and buried roots that they infect during the parasitic phase, as generally seen in case of root rot fungi like *Fomes annosus* (Fr.) Cke., *Ganoderma lucidum* (Leyss.) Karst., *Armillaria mellea* (Vahl.) Quel., *Fomes lignosus* (Klotzsch) Bres. and others. Furthermore, these fungi which parasitised roots of trees when living are able to spread quickly into healthy root tissues once the tree is felled and host resistance lost. However, some of these fungi have little free mycelial spread in the soil, possibly due to antagonism of soil microorganisms. Braun (1958) presented results of his experiments in the laboratory where he showed that *F. annosus* is unable to grow in non-sterile soils fortified with nutrients and with an adequate food base. The fungus, however, was able to spread in sterile needle humus. Braun concluded that the inability of *F. annosus* to grow in non-sterile soils was due to antagonistic effect of soil microorganisms rather than deficiency of nutrients. In acid soils, the dead pine roots are colonised by *Trichoderma viride* Pers. ex Fries and other root surface microflora preventing colonisation by *F. annosus,* while in alkaline soils, the roots are covered with the mycelium of *F. annosus*. This antagonistic effect of soil saprophytes inhibiting growth of *F. annosus* in pine roots, however, does not occur in California's forest soils which are acidic (Bega 1963).

PARASITIC PHASE

Pathogenic soil-inhabiting fungi

There is a fundamental difference in parasitic behaviour of the fungi belonging to soil-inhabiting and root-inhabiting groups. The degree of parasitism depends upon the tolerance of a parasite to host resistance. Based on this, Garrett (1960) built up a possible evolutionary sequence for soil-inhabiting parasites beginning with (1) primitive parasites that destroy seedlings and juvenile tissues of a wide host range but become restricted by mature host tissue resistance and can live saprophytically e.g., damping-off fungi. Such fungi kill the host tissue by their toxic excretions well ahead of actual penetration ; (2) less primitive parasites that are less restricted by effective host resistance ; (3) specialised parasites that attack host tissues and do not exhibit saprophytic activity other than passive survival in tissues invaded by them as parasites ; and finally (4) parasites, whose relationship with the hosts is one of symbiosis, e.g., mycorrhizal fungi. This evolutionary sequence does not necessarily imply inverse relationship between parasitic specialisation and saprophytic ability. A facultative parasite which has acquired the ability to parasitise a host plant is not less of a saprophyte than the so-called obligate saprophytes (Sadasivan and Subramanian 1960). The fungi causing vascular wilts restrict themselves in the vascular tract where they survive and multiply. With the possible exception of the penetration phase, the fungi lead a semi-saprophytic existence and are, therefore, primitive parasites than the root rot fungi. The vascular fungi do not emerge from the vascular cylinder until the disease is in an advanced stage or until the death of the host plant is imminent. The vascular parasites such as species of *Verticillium* and *Fusarium* causing wilt belong to this group.

Vascular wilts: The soil-inhabiting fungi causing vascular wilts belong to the group of primitive parasites and in their parasitic phase they enter their hosts through young roots and rootlets and then restrict themselves in the vascular tract where they survive and multiply. They do not attack the living cells surrounding the vascular tracts until the death of the host. The pathogen impairs water conductivity of the conducting tissues.

Wilt is a systemic disease where the entire individual or its part exhibits wilting of the foliage in acropetal succession up the shoot. The leaves become yellow, lose turgidity and fall off. The affected plant or the branch soon dies. Dwarfing, stunting and necrosis may also occur. In all cases, however, a growing imbalance of water economy in affected plants is common to all wilt diseases. A characteristic symptom of several wilts is the vascular discolouration in the outer layers of the sap-wood, green streaks in *Verticillium* wilt of elms and also many hardwoods and pink stain in sissoo wilt. The vascular browning is ascribed as due to enzymatic action of the wilt fungus. Such discolouration may not be, however, diagnostic and cultural studies are necessary to diagnose the cause of wilt.

Two theories are advanced to explain the mechanism of wilt: (i) occlusion theory, which assumes plugging of vessels by the mycelium of the wilt fungus interfering with translocation of sap. Such physical plugging which though normally associated with the disease is not generally considered as a major cause of wilt; and (ii) toxin theory, which postulates production of plasma poison by the wilt fungus and these toxins destroy the osmotic function of living cells (Gaumann 1954, Dimond 1955), particularly of the foliage, due to which there is an excessive transpiration resulting in loss of turgor. The evidence of vascular occlusion in tomato wilt due to *Fusarium oxysporum* Fr. f. *lycopersici* (Sacc.) Snyder and Hansen was given by Ludwig (1952) who demonstrated that it is not the actual mycelium but a viscous transparent substance in the vessels that was responsible for the development of wilt. It is believed that in the xylem vessels, occluding blocks of pectin material form at the expense of the cell wall (Dimond 1955). Fusaric acid, alternic acid and lycomarasmin are some of the important wilt toxins that have been investigated by many workers (See reviews: Ludwig 1960, Sadasivan 1961).

Fusaric acid, a vivotoxin, is detected in soils and in the rhizosphere. A vivotoxin may be produced in the infected host by the pathogen or host or by both. In general, wilt diseases are favoured by high temperature (25°–30°C) and by acid soil conditions with a high nitrogen and a low potassium content. Occurrence of nematodes also increases the incidence of wilt disease. Soil microflora, especially in the rhizosphere of the host plants, exert a dynamic influence on root-infecting pathogens and this in turn is influenced by the host (Sadasivan 1961).

Pathogenic root-inhabiting fungi

Pathogenic root-inhabiting fungi are generally characterised by a continuous external mycelial spread over the host roots. This continuous external spread is named by Garrett (1960) as ectotrophic growth habit of root-inhabiting pathogenic fungi.

While invading the internal tissues of the host roots, the fungi of this group attack living tissues and spread externally over the underground parts of the host plant and often in advance of the internal spread and may sometimes spread freely through the soil as hyphae or hyphal strands or rhizomorphs. These fungi belong to specialised root-inhabiting parasites and root rots belong to this group.

Root rots : Root rot fungi generally possess a wide host range. Once the infection has established in the roots, its further spread depends on the host resistance. In the susceptible hosts, the pathogen may spread quickly into roots and cause disease. The symptoms of root rot attack, however, do not manifest till a large percentage of roots is attacked. In the less susceptible hosts, the fungi may colonise roots, but then the progress inside the roots is slow because a large mass of healthy roots react strongly against spread of infection. Such hosts, therefore, generally do not show visible symptoms of infection till most of the roots become attacked. In the natural forests, therefore, where resistant and susceptible species are in mixture, the fungi remain in an endemic state. When such natural forests are clearfelled for raising plantations, the host resistance is lost and the fungi on roots of susceptible hosts are able to spread quickly in dying and dead roots building a large inoculum potential. When trees are felled, the stumps and their roots remain alive for a considerable period. Gradual lowering of host resistance paves the way for colonisation of roots first for specialised parasites like root rot fungi followed by weak secondary parasites and finally for saprophytes. *Armillaria mellea* which is able to spread quickly on roots of felled trees becomes greatly restricted or is absent if the roots are killed more quickly by ring barking the trees when green (Leach 1937, 1939).

While the root rot fungi are considered to be specialised parasites, two groups may be distinguished: in one, the fungi cause typical root rot and do not attack the heartwood as in *Ganoderma lucidum*; in the other, the fungi may cause root and butt rot as in *Fomes annosus*, *Armillaria mellea* and *Polyporus schweinitzii* Fr. Due to strict parasitic habit of the former group, the fungi are unable to freshly colonise dead roots or stumps and are also unable to make any free mycelial spread in the soil, except on root surface or outside the root when in contact with a solid surface like roots of adjoining plant or stones. In forests, therefore, healthy trees become infected when their roots come in contact with a decayed root. In the second group, the fungi not only attack the living sapwood but also the heartwood as in case of *F. annosus*, *A. mellea*, *Peniophora rhizomorpho-sulphurea* Bakshi and Sujan Singh. In the last instance, the fungus is an active saprophytic coloniser of dead roots and stumps. Some fungi are able to colonise freshly cut stumps through basidiospores as demonstrated in *F. annosus*. In *F. lignosus*, stumps were believed to be colonised by basidiospores (Petch 1921) though Napper (1932) did not hold this view. This ability for colonisation, however, decreases markedly with time because of competition with other fungi, as shown for *F. annosus* (Rishbeth 1951). These fungi exhibit vigorous ectotrophic growth. The mycelium of *F. lignosus* can travel independently through the soil unattached to any root or dead wood, except of course, at starting point (Petch 1921), though such mycelium is unlikely to survive for long unless it soon meets fresh material from which it can derive nutrition. *Armillaria mellea* also spreads through the soil by subterranean rhizomorphs which are able to withstand adverse conditions, particularly desiccation, and survive for a long period. *Peniophora rhizomorpho-sulphurea* is able to spread in soil by long, spreading strands (Bakshi, Sujan Singh and Ujagar Singh 1966). These pathogens have, therefore, evolved a successful means of vegetative propagation by means of rhizomorphs, through which the fungi are able to spread from infected trees and stumps to adjoining healthy trees.

For both the groups as discussed above, the food base or the wood mass colonised by the fungus is important in root disease development in the forest. However, root rot is not exclusively due to the absolute value of infection foci mass but to the ratio between infection foci mass on one side, and the host plant root mass, the plant's vital reaction and the distance between parasite and host plant on the other (Fassi 1964).

CONTROL

The control of root diseases generally is more difficult because of the difficulty of combating the disease on underground plant parts which are not easily accessible as compared to diseases attacking aerial plant parts. In the latter case, the control may be possible through use of chemical fungicides or by variation in the planting time to avoid epiphytotic conditions or through sanitation measures. However, chemical control of root diseases with fungicides applied to soil may not be effective as these chemicals may not reach the affected parts and also they may soon become inactivated due to reaction with other chemicals present in the soil or through leaching. Any soil fumigant intended to kill the ectotrophic growth of root-inhabiting fungi may be toxic generally to the host also, though fumigation with methyl bromide and steam are commonly practised under glass house conditions, prior to sowing to control damping-off and other root diseases.

Crop rotation is an old and effective method of root disease control in agricultural crops particularly for diseases which have got limited host range. In absence of the host, the population of such fungi tends to decrease below the limit at which the disease is economically dangerous. Therefore, keeping such land free from particular susceptible host for that period is an effective method of control of root diseases. Other effective methods of root disease control are plant sanitation and through various soil amendments.

Difficulty in the control of root diseases of forest crop arises because of the long rotation age of such crops. Crop rotation as practised for agricultural crops is, therefore, not generally feasible in forestry practice. In the natural forest where the species susceptible to a root disease fungus is growing with other species which are resistant, the disease does not normally assume alarming proportion. This phenomenon is observed in mixed plantations of susceptible and resistant species, where root disease remains very much in check. However, in pure plantations, once the root disease breaks out, the spread is very rapid from tree to tree and, therefore, within a short time the disease becomes alarming. In the present trend in forestry, preferred species are grown as a pure crop in plantations. If a root disease is anticipated, all care must be taken to avoid the disease appearing in the plantation for which a knowledge of the etiology and epidemiology of the disease is important. Forest root disease control is best achieved through a proper understanding of the predisposing factors which lead to the disease outbreak and eliminating such factors prior to raising forests or through proper management practices while the forest is growing.

SILVICULTURE AND MANAGEMENT

Suitable site

For devising successful control measures for root diseases, a thorough knowledge of silviculture of the species is essential. Tree vigour is important in determining the susceptibility or otherwise of a species to root diseases and this may be obtained by raising forest species on sites favourable for their healthy growth. For example, sissoo, a plantation species, grows healthy free from root diseases on sandy loam soils with good drainage but becomes diseased due to various root-infecting fungi when raised in stiff soils of clay or clay loam or in areas where water-logging occurs even for a short duration during the year (Streets 1962). Therefore, the bhabar tract where water-table is low is generally suitable for raising sissoo plantations, while the tarai region which becomes waterlogged during the rains generally appears to be unsuitable for raising sissoo (see Chapter 14 for description of bhabar and tarai).

Control burning

Control burning may inhibit development of root diseases. *Polyporus shoreae* Wakef., the cause of root rot of sal, is particularly damaging in wet sal forests under fire protection measures in early age. High rainfall favours weed growth and high soil moisture conditions creating ideal predisposing factors for the root rot fungus to become epidemic. Control burning, a necessary silvicultural measure in wet sal, keeps the root rot in check (Bakshi and Boyce 1959). In the southern United States, the common management practice of burning before thinning has been shown to reduce the incidence of *F. annosus* in the residual stand (Anon. 1970). Burning results in the increase of *Trichoderma* in the soil.

Site preparation

In India, large areas of derelict miscellaneous natural hardwood forests are clear-felled to raise plantations of economic species. Preparation of the site prior to reforesting the area varies from removal of residual roots and stumps over the entire area with simultaneous soil working as in mechanised plantations to their removal along lines in line sowings by mechanical extraction of stumps and larger roots. In the latter case, roots of stumps lying deep in the soil or in between lines may not be removed. In other cases, it may not be possible to raise mechanised plantations on a bouldery site or a hilly terrain. *Ganoderma lucidum* which remains endemic in the natural forests builds up a high inoculum potential on residual roots and stumps when such forests are clear-felled and attacks a susceptible species planted on the site. In mechanised plantations, however, the infected roots are mostly removed due to which the reforested stands grow healthy free from root rot. In afforested stands, root rots are not important in the first rotation or in case of *Fomes annosus* until thinning, when the pathogen colonises the stumps and gradually builds up inoculum potential to subsequently infect the standing crop. Root rots due to *Fomes annosus, Armillaria mellea, Poria weirii* Murr., however, become important in second growth stands, where the stumps and roots of the previous stand are not cleared.

Resistant species and mixed stands

In reforested stands, if threat due to root disease is suspected, planting resistant species in the first rotation is advisable. By the end of the rotation, the decay fungus present in old stumps and roots is expected to decline because the available food base being limited, will be utilised and disintegrated. In the second rotation, the susceptible hosts could, therefore, be planted. In case a species susceptible to root rot is raised in the first rotation, it should be planted in intimate mixture with resistant species which will serve as barriers to rapid spread of the disease.

Isolation trenches

Root diseases generally spread in plantations from root to root by contact or root graft resulting in diseased plants occurring in groups. Isolation trenches may prove effective in containing the disease in between the trenches thus preventing its spread beyond the trenches.

Other management practices recommended generally for root disease control include wider spacing to minimise the chances of root contact. For control of *Fomes annosus* root rot, delayed thinning or its exclusion or adjusting time of thinning during a period when temperature at the stump surface reaches levels lethal to *F. annosus* are suggested (Anon. 1970).

CHEMICAL CONTROL

Colonisation of virgin woody substrates by root rot fungi may be prevented by use of chemicals. An ideal stump treatment should be effective not only in preventing infection at the stump surface, but in killing the stumps rapidly allowing colonisation by organisms antagonistic to the root rot pathogen. The chemical should be easily available, cheap, easy to apply, non-poisonous, particularly to live stock and wild life. Stump treatment has been successfully evolved to control colonisation by *F. annosus*. Among chemicals tested and recommended for use are creosote, ammonium fluoride, sodium nitrite, borax, urea and ammonium sulphamate. In Great Britain, it has been the practice to paint the surface of stumps with a protectant to prevent stump colonisation by *F. annosus*. Creosote which was earlier used (Rishbeth 1952) is variable in performance and is liable to breakdown. This is now replaced by sodium nitrite in Forestry Commission plantations (Greig and Burdekin 1970), though it is somewhat toxic to animals.

BIOLOGICAL CONTROL

Biological control of soil-borne pathogens has not been promising because the success depends upon complex interactions in a medium which is already colonised by a variety of microorganisms. The wood-inhabiting pathogens, on the other hand, colonise woody tissues which are virgin substrates, at any rate during the early stages, and also are highly selective and therefore, opportunities for biological control may be greater. This is exemplified in the biological control of *Armillaria mellea* (Leach 1939) by ring barking of trees sometime prior to felling after which the roots are not colonised by *A. mellea* but by fungi innocuous to plantation crops. Root rot due to *Fomes annosus* is severe in plantations soon after thinning because the pathogen is able to colonise freshly cut stumps. *Peniophora gigantea* (Fr.) Massee is a vigorous competitor of *F. annosus* for initial colonisation of stumps and may also replace *F. annosus* in stumps (Meredith 1960, Rishbeth 1963). Inoculation of pine stumps with *P. gigantea* soon after felling by oidia of the fungus gives satisfactory protection against *F. annosus*. A method is developed (Rishbeth 1963) where tablets containing oidia of *P. gigantea* are dissolved in water and spread over freshly cut pine stumps. *P. gigantea*, however, does not colonise spruce stumps well.

7. HEART ROTS

Among the different causes responsible for loss to the standing crop in forests, decay is by far the most important. In the U.S.A. timber trend study (Hepting and Jemison 1958), decay accounted for nearly 80 per cent of the loss through diseases. In an assessment of sal (*Shorea robusta* Gaertn. f.) forests in India, about 73 per cent trees exhibited decay resulting in 10 per cent loss of timber (Bakshi, Rehill and Choudhury 1963). The actual loss may be greater because decay in the tree, particularly at short intervals along the bole, will prevent sampling of requisite lengths of timber for the purpose of beams and railway sleepers for which sal timber is preferred.

In the tree, decay does not set in till the heartwood is formed and this usually occurs between 15-30 years of age depending on the species. The central core of dead heartwood is protected from attack by decay fungi by the outer living sapwood and the bark. However, when the heartwood becomes exposed through openings in the bark and sapwood, decay may establish in the heartwood as heart rot which becomes progressive with age. Since decay occurs in the heartwood which is dead and the living wood remains mostly unaffected, trees are not killed but continue to grow and may present all the outward appearance of a healthy tree and often complete the rotation age. Literature on heart rots in trees has been reviewed by Wagener and Davidson (1954) and recently by Bakshi and Sujan Singh (1970).

DECAY FUNGI

There are many fungi causing decay in the heartwood of standing trees. These fungi mostly belong to the Hymenomycetes of which the genera under Polyporaceae are important. Decay fungi belonging to Polyporaceae attacking different forest trees in India are described by Bakshi (1971) who also listed important decay fungi around the world (Bakshi 1964 a). Similar contributions from other countries include those of Baxter (1943), Boyce (1961) and Overholts (1953) in America; Cartwright and Findlay (1958) in the U.K.; Pinto Lopes (1952) in

Portugal; Benito Martinez (1952) in Spain; Bondartsev (1953) in Russia; Imazeki and Hongo (1957) in Japan; Cunningham (1965) in New Zealand; Cleland (1935) in Australia among others. Sporophores develop diagnostic characters which aid in identification of species. In absence of sporophores, the fungi may also be identified from the culture (Bakshi, Sehgal and Balwant Singh 1969, Bakshi, Sen and Balwant Singh 1970, Sen 1973, Cartwright and Findlay 1958, Nobles 1948, 1965). The types of decay in the wood, whether white rot or brown rot, are also characteristic of the species and may be determined in culture from chemical tests (Bavendamm 1928, Nobles 1958, Preston and Mclennan 1948). Cardinal temperatures for growth have been determined for decay fungi in culture (Sehgal, Sen and Bakshi 1966).

INFECTION COURTS AND ESTABLISHMENT OF DECAY FUNGI

Tree wounds due to various causes are the principal infection courts for trunk and basal rots. In root and butt rots, fungi enter through either of the two regions and spread from one into the other. At the infection court, moisture, temperature, nutrients, microflora and natural extractives determine the establishment of decay fungi (Merill 1970). High moisture in the sapwood prevents heart rot fungi to progress in the sapwood, at any rate, to any appreciable degree. In the heartwood, the moisture is such that decay fungi are able to establish and progress. The cardinal temperatures for growth of fungi in the wood are approximately the same as those on agar (Cartwright and Findlay 1958).

When the wood of a living tree is freshly exposed following an injury, chemical and biological changes occur. Chemical changes result in formation of substances, especially phenolic compounds that afford protection against invading microorganisms. The tissues around wounds containing phenolic compounds are termed protection wood (Jorgensen 1962). Only a limited number of microorganisms can tolerate the active inhibitory principles. In successful infections, a succession of micro-organisms occurs. Initially, bacteria and non-decay fungi become active on the wound usually causing discolouration which may advance inward towards the pith and around the tree but seldom affects new tissues formed after wounding. The primary colonisers suitably change the wood substrate for further colonisation by other microoganisms or decay fungi. The succession of organisms continues till the tissues are completely digested. However, the succession may not go through to completion and may stop at any stage depending on the healing of wound, colonisation by antagonistic microorganisms that preclude infection of heart rot fungi and unfavourable environment. In a typical attack, the discoloured zone of wood forms a column extending up and down the tree. Each such column is associated with one injury and therefore, when injuries occur on the tree at different times, multiple columns of discoloured wood develop. Decay occurs within the discoloured columns. Discolouration and decay are compartmentalised within the tissues present at the time of wounding (Shigo 1972) and seldom move into the new wood formed after the injury. The decay columns are separated by a narrow band of discoloured wood which remains as a hard black rim lining the core of healthy wood in advanced stages of degradation when decayed wood becomes digested resulting in hollow cores.

Toxic natural extractives which impart resistance to decay are formed during development of the heartwood and these substances are mostly deposited in the cell wall. There is a great variation in decay resistance of wood in different timber species. Within a species, individual trees may vary considerably in decay resistance. This variation is mainly controlled genetically and is also due to difference in growth rate, age and size of the tree. In the same tree, there is radial gradient in decay resistance which decreases progressively from the outer heartwood to the inner. Vertically, the decay resistance of the outer heartwood becomes progressively less from base of the tree upwards. The reverse is true of the inner heartwood. The chemical nature of extractives, most of which are phenols, is reviewed by Scheffer and Cowling (1966) who also listed different heartwood extractives occurring in different tree species. Some important extractives are thujaplicin in western red cedar, lepachonon occurring in several species, tectol in teak, sequirins in Californian redwood and quinones, coumarins and phenols in other species. These toxic chemicals do not, however, provide protection indefinitely but become inactivated, possibly catalysed by oxidative enzymes through auto-oxidation, or due to slow chemical changes such as polymerisation, oxidation or acid hydrolysis (Scheffer and Hopp 1949). Degradation of heartwood extractives can also occur through action of microorganisms as stated earlier.

PROGRESS OF DECAY IN TREES

IN INDIVIDUAL TREES

Once the decay is established, it becomes progressive in the tree. The rate of spread of the decay depends on the host and the fungus. The characteristics of the host include wood properties and internal conditions of the tree such as composition of gases, moisture and temperature (Wagener and Davidson 1954). The rate may be determined either from natural infections or through inoculations.

IN STANDS

Progress of heart rot in a stand may be affected by the age of the stand, environmental conditions and stand vigour.

Age

While the tree is vulnerable to attack by heart rot fungi as soon as the heartwood develops, the establishment of infection depends on the age at which infection courts develop. Most infection courts, like those caused by fire, frost and pruning develop at particular times throughout the stand. For example, sal is frost tender when young and frost causes cankers through which *Fomes caryophylli* (Rac.) Bres. establishes. Incidence of infection due to this fungus is, therefore, high in young stands and sharply declines with age. On the other hand, the incidence of infection in sal due to *Hymenochaete rubiginosa* (Schrad.) Lev. is low at young age but increases with age of stand when fire wounds, the most important infection court for the fungus, are common (Bakshi, Rehill and Choudhury 1963). In any case, the volume of decay in a stand generally increases with age. In an overmature stand where no new wood is normally added and decay, if present, becomes progressive, the total decay becomes high and the retention of such a stand becomes unprofitable.

Site

There is no direct correlation between site and decay in the tree. Site may indirectly affect the incidence of infection by increasing the period of susceptibility through a prolonged persistence of infection courts on the tree, like the slow healing of wounds and branch stubs on the less vigorous trees on poor sites. Site has a direct influence on infections due to root and butt rot fungi, like *Fomes annosus* which is not serious on former moorland and pasture land (Peace 1938) but often becomes serious in plantations on land that has been previously cultivated or has carried a forest crop (Peace 1938). Infection due to *Armillaria mellea* (Vahl.) Quel. is also serious in plantations on sites previously occupied by broadleaved species (Ono 1965), but development of *Fomes annosus* on such sites is slower than in second rotation conifers. In general, diseases due to *A. mellea* and *F. annosus* develop from the unimportant latent stage in natural forests to dangerous levels under intensive management (Van Vloten 1962).

The decay volume per cent in the tree increases more rapidly on poor sites than on good sites. In the latter, trees may become infected more readily and may have higher decay cull than those on poor sites, but higher productivity on better sites more than compensates for the higher decay volume on such sites.

Soil moisture. The site factor expressed as available soil moisture has a greater effect on cull in conifers than in hardwoods (Morawski, Basham and Turner 1958). Generally, dry sites are more favourable for progress of heart rots than wet sites (Boyce and Wagg 1953, Basham 1958, Etheridge 1958), though higher incidence of heart rot on mesic sites is reported in *Populus tremuloides* Mickx. (Thomas, Etheridge and Paul 1960) and *Picea* spp. (Etheridge 1956).

Soil texture. Soil texture may also influence the incidence of heart rot. In *Picea*, decay is greater on clayey soils and less on sandy soils, whereas the reverse is reported in *Larix* (Peace 1938). Stands of *Populus tremuloides* on deep sandy silt or loam or on shallow sandy loam over impervious material were generally less defective than those on drier or wetter sites (Basham 1958).

Soil reaction. Infection of *Pinus elliottii* Engelm. and *P. taeda* Linn. due to *F. annosus* increased with increasing pH and sand or clay content of the soil and decreased with increasing organic matter content and grass cover (Froelich, Dell and Walkinshaw 1965). In *Picea*, the damage due to *Fomes annosus* was greater on limed than unlimed plots (Dimitri 1966). Mortality of pines in East Anglia due to *F. annosus* could be closely correlated with soil reaction in the major lateral rooting zone, extensive damage occurring only on pH 6 (Wallis 1960). Losses were heavy only on former agricultural land where liming had raised the pH above 6 or in calcareous grass heath with a naturally high pH (Dimitri 1963) though in spruce, attack of *F. annosus* was favoured by acid soil rich in humus (Comic 1957).

Stand vigour

Within the same age class, tree vigour may affect decay volume. The incidence of infection may not be affected by tree vigour, although slow growth may increase susceptibility (Gruschow and Trousdell 1958). In sal, the volume of decay was greater in slow-growing trees than in fast-growing ones (Bakshi, Rehill and Choudhury 1963). In *Pinus strobus* Linn., the incidence of decay in vigorous and non-vigorous trees was about the same, but percentage of volume lost through decay was considerably higher in slow-growing and less vigorous trees (White 1953). Such a difference was, however, not observed in *Picea sitchensis* (Bong.) Carr. (Bier, Foster and Salisbury 1946) and in *Abies* spp. (Bier, Salisbury and Waldie 1948).

DETECTION OF HEART ROTS

Heart rot causes progressive decay and, therefore, the volume of decay increases in the tree and in the stand with age. Also, decayed trees occupy a space that could more profitably be utilised by adjoining healthy trees or by replacements. Furthermore, diseased trees constitute a source of infecton to healthy trees. Early detection of decayed trees in the forest is, therefore, necessary so that the maximum timber may be salvaged from them and the forest may be protected from excessive decay losses. Heart rots in trees may be detected either from external symptoms manifested by diseased trees or through direct probing into the heartwood.

EXTERNAL INDICATORS

Sporophores

The presence of sporophores of heart rot fungi on the tree is a sure indication of decay. Some fungi develop sporophores abundantly soon after the decay is established in the heartwood, while others develop sporophores, less commonly or very rarely, even when decay is extensive. Detection of such trees may become difficult unless they develop other external symptoms.

Punk knots

Punk knots are cylindrical cores of soft punky tissue of the wood decayed by the fungus, embedded in the sapwood upto different depths and opening up on the bark as round or oval hard structures. Their presence indicates heart rot in all cases. Decays due to *Fomes pini* (Thore ex Pers.) Lloyd and *Fomes caryophylli* are characterised by punk knots.

Swollen knots

Ordinarily, a knot indicates the position of the base of a branch that has become embedded in the wood during growth of the tree. Swollen knots are formed when living wood tries to grow over a knot in which a sporophore is forming (Boyce 1961). Such knots are reliable indicators of decay (Thomas and Thomas 1954, Foster 1954, Boyce 1961). In sal, a protruding branch stub may be callused over leaving a slight but perceptible swelling on the bark, like a swollen knot. Heart rot fungi may enter through such branch stubs before they are callused over. However, if such knots are callused before infection can establish, the knots remain free from infection. Bakshi, Rehill and Choudhury (1963) observed that such callused knots are reliable indicators of decay at 0.1% level of probability and recommended that callused knots be used as indicators of heart rot in sal only on a percentage basis.

Swollen boles

Advanced decay in the tree is usually evident as perceptible swelling of the bole in the region of decay. Swollen bole is a reliable indicator of decay. In sal, it indicates decay in more than 80 per cent cases (Bakshi, Rehill and Choudhury 1963). In bottomland hardwoods in the southern United States, swollen bole indicates decay in trees (Toole 1959 a).

Other indicators include resin flow, crown dying, forking, and cracks in the bark and protuberances from it (Yokota 1957, Foster, Craig and Wallis 1954, Foster 1954). However, these symptoms do not indicate decay in all cases. Some heart rot fungi produce cankers prior to establishment of heart rots (Toole 1956, 1959).

External injuries

Openings and scars may indicate former infection courts and may, therefore, serve as indicators. Among these can be included broken tops, basal and trunk scars due to logging or fire injuries and wounds caused by lightning. All these can be used on a percentage basis to detect heart rot.

Direct probing

Heart rot may be detected by taking increment cores from trees suspected to contain decay and isolate the fungus in culture from the decayed wood.

In advanced decay, trees may give a hollow sound if struck on the bark with a hammer or back of an axe at the region of decay. It may be necessary to remove the bark, particularly if thick, in order to obtain the correct sound.

Zycha and Dimitri (1962) used a device for determining hidden decay in trees by forcing a steel needle into the stem by means of a spring. The ease of needle penetration in relation to depth indicates decay values which are recorded on a chart. Iizuka (1956) devised an apparatus using the radioactive isotope Co^{60} and a portable counter to detect the existence and extent of hidden decay. Waid and Woodman (1957) used an ultrasonic equipment for detecting rot.

Eslyn (1959) reviewed the use of X-ray photography to determine the extent of decay and successfully used a thulium X-ray unit to detect the presence of cavities and advanced decay in trees. Densitometric measurements of film are more objective than methods requiring visual examination. However, in the X-ray method, decayed areas, if present in the vicinity of involutions and flat faces, are masked on the film. Also, decayed areas saturated with water are difficult to determine due to greater absorption of X-rays by rots. Lange (1959) reviewed methods of diagnosing stem rot in standing trees. Cervinkova and Temmlova (1966) described a method for diagnosing decay on the basis of the coefficient of thickening of the butt which has a positive correlation with decay. Shigo (1973) described an electric device for quantifying progressive stages of discolouration and decay in wood by measuring the resistance to a pulsed current which decreases with increase in concentration of the cation due to discolouration and decay in wood.

ESTIMATION OF DECAY IN TREES

Estimation of decay in individual trees and in the stand is important for determining the net volume of merchantable timber. Such data is useful for the timber industry because decay not only reduces the volume of sound wood but also increases logging and milling costs. Therefore, an accurate estimate of decay in a tree is important to assign it to the correct quality class during marking prior to felling. This is also necessary for the proper management of the forest and to fair and honest deal in timber sale contracts.

A tree or its part which is rendered unmerchantable because of defects is termed as cull which includes all defects such as fire scars, shakes and others, in addition to decay which is the major contributory factor for cull. The amount of cull expressed as a percentage of the gross volume of a tree or stand is termed as cull factor. When the cull factor is applied to individual tree volumes where the cull is associated with specific indicators, it is called indicator factor. For this factor to be most useful, the average incidence and volume of cull associated with any indicator must be known from previous estimates. Estimation of cull based on indicator factors applied to individual tree volume is more accurate particularly when it is done over greater part of the stand rather than for single trees or groups of trees. Cull factor applied to groups of trees or stand volume regardless of indicators is called a flat or general factor (Kimmey 1950, Kimmey and Hornibrook 1952). Flat factors are the simplest to apply but are likely to be least accurate especially on small areas. However, flat factors are satisfactory for tree species having small percentages of cull and those which do not possess suitable indicators upon which to base indicator factors. Even where indicator factors are used to estimate cull in a species, supplementary flat factors are usually applied for those trees without indications of cull.

Cull factors are invaluable in forest inventories because accuracy in estimate of merchantable timber determines the usefulness of the inventory. The success of obtaining sustained yield from a timber stand depends upon the accuracy of the estimate of net volume of merchantable timber that may be harvested annually. An accurate estimate of the present and also the future cull will influence the allowable cut.

CONTROL OF HEART ROTS

In heart rots, the decay volume increases with age. Therefore, the greatest damage due to heart rots occurs in mature and overmature virgin forests. Major losses from heart rots in managed stands may be prevented or reduced to an acceptable level through management practices and silvicultural measures. However, some decay will occur in any timber stand because some injuries, such as those from natural causes, are unavoidable. A forest stand has too low a value per unit area to justify treatment of individual trees. Control measures should, therefore, be applied on a stand basis. Shade and ornamental trees in parks and avenues are valuable individually and, therefore, measures to minimise damage in any tree due to heart rot are economically justifiable in such cases.

The management and silvicultural systems under which a forest crop is grown largely determine the nature of injury to trees, and development of infection courts through which heart rots may establish. Control of heart rots is effected by regulating the silvicultural systems and management practices as discussed below.

IMPROVEMENT IN STAND MANAGEMENT

Any recommendation concerning management of forests in order to minimise damage due to heart rot must take into consideration the silvicultural characteristics of the tree species. Fire is the most important cause of tree wounds and associated infection courts for decay fungi and should, therefore, be kept out of forests unless necessitated in the management of a species, as in moist and wet sal forests (Champion 1933). In dry sal forests, early control burning minimises the risk of severe fire injury in fire season (Bakshi 1957 c). In selection forests, suppression may occur in a dense stand leading to dead tops and branches and branch stubs, all forming infection courts for decay fungi. In open stands in frosty locations, frost-tender species may suffer from frost damage and decay fungi may become established through frost injury (Bagchee 1961). Adjustment of the stand canopy and density is, therefore, necessary to minimise both suppression and frost damage.

Thinnings and improvement fellings remove diseased, forked, malformed and injured trees and also those bearing sporophores of decay fungi thus leading to stand improvement. No useful purpose is served in retaining decayed trees in the stand because (1) more and more timber becomes decayed with age; (2) decayed trees occupy space that may profitably be taken up by healthy trees ; and (3) the sporophores developing on decayed trees release spores which serve as inoculum for new infections. Based on these considerations, Bakshi (1957 a) recommended removal of green but decayed trees of khair, *Acacia catechu* Willd., below the exploitable diameter and also removal of sporophores from trees.

Fomes annosus, the cause of butt rot (and also root rot) in conifers, becomes established through colonisation of stumps. This may be prevented either by stump treatment with chemicals or through prior colonisation of stumps by some innocuous fungus (Chapter 6).

During marking, the two stems of a forked tree are considered as two trees if forking occurs below about 1 m. Above this height, they are considered as one tree. In the former case, it is usual to mark one of them for felling, leaving the other for future working. If decay is present at the basal region, it is likely to be present in both the stems; else, such decay is likely to progress into the tree that is left uncut in course of time. Therefore, if any rot is detected when one stem is cut, the companion stem should also be felled.

DECAY IN RELATION TO SILVICULTURAL SYSTEMS

Even-aged stands are more susceptible to damage by snow, frost, and high winds than uneven-aged stands. However, in the former, the trees do not normally suffer from suppression, and they develop straight clean boles from which the lower branches become pruned naturally. Even-aged stands may be cut before decay becomes serious. Many plantations, particularly industrial ones, are raised over short rotations during which period little or no heartwood is developed. This, together with control over rotation periods, and the fact that managed even-aged stands are less subject to injuries through which heart rot fungi could establish, accounts for considerably smaller losses due to heart rot in man-made forests as compared to natural stands (Bakshi 1967).

In uneven-aged forests, as in the selection system, trees of the lower storey may suffer from suppression and associated deformities in stems, and may have dead tops, dead branches, and branch stubs that provide infection courts for heart rot fungi. Such forests, however, are subject to thinnings and improvement fellings when trees having abnormalities or deformities, and also those bearing sporophores of decay fungi may be removed. However, injury to residual trees in the stand may occur during selection cuttings. Utmost care should be taken to minimise such injury. Wounds on residual trees may be protected by wound dressing chemicals. If trees have large, thick branches, these should be lopped before felling to minimise damage to the residual crop. All trees seriously damaged during felling operations should be marked and removed. This should include secondary species in the stand which may be uneconomic to extract but which serve as additional hosts for heart rot fungi.

In both selection and clear cuttings, a great deal of slash is left in the forest. Decay fungi may develop on slash and produce sporophores, thus increasing the supply of inoculum for infection of the standing crop. Slash should be burnt after clear cutting.

The progress of decay in the initial stages is slow whereas the formation of new heartwood is relatively rapid. Therefore, the net increment of sound wood is greater than the loss through decay. As the trees approach the normal rotation age, the volume of new wood added every year slows down. However, if heart rots are present, the volume of wood decayed increases rapidly with age and surpasses the yearly increment of new wood. Thus, net mean annual increment of sound wood becomes negative and the stand becomes increasingly uneconomic. Therefore, if there is a high incidence of heart rots, it is inadvisable to allow the stand to complete its normal rotation age. Instead, a pathological rotation, as indicated by net mean annual increment, should be prescribed when the maximum volume of sound wood may be salvaged profitably.

In coppice forests, decay is generally present in the stool due mainly to injury during coppicing. Coppice shoots may originate in two ways, those arising out of callus growth at the cut end of the stool (callus shoots) and others developing from buds on the sides at various levels, either above or below ground. Callus shoots are organically connected with the stool and, therefore, decay present in the latter eventually migrates into callus shoots when heartwood develops in them. Coppice shoots arising from buds, termed side shoots, develop their own root system in course of time and may grow into independent plants free from heart rot even though decay may be present in the stool. Management of coppice forests where there is a high incidence of decay in the stools may be practised by gradual change over to seedling crop and preventing further coppicing of stumps after the seedling crop is established (Bakshi 1957 c), by girdling of stumps and killing of tissues from which sprouts arise with biocides (Bakshi 1957 c), or if this is not feasible, by mechanical extraction of stumps where such a method is economically practicable, or by raising coppice crop of shoots arising out of buds at or below ground level as demonstrated for teak (Chapter16).

8. PATHOLOGICAL PROBLEMS IN URBAN AND FARM FORESTS

TREES IN URBAN AREAS

With increase in urbanization throughout the world, there is an emerging awareness of the value of urban forests for improved environmental conditions. Trees located near human habitations provide shade and shelter against sun and wind, yield timber and fruit in orchards, serve as a screen for privacy and camouflaging, provide an aesthetic beauty to the landscape, play a role of a natural filter purifying the air from dust and in effectively changing the concentration of harmful gases. Tree species under forestry and urban conditions are the same but those in the latter attain high economic and aesthetic values.

In urban areas, the original ecosystem near large human populations is so disrupted that it may be considered degraded (Wilson 1975). In an urban ecosystem, recycling of essential nutrients and energy obtainable in a balanced ecosystem is curtailed. Aeration and water supply to roots are severely restricted due to buildings, pavements and soil compactness. Leaf litter is removed instead of allowing it to decay and replenish the soil. In heavily industrialized areas, damage to trees occurs from various air pollutants. Physiology of the tree becomes seriously disturbed. Mechanical damage to trees occurs during construction work, movement of vehicles, browsing by animals or deliberate wounding by man. All these causes result in physiological disorders or in infection by pathogens causing death of branches or the tree.

TREES IN AVENUES

In urban areas, plants need protection at all stages of growth from human interference and browsing by cattle, the two factors that are common to trees along rail roads or canal banks. Injury to roots and stem bases results in infection by root-infecting fungi like *Ganoderma lucidum* (Leyss.) Karst. In India, avenue trees are usually numbered after blazing the bark (Pl. 5, 1). Eventually the blaze is overgrown by callus growth and a fresh blaze is made. Apart from disfiguring the tree, the injury exposes it to attack by decay fungi like *Fomes senex* (Nees and Mont.) Cke., *Ganoderma applanatum* (Pers. ex. Wallr.) Pat., and *Polyporus sulphureus* Bull. ex. Fr. on toon (*Toona ciliata* Roem.) in avenues. As the trees grow, lower branches are generally lopped to obtain a clear avenue or for making way for overhead communication wires. Decay fungi establish through resultant injury. Decay in the heartwood does not kill the tree but results in snapping of branches and stems (Pl. 5, 2) and uprooting of trees during high wind involving accidents and risks to life. Numbering should be done on metal plates nailed on the tree at a height at which they cannot be tampered. In lopping branches, clean cuts close to the stem should be given, without leaving splinters, to permit rapid healing. The cut surface may be immediately coated with an antiseptic dressing as a safeguard against infection by decay fungi (Bakshi 1964).

TREES IN FARM LAND

In the rural economy, trees provide fruits, fodder and various minor forest produce and act as shelter belts besides other beneficial attributes of trees in urban forests. On maturity, trees may be felled for timber and fuel wood. Illegal encroachment of reserved and protected forests, situated particularly in areas close to villages and towns, by the local inhabitants to meet their need from forests, is inevitable if they are not provided with an alternative. Therefore, there is an imperative need for farm forestry or growing trees in or around farm lands for the benefit of farm and the farmer in order that reserved forests are not subjected to maltreatments (Bakshi 1958). Proper land use, choice of species and method of working farm forests need due consideration in this respect. In farm lands, injury to trees which are raised along with the crop becomes inevitable during different stages of farming practices leading to infection by root disease fungi.

In parts of western Himalayas, villagers are granted the right of controlled lopping of blue pine (*Pinus wallichiana* A.B. Jackson), whose soft needles are used for spreading on the floor of cattle sheds to be later removed along with animal refuse for making compost. Blue pine growing on farm land is heavily lopped to clear the surrounding land from shade. The concession of controlled lopping of blue pine is

PLATE 5

DECAY IN TREES IN AVENUES AND PARKS

1. Avenue of *Toona ciliata* with sporophores of *Fomes senex* on blazed bark and lopped branches. 2. *Cassia nodosa*, tree snapped due to decay by *Polyporus palustris*.

PHYSIOLOGICAL DISORDERS

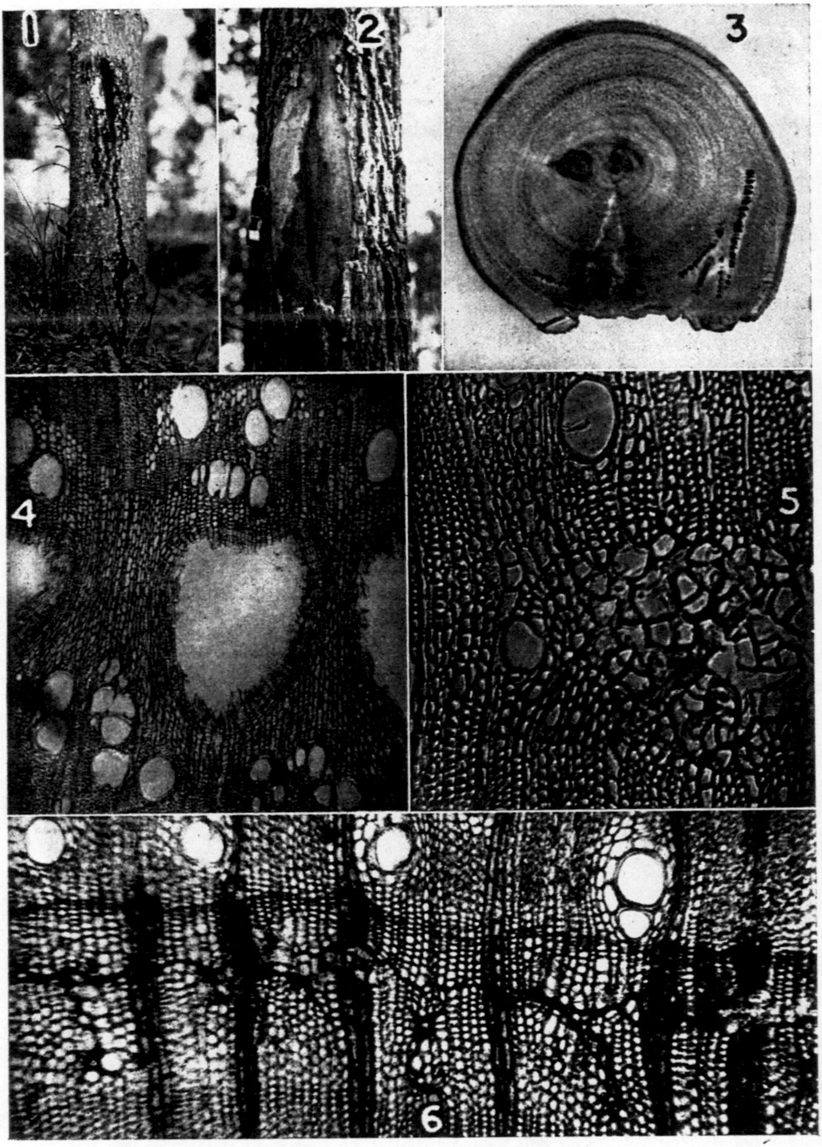

1–2. *Populus deltoides* : 1. Bark burst. 2. Canker on bark burst due to *Botryodiplodia palmarum*. 3–5. *Eucalyptus citriodora*, frost canker in wood : 3. Cankered wood with kino ducts. 4. Undifferentiated parenchyma and gum veins in wood. 5. False ring showing undifferentiated parenchyma and deflected medullary rays in wood 6. *Acacia mearnsii* showing healthy autumn wood and frost ring with disorganised parenchyma in spring wood.

misused in most cases with the result that trees stand like poles. (Pl. 17, 3 What is more disastrous is that *Fomes pini* (Thore ex Pers.) Lloyd enters through lopping injury. Such lopping of branches and leaves for use as green manure and also as fodder for cattle is practised in many other parts of India. This results in extensive damage to trees and makes them liable to attack by diseases. Such practices can be restricted if grazing pastures are provided for the cattle and growing farm forests is encouraged for providing fuel wood (Bakshi 1964).

TREES IN INDUSTRIAL AREAS

Besides pollution of water and soil, air pollution with dusts and gases emitted by factories and automobiles is becoming increasingly important with growing industrialization. Pollution damage to trees along railway lines occurs from smoke emanating from locomotives. While trees can absorb pollutants and thereby improve the environment, they in turn may become affected by the pollutants. Dusts of various chemical compositions accumulate on leaves as a thick plaster, thus making penetration of sunlight and exchange of gases difficult. Physiological processes are hampered resulting in premature leaf fall. Sulphur dioxide, the main pollutant of air formed during burning of coal in the industry, is absorbed in leaves in the form of sulphates and in this process, causes damage to tissues and mortality in trees. While industrial pollution may be minimised through improved technology, selection of suitable species resistant to such damage will enable raising a green tree belt around industrial centres for reducing pollution hazards.

Forest pathology problems in urban and amenity forests where trees are growing under continuous stress deserve special attention. Maltreatment of trees in urban areas is not always done with any malicious intent but due to lack of tree sense which can be aroused in the people through education for the larger interest of the community. The *Vana-mahotsava* or the annual national tree festival in India helps to arouse tree sense in the people.

9. PHYSIOLOGICAL DISORDERS

Diseases due to non-parasitic or physiological agencies do not normally cause damage of great magnitude to forests as compared to parasitic diseases. However, over limited areas, they may cause losses in forests. Physiological disorders of forest trees in India have not been studied in any great detail. Some have been studied and are reviewed below. A comprehensive review on the subject may be obtained from treatises by Boyce (1961) and Peace (1962).

FIRE

Fire is a common feature in forests in India. The intensity of a forest fire will depend on the amount of ground litter, inflammable material on the ground and on trees and temperature and humidity conditions. If the heat from the fire reaches a lethal temperature, it may adversely affect the useful flora and fauna. The beneficial microorganisms like mycorrhizal fungi which are more frequent in the top layers of soil may be killed after an intense fire. Young regeneration in a heavily burnt area may show initial poor growth with pale foliage due to lack of mycorrhizal roots. Health of plants is restored with return of beneficial fungal symbionts (Bakshi 1957d). Young plantations of fire-tender species, like *Eucalyptus*, may be killed outright by ground fires.

The extent of fire damage to trees depends on the resistance of the species to fire and the heat generated by it. Thick-barked species are comparatively more fire-resistant than thin-barked ones ; conifers are generally more inflammable than broadleaved species. Fire may kill back young regeneration though the underground parts usually remain unaffected. If the species is a good coppicer e.g., sal (*Shorea robusta* Gaertn. f.), it can sprout giving rise to coppice shoots and thus survive the effects of fire. In fire-tender species, the bark and the cambium may be killed all around the tree resulting in its death. If, however, the injury occurs on one side, that face only is killed resulting in an open wound. If the wound is small, it may be healed up and the injured wood becomes included in the tree. Such unsoundness due to fire is a common phenomenon in sal, a fire-hardy species. If, however, the wound is large, it remains open. Basal injury caused by fire may form an infection court for decay fungi to establish in the tree.

SUN SCORCH

The surface of the stem receiving full insolation can reach a temperature far above that of the surrounding air. Death of bark leading to open wounds occurs on this surface in thin-barked species. Sun scald is always oriented on the sunny side. In even-aged stands, sun scorch may occur on fringe trees exposed to sun or on trees which become exposed after felling neighbouring trees or after thinning or when trees become leafless during a certain period of the year. In teak (*Tectona grandis* Linn. f.) stands, long open scars extending from near the ground level to a distance of 3-6 m of the trunk may develop. Such scars, always oriented on the southern half of the stem, are caused by excessive heat from the sun in these open stands to which the thin-barked teak is susceptible (Boyce 1959). In Tarai and Bhabar Plantation Div., U.P., some clones of *Populus 'euramericana'* (Dode) Guinier were affected by sun scorch resulting in opening of the bark lengthwise for 2-4 m on the southern face. The wound subsequently enlarged due to colonisation by *Botryodiplodia palmarum* (Cke.) Petr. and Syd. Occasionally *Trametes corrugata* (Pers.) Bres. established and caused decay in wood (Bakshi, Reddy, Puri and Sujan Singh 1972).

FROST

Trees can, to a certain extent, adjust to climatic fluctuations but may suffer from injury under abnormally low temperatures. Winter frost which occurs when trees are dormant is not so injurious as early frost and particularly late frost in the spring when trees produce tender succulent tissues which are most susceptible to frost injury. The common symptoms of frost damage include discolouration, shrivelling and death of tender leaves and current year's shoots. Frost causes injury to the cambium. In severe frost, the cambium may be killed all around the stem thus girdling it or it may be killed in localised areas, causing cankers as in sal which is frost-tender at the young age or *Eucalyptus citriodora* Hook., which cannot stand even a slight frost (Bakshi and Sujan Singh 1964). If the injury is not severe so as to kill the cambium, undifferentiated parenchyma is formed in place of normal wood tissues and this is accompanied by formation of traumatic canals and gummosis in *E. citriodora* (Pl. **6**, 3-5) but most invariably by frost rings as in *E. citriodora* (lit. cit.) and black wattle (*Acacia mearnsii* De Wild.) (Pl. **6**, 6) (Bakshi, Reddy, Puri and Sujan Singh 1972). In a frost ring, abnormal tissues develop consisting of undifferentiated parenchyma instead of normal wood. Rays may become discontinued or laterally displaced bordering the affected region (Pl. **6**, 5).

Control of frost damage depends on sound silviculture and choice of species and site for planted stands. Protection of young sal regeneration can be obtained by suitable adjustment of the overhead canopy during management under Selection system or by gradual opening of the canopy during seeding and intermediate fellings under Uniform system. In raising plantations particularly of exotics far removed from their natural homes, site conditions of the species in their native homes should be given due consideration in raising the species in exotic locations. In its native home in Australia, *Eucalyptus citriodora* cannot withstand even slight frost (Anon. 1955) and, therefore, frosty locations should be avoided in raising plantations of the species. Black wattle (*Acacia mearnsii*) in its indigenous home in South Australia and Tasmania occurs in cool and moist locations where winter temperature does not fall below 4.4°C. The species is widely planted in the Nilgiris in South India. Frost injury occurs in plantations raised on former grassland sites which are believed to have been formed mainly due to firm retrograde effect of frost. In grasslands, the frequency and intensity of frosts immediately over the ground are greater than normal (Day 1928). In a hilly terrain, cool air settles in valleys and a frost-tender species like black wattle suffers in such frost pockets which should be avoided for raising the species. Ideal conditions for growth of black wattle are an altitudinal range of 1830–2440 m, rainfall of 125–200 cm with over 100 rainy days in a year, a temperature range from a mean maximum of 24°C to a mean minimum of about 2°C and sites of deep clay soils of gneiss origin (Streets 1962). The species is now successfully raised in Kodaikanal hills (Tamil Nadu) on hill tops, while in the valleys, *Pinus patula* Schlecht. and Cham. is showing promise.

SOIL MOISTURE

Under high soil moisture, soil aeration is poor. Also, carbon dioxide accumulates in the soil due to poor diffusion. Under these conditions, death of roots occurs whose effect may be noticed in the dying of aerial parts and in extreme cases, death of the tree. Wet soil with poor drainage results in poor soil aeration under which condition, plants may die e.g., sissoo (*Dalbergia sissoo* Roxb.) which cannot withstand water-logging even for a short period (Streets 1962). Under permanent water-logged conditions, most tree species develop poorly and may die. However, sissoo grows healthy when raised in sandy wastes and under irrigated conditions, when flowing water carries dissolved oxygen to roots and also adequate aeration is provided to roots growing in a loose textured soil. In soils of high moisture under good drainage, adequate aeration to roots occurs and plants grow healthy e.g., poplars which can thrive under irrigated conditions. However in water-logged soils in the tarai region, bark bursts occur on the stems of *Populus deltoides* Marsh. and *P.* '*euramericana*' (Pl. **6**, 1). *Botryodiplodia palmarum* colonises the affected region of the blister causing flaking of the bark and formation of canker (Pl. **6**, 2). Bark bursts are recorded on *Populus deltoides* in wet hollows on shallow soil overlying heavy clay (Streets 1962).

Insufficient soil moisture can arise from many causes. Trampling of the forest floor by the grazing cattle causes soil compactness due to which water does not percolate in the soil but runs off. Lessened soil moisture resulted in large scale dying of sal in Bihar (Boyce and Bakshi 1959). Other contributory factors for inadequate soil moisture include low atmospheric precipitation and low water-table. If the water requirement of the tree falls below the level necessary for the plant to carry out vital functions, drought conditions occur as in some coastal plantations of *Casuarina equisetifolia* Forst. Though the species is fairly tolerant to drought, it cannot withstand long periods of atmospheric drought. The capillary force in a coarse sandy substrate is insufficient to raise water from the water-table to the surface where most roots lie. As a result, plantations fail. Deep planting and watering during the first two years enable plants to develop deeper root system in the capillary zone (Chapter 17).

10. SEED PATHOLOGY

Plants are normally propagated through seeds. However, seeds may carry with them harmful pathogens which may affect the future health of the crop. Seeds may be attacked by pathogens while still borne on the plant in the field. However, infection may not occur on all the seeds. Also, seeds may be attacked by different pathogens. On sowing, therefore, infected and healthy seedlings show a random pattern of distribution in the field. Seeds may also become infected during storage and subsequent handling prior to planting. Deterioration of seeds in storage is governed by moisture content of the seed and by storage conditions, mainly humidity and temperature. Fungi attacking seed in storage usually belong to *Aspergillus* and *Penicillium* which normally invade seeds after harvest and may cause loss of viability by invading the embryo (Christensen 1973). The only effective method of preventing damage due to storage fungi is to store seeds under dry conditions and at a temperature which will limit fungus growth on the seed.

Seed-borne infections may behave in different ways depending on the nature of the pathogen and also on sowing and growing conditions in the field. Severely infected seeds in which the embryo is affected will not germinate. In the field, such an infection is the culmination point and the disease may disappear along with the seed. More superficial infections as those caused by Ascomycetes and Fungi Imperfecti mainly cause pre-emergence blight but do not manifest on the older crop. Slightly infected seeds will, however, germinate to serve as infection centres from where the disease will spread in the field causing significant loss.

SEED AS CARRIER OF PATHOGEN

The pathogens may be carried within or on the seed, on fruit and cones or on the inert matter like chaff and straw accompanying the seed. However, the presence of a fungus on a seed does not invariably mean that it is harmful nor that it is necessarily transmitted through seeds.

An annotated list of seed-borne plant diseases including those of forest trees is compiled by Noble and Richardson (1968). In addition to common moulds which may infect seeds in storage, other important pathogens recorded include different species of *Ciboria* causing sclerotinia disease on seeds of alder, birch and chestnut, *Cytospora longispora* Servazzi on *Araucaria heterophylla* (Salisbury) Franco, *Ceratocystis fagacearum* (Bretz) Hunt on oaks and chestnuts, *Deuterophoma tracheiphila* Petri on *Citrus* spp., *Cercospora coffeicola* Berk. and Cke. and *Colletotrichum coffeanum* Noack on *Coffea* spp., *Endothia parasitica* (Murr.) P.J. & H.W. And. on chestnut, *Sclerotinia laxa* Aderh. & Ruhl. on hazel, *Botryodiplodia theobromae* Pat. on rubber, *Gnomonia leptostyla* (Fr.) Ces. & de Not and *Xanthomonas juglandis* (Pierce) Dowson on walnuts, *Diplodia pinea* (Desm.) Kickx and *Lophodermium pinastri* (Schrad.) Chev. on pines, ring spot virus on *Prunus*, *Guignardia aesculi* (Pk.) Stew. on horse chestnut, *Alternaria ricini* (Yoshii) Hansford on castor bean and elm mosaic virus on elms. During recent years, there have been many records of new outbreaks of diseases in many countries (Chapter 10). In most cases, there is no evidence whether whole live plants were imported into those countries. It is a matter of speculation whether the diseases could have been transferred through seeds. Quarantine safeguards are, therefore, necessary in the import of seeds into any country.

SEED HEALTH TESTING

Standard procedures for seed testing, including testing for their health, have been laid down by the International Seed Testing Association (Anon. 1966). Seed health testing for the presence of disease organisms and insects is important for the following reasons : (1) seed-borne organisms may give rise to progressive development of diseases and insect pests in the field and thereby reduce the commercial value of the crop ; (2) seed health testing may supplement germination test in order to obtain a better picture of the performance of a seed lot in the field ; (3) through infected seeds, diseases and insects may get established into new regions and may cause significant losses : this is important in plant quarantine work both within a country and also in international seed trade necessitating in certification of seeds ; (4) testing seeds for health is useful for determining the necessity or otherwise of seed treatments for controlling seed-borne pathogens. Such tests may also be used for evaluation of fungicides in seed treatments.

The usual methods of seed health testing are based on examination of a limited number of seeds out of a lot and may not assure complete freedom from an injurious pathogen. In this regard, it may not be a fool-proof method for certification under quarantine testing.

In testing seeds for health, the methods in use are not yet perfect for determining the different kinds of seed-borne diseases. An important feature for testing seeds for health should be the reproduceability of results. Seeds normally carry a complex microflora of which the pathogens constitute a small part. Seed health testing should clearly project the full disease potential harboured by the seed. Such tests determine the kind and the amount of inoculum, distribution of inoculum within a seed lot and the vigour of inoculum at the time of testing (Neergaard 1973).

Preliminary inspection of dry seeds may be done by observing fruiting bodies of fungi on seeds. Spores and hyphae adhering to seeds may be observed by examining the washings of seeds in water. A closer examination of seed-borne pathogens may be made from agar test to determine growth of pathogens and incidence of fungi on seeds. Plating seeds on agar after surface disinfection permits ready detection of most storage fungi. In blotter test, growth of the seedling and the pathogen is simultaneously observed and the growth rate and virulence of inoculum studied. Seeds are exposed to fluorescent light or near-ultraviolet light alternating with a period of darkness for stimulating sporulation of fungi (Leach 1967). The blotters are kept adequately moist during incubation but not too wet to stimulate bacteria that may be antagonistic to pathogenic fungi. In seedling symptom test, the vigour of germination of the seed as well as the total inoculum potentials of seedling pathogens are determined. In growing-on test, plants are grown beyond the seedling stage to allow development of a disease that appears on plants of advanced age only. The method is adopted for post-entry quarantine control to intercept dangerous pathogens. It is applicable for detection of those diseases which cannot be detected in the seedling stage, e.g., downy mildews, certain systemic diseases including viruses. In indicator test, indicator plants are inoculated by inoculum harboured in the seed being tested and is used for detection and identification of seed-borne viruses and for checking the viability of pathogens (Neergaard 1973).

SEED TREATMENT

Treatment of seeds to control seed-borne diseases is a common practice usually with agricultural seeds. Seed dressing with certain protectant chemical has the effect of spot treatment of the soil around the sown seed and frequently controls seed- and soil-borne pathogens to a degree giving valuable increase in emergence. It may also control the disease sometime after emergence. In conifers, however, control of pre-emergence damping-off by fungicides is not encouraging possibly because of the long susceptibility of conifers to these diseases and also because the seed coat on which the fungicide is applied normally comes above the ground during germination. To obtain lasting control, pelleting of seeds with a toxicant and methyl cullulose as sticker is suggested when the fungicide is released slowly in the soil (Berbee, Berbee and Brener 1953). Seed disinfection is done to kill spores on seeds and also against excessive moulding during certification and germination tests.

Seed disinfection developed with the use of organic mercury compounds which possess a comprehensive efficacy against most seed-borne diseases. Disinfection of seeds is also possible by using toxic vapours and other chemicals (Vaartaja 1964). Systemic fungicides, like vitavar, are also being used to effectively control seed-borne diseases (Hansing 1970).

11. FOREST DISEASE CONTROL : GENERAL PRINCIPLES

One prime objective of forestry is to grow forest crops free from diseases and obtain a profitable harvest. In the event of an outbreak of a disease, control measures that have to be adopted should be economically justifiable. It has to be ensured that substantial benefits are gained by adopting any control measure. For this, precise information is needed on the monetary losses due to diseases. The impact of forest diseases is assessed not only by the actual damage they cause but also by the cost of preventive and control measures, limitations in raising crops in areas of high disease hazard and in the upsetting of forest industrial developmental plans of the country for which the raw material is obtained from forests. Forest disease control is circumscribed by the fact that forest crops are perennial, usually with a rotation age of many decades and have a relatively low value per unit area as compared to agricultural crops. For these reasons, the intensive methods of disease control, particularly through use of chemicals, crop rotation and breeding resistant stock, as used in agricultural crops, are not generally economically feasible in forestry practice. Control measures are limited, to what can be achieved through silvicultural and management practices, to create growing conditions ideal for growth of the crop but unfavourable to development of the disease. However, in plantations where crops are intensively managed, the intrinsic value of the crop is comparatively high and, therefore, intensive control measures may become applicable. This is also true in case of forest nurseries. The principles of forest disease control are broadly discussed under direct and indirect measures. Quarantine is a measure of exclusion of diseases (and also insects) and is discussed separately (Chapter 12)

DIRECT MEASURES

SANITATION

. Sanitation is the process that reduces or eliminates the initial inoculum from which the disease develops. Such inoculum may be present either on principal hosts or on collateral or alternate hosts. *Corticium salmonicolor* Berk., & Br., the cause of pink disease, occurs on rubber (*Hevea brasiliensis* (H. B. K.) Muel.-Arg., teak (*Tectona grandis* Linn. f.) and many other hosts and attacks *Eucalyptus* planted in sites under suitable conditions of development of the disease. The presence of the inoculum thus limits healthy growth of *Eucalyptus*. Such locations should, therefore, be avoided. Debris as lops and tops from thinnings, brashed branches, fallen leaves are normally allowed to remain on the forest floor to decay and add to the humus. In a few instances, such as Lophodermium needle-cast disease, the fungus thrives on infected fallen needles and spreads. Mulching soil effectively controls dispersal of spores of the fungus. Control by sanitation requires an early detection of the disease in the forest and taking suitable steps to remove the disease inoculum before it can spread the disease. Diseased trees should be removed at the earliest opportunity through sanitation cuttings made during thinnings and improvement fellings. Sanitation may also be carried out by removal of fruiting bodies of fungi, particularly those belonging to Basidiomycetes, to reduce disease inoculum.

ERADICATION

Heteroecious rusts require two hosts to complete their life-cycles. These diseases can, therefore, be controlled by eradication of one of the hosts. The alternate host may be a weed, or may be a valuable species in which case it has to be eradicated in favour of the more valuable crop. Eradication may be done mechanically or through use of weedicides, thereby keeping the other host plant of economic importance free from the disease. Thus eradication of *Ribes*, the alternate host for the white pine blister rust *Cronartium ribicola* Fischer, protects the susceptible 5-needle pines from attack by the rust. Similarly, eradication of insect vectors may control diseases transmitted by them.

In natural forests, fungi normally remain endemic. Root and butt rot fungi that may occur on trees usually have restricted growth on roots but quickly colonise residual roots and stumps after clear-felling. The large inoculum potential thus built up becomes a potential threat to susceptible hosts later planted in such areas. Control of root rot may be obtained by removal of residual stumps and roots prior to planting either by raising mechanised plantations or through extraction by physical means.

ISOLATION TRENCHES

Once a root disease establishes in the forest, it spreads centrifugally from the infection centre to adjacent trees. For pathogens which do not have a free spread through the soil, the disease is mainly communicated to adjacent healthy plants through root contact or root graft. In such cases, the spread of the disease can be checked by isolating diseased plants by trenches.

CHEMICAL CONTROL

Chemical control of forest diseases is generally prohibitively expensive and, therefore, not practicable except probably in nurseries and plantations where the incidence of the disease and the damage are high. Control of Dothistroma blight in radiata pine in New Zealand is an instance to this. Control of pre-emergence blight and damping-off in the nursery is possible through application of chemicals in the soil and as seed dressings to destroy any pathogenic organism carried with the seed and to protect the seed from attack of soil-borne pathogens during germination. In *Rauvolfia serpentina* (Linn.) Benth. ex Kurz. raised as minor forest produce which yields alkaloids of medicinal importance from roots, captan, a copper-based fungicide not only controlled defoliation caused by the target leaf spot pathogen, *Corynespora cassiicola* (Berk. & Curt.) Wei but also resulted in a net increase of about 140 kg of roots per acre over unsprayed plants at the end of a 3-year rotation period. The price of the increased yield amounted to about Rs. 2500/- per acre at the then prevailing price for which the cost of the fungicide and its application worked out at Rs. 300/-, thus resulting in a net gain of Rs. 2200/- per acre (Reddy, Tewari and Bakshi 1971).

Fungicidal chemicals may act as prophylactics in preventing infection. For this, the chemical should be applied on the surface prior to the arrival of the inoculum or at least before it infects the host. Once the infection occurs and the pathogen establishes within the host, a surface protectant chemical will not be able to check further progress of the disease in the plant. In such cases, a systemic fungicide may be used when the chemical is distributed throughout the system of the host. Such a chemical is curative in action.

A fungicide will have its desirable effect if it attacks the pathogen at the weakest point in its life-cycle and at the same time proves non-toxic to the host at the dosage used. Furthermore, the chemical should not cause any pollution hazards to wild-life. The aerial plant part sprayed with a fungicide is expected to get protection over a period of time when it is exposed to rain. This necessitates that the fungicide must not be washed away during rains and should be insoluble in water but released slowly and at the same time act as a protectant. The fungicide must adhere well to produce a lasting effect. For this, a sticker as well as a spreader are added to the fungicide. The latter is meant to improve contact between the fungicide and the surface to be protected.

To control a disease of aerial plant parts by a fungicide, the timing of application is important. For this, a knowledge of the biology of the pathogen is necessary including the infective stage, vulnerable stage in the life-cycle of the pathogen, vulnerable period of infection of the the host and the mode of action of the chemical. All air-borne pathogens have at least one active and one passive period during one year (Evans 1968). In the temperate climatic zones, the active period coincides with the summer while in the tropics, with the wet season. If there are frequent rainy days, the spray may have to be repeated and will prove more effective if a dry period occurs or could be predicted immediately after an application.

INDIRECT MEASURES

CHOICE AND IMPROVEMENT OF SITE

Natural stands are adapted to the site they occupy. Such forests, therefore, generally grow healthy as long as the stands are properly managed. Large scale dying of dry type of sal (*Shorea robusta* Gaertn. f.) forests in Bihar State is due to deteriorating site conditions, the most important single factor being lessened amount of soil moisture. The contributory factors for this are repeated fires and heavy grazing resulting in compactness of the soil. Soil erosion is heavy because the ground litter and undergrowth are nearly absent. It is, therefore, necessary to improve the site through management practices, like fire protection measures, closure to grazing and trenching to hold water run-off to restore conditions for healthy growth of sal.

Suitable selection of site is important for plantations, particularly of exotics, which are far removed from their natural range of distribution.

There are many instances around the world where the attempts to raise preferred species in plantations have failed when they are grown under adverse conditions or in localities where diseases are present. Good site is necessary for maintaining stand vigour to resist attack due to diseases. Afforestation projects may sometimes be limited to sites which are unsuitable or at best marginal for agriculture. Some exotics may be more exacting in their site requirements. It may be necessary to establish forest conditions by growing hardy species prior to planting exotics. Fast growing species cause a severe drain on soil nutrients which become more and more depleted with years. Reforestation involving replacement of the natural vegetation by a new and essentially different species should be approached with care because the success of the exotic cannot be predicted unless it has completed at least one rotation successfully. An extreme case of species conversion may be seen in the high rainfall areas in Karnataka and Kerala States in South India where semi-evergreen and evergreen natural forests are clear-felled and planted with *Eucalyptus*. The soil which used to be protected by the lush evergreen vegetation becomes exposed to erosion upon clear-felling and also thereafter under *Eucalyptus* which fails to protect the soil with its light canopy. Deforestation results in invasion by *Eupatorium*, an impenetrable weed, which creates problem in plantation work and poses risk of fire. The situation becomes aggravated with large-scale mortality in *Eucalyptus* due to pink disease resulting in large gaps in plantations. Heavy soil erosion and site deterioration will make it difficult to reclaim these areas for raising forests.

CHOICE OF SPECIES

Natural forests have inherent slow growth and low productivity. Also, such forests are of mixed composition where only some species are economically valuable. In India, forests consist mostly of broadleaved species which are mostly short-fibred. The pattern of demand is changing in favour of industrial wood for which long-fibred conifers are most suitable. Indigenous species are often not capable of making full economic use of the sites. For all these reasons, exotics, by virtue of their fast growth, are introduced with the prime objective of fulfilling a definite need not met with by indigenous species. Exotics should possess the ability to thrive in their new environment for which a great deal of care is needed in selection of suitable species, provenance and site to verify the performance of a species in trial plantings before taking up large scale plantation programmes.

Plantations may be raised either pure or in mixture. A pure crop is preferred because the management is simplified. Also, the area can be easily replanted following harvest. Furthermore, a sufficient and sustained supply of wood of one species becomes available, thus facilitating its industrial utilisation. A pure stand is, however, ideal for a pathogen to build up to epidemic proportions since infection is direct and rapid from tree to tree. Parasites are often damaging at a particular age in the development of the tree so that in an even-aged crop, the entire plantation becomes susceptible in the event of an epidemic outbreak of a disease. For these reasons, mixed stands may be raised in plantations where the species can be in intimate, strip, group or block mixtures. In selecting species in a mixture, it is always advisable to plant species, one or more of which are resistant to the potentially dangerous pathogens in the area. Monoclonal plantation that lacks the genetic variability of a natural seeding population is unjustified on pathological grounds. The disease hazards become high in such plantings as may happen in poplars. In using clonal stocks in planting, it is advisable to plant a number of clones in mixture and limit the area occupied by each clone (Hartley 1939).

Plantations may be raised using indigenous or exotic species. In the former case, the disease risk is usually no more than what the species is exposed to in its natural habitat and the diseases caused by native fungi do not usually cause serious damage when the species is raised in pure stands. An exception to this is seen in rubber trees which grow healthy in natural forests in South America but failed when raised in plantations due to serious defoliation by an indigenous fungus, *Dothidella ulei* P. Henn. Exotics, on the other hand, may be exposed to two hazards. They may be threatened by an introduced parasite. If the local climate is congenial to the development of the foreign fungus, it finds the exotic as a suitable host which becomes diseased. Secondly, the exotic may encounter a native parasite to which it has no resistance and, therefore, becomes diseased.

Pathogens on stem and leaf have a limited host range usually confined to a genus within which some species are susceptible and others resistant. When *Eucalyptus* is planted outside Australia, the natural home of the genus, or pines are planted in southern hemisphere where no native pines occur, the endemic stem and leaf fungi are unlikely to become major pathogens on exotics as these have no near relatives in indigenous genera. Diseases of stem and leaf which may occur on such exotics are caused by pathogens inadvertently introduced. However, when exotics do have indigenous near relatives in their new home, the native parasites may become virulent on the exotic as happened in East Africa where the Monochaetia canker disease of the native *Juniperus procera* Hochst. got transferred to *Cupressus macrocarpa* Harw. (Jones 1953). Root disease fungi, on the other hand, generally have a wide host range so that native pathogens may cause root diseases of exotics (Bakshi 1967).

SILVICULTURE AND MANAGEMENT PRACTICES

Natural forests which are worked on sound silvicultural practices and are properly managed, do not normally suffer from any serious disease. However, such forests may become diseased if profoundly disturbed as happened in wet sal forests where intensive fire protection measures carried out at the early age of the crop resulted in a serious outbreak of *Polyporus shoreae* Wakef. root disease which was brought under check through control burning which is thus found to be a necessary silvicultural measure in wet sal forests (Bakshi and Boyce 1959).

Fire, which is a regular feature in forests, is the principal cause of injury to trees through which decay fungi establish in the tree. Fires normally occur between April and June when weather is dry and the dry undergrowth easily takes fire. Early control burning reduces the risk of fire damage to trees.

In an uneven-aged stand, suppression may occur to a certain degree to the understorey crop under the shade of dominants, particularly with a thick overhead canopy. Suppression causes death of leaders and branches which form favourable infection courts for fungi.

In forests raised for timber, a high incidence of decay in the heartwood causes serious loss in yield of timber. At a young age, progress of decay is slow. With increasing age, the volume of wood decayed increases and overtakes the new wood added to the tree. From this stage, the tree continues to grow at an unprofitable rate. In stands with high incidence of decay, it is, therefore, advisable to lower the rotation age when it will be possible to salvage a greater quantity of sound timber than at a later age. Industrial plantations are usually raised for a short rotation when development of heartwood is little. This, together with the fact that managed even-aged stands suffer much less from injury, accounts for considerable reduction in decay in such plantations in contrast to natural stands.

Infestation by dwarf mistletoe is favoured by selective cutting, while in clear cutting, the infestation is less. Heavily infested stands are, therefore, clear-felled leaving healthy standards free from the parasite as seed trees. The regeneration is thus free from mistletoe.

A pure even-aged stand may suffer from injury due to frost and from excessive insolation resulting in sun scorch. Pathogenic fungi may establish on the wounds. Drought favours many diseases including *Diplodia pinea* die-back of pines, *Dothichiza populea* Sacc. and Briard and Cytospora cankers on poplars ; high moisture favours Rhabdocline needle-cast of Douglas fir and cedar leaf blight due to *Didymascella thujina* (Durand) Maire. Damage by hail and snow may be extensive in an even-aged crop. Hail damage may form infection courts for fungi like *Diplodia pinea* on *Pinus radiata.* Branches may break under weight of snow and decay fungi may establish through wounds. An even-aged crop is also liable to wind damage which is accentuated by thinning and pruning.

CULTURAL PRACTICES

Diseases may be controlled through cultural practices. In plantations, the crop is initially raised at a close spacing to suppress the competing vegetation and also allow early closing of the canopy. Timely thinning with a view to provide spacing is, therefore, important as it increases plant vigour which in turn reduces damage from diseases. Thinning also provides a tool for removing diseased individuals from the stand for which the pathological condition of the stand should be determined before marking for thinning. Diseased trees are likely to be randomly distributed for which reason thinning has to be irregular. All the same, it will serve the desired purpose of removing sources of infection. On the contrary, stumps left after thinning in conifer stands may be colonised by *Fomes annosus.* Such stumps serve as infection centres for spread of the disease in plantations. Chemical treatment of the fresh stump surface will prevent colonisation of stumps by the pathogen.

Pruning of trees has the primary objective of improving timber quality by elimination of included branches. It is also done for trees in parks and avenues to remove dead and diseased limbs, create openings for sun light or make clearance for overhead communication lines.

Pruning cuts should be clear and made flush to the main bole for rapid healing of pruning wounds through callus formation. If branch stubs are left, they form courts of entry for decay fungi. Large pruning wounds, where healing is delayed, may be treated with a wound dressing chemical to prevent colonisation by parasitic fungi. Such treatment can, however, be feasible for trees which have a high intrinsic value.

To meet the demand of pole size timber, broadleaved forests may be worked on short rotations by cutting trees and allowing the stool to sprout either from callus growth developing on the cut surface of the stump or from dormant buds that may occur on the side of the stump. Control of decay in sprouts may be possible by retaining low origin sprouts instead of those of high origin.

BIOLOGICAL CONTROL

Biological control involves the use of an innocuous organism to prevent attack by a pathogenic organism. Control of *Armillaria mellea* root rot in tea bushes is possible by girdling original forest growth before planting (Leach 1939). In girdled trees, roots are depleted of carbohydrates to an extent that they cannot be invaded by *Armillaria mellea* but colonised by innocuous fungi. By this process, the disease inoculum is reduced to the minimum and the future crop grows healthy free from *Armillaria* attack. Likewise, colonisation of fresh conifer stumps soon after thinning by *Peniophora gigantea*, an innocuous fungus, prevents stump colonisation by the pathogen, *Fomes annosus*. This principle is taken advantage of in inoculating fresh stumps with *P. gigantea*.

RESISTANCE BREEDING

Resistance is the ability of an organism to hinder the development of a disease. There may be active resistance where the host reacts to the attack by a pathogen or passive resistance due to qualities inherent in the host.

A pathodeme is a host population in which all individuals have a given resistance in common, while a pathotype is a pathogen population in which all individuals have a given pathogenicity in common. If in a series of inoculations, the amount of disease shows a differential interaction between pathodemes and pathotypes, the resistance and pathogenicity are called vertical, which involves mechanisms which are within pathogen's capacity for change. In the absence of any such interaction between pathodemes and pathotypes, the resistance is termed horizontal, where the mechanisms involved are beyond pathogen's capacity for change (Van der Plank 1963). In vertical resistance, the inheritance is governed by a single gene and, therefore, easy to manipulate in a breeding programme. Horizontal resistance is controlled by polygenes and, therefore, difficult to manipulate genetically. The effectiveness of horizontal resistance does not break down. Vertical resistance, however, is liable to breakdown. For example, varieties of wheat used are normally resistant against all prevalent races of stem rust. As long as the races of the rust do not change, the resistance remains unchanged. The fungus is mutable and new races may arise against which the earlier host resistance is inoperative. Unless, therefore, the host population is also changed, the breakdown is permanent. In case of a breakdown in a vertical resistance, the vertical pathodeme in question must be replaced. Replacement is easy with annual crops, difficult with short term perennials and impracticable with perennials like trees. In horizontal resistance, there is a constant ranking of pathodemes and pathotypes in the absence of a differential interaction. The ranking in pathogenicity of horizontal pathotypes is constant irrespective of the pathodemes tested (Robinson 1973).

The development of a genetic resistance is the objective of all resistance breeding which, though slow and needs considerable initial outlay and therefore costly, is potentially the most effective long range solution to the control of some forest diseases. The objective of disease resistance improvement is to select or breed trees to establish a host-parasite relationship to yield an economic harvest. Forest trees provide a wide range of genotypes for selection and breeding for disease resistance, unlike agricultural and horticultural crop varieties which are usually more homogeneous and comprise few biotypes than the old indigenous breed of crops. Species, varietal and individual variation in the inherent resistance of forest trees, constitute the basis for developing disease resistance. Vertical resistance is unlikely to be valuable in a perennial crop or in a crop which is difficult to breed (Robinson 1971). In a heavily diseased area, it is not unusual to find a few disease-free individuals. The freedom from infection may either be due to escape or to inherent character in the presence of genes for resistance in the host or absence of gene for infectivity in the parasite. Intra-specific genotypic variation in resistance in many host species and to many pathogens is recognised in *Castanea dentata* (Marsh.) Borkh. with respect to *Endothia parasitica* (Murr.) P.J. & H.W. And., in

Pinus monticola Dougl. and *P. strobus* Linn. to *Cronartium ribicola* Fischer, in *Pseudotsuga menziesii* (Mirb.) Franco to *Rhabdocline pseudotsugae* Syd. and *Phaeocryptopus gaumannii* (Rohde) Petr., in *Ulmus* to *Ceratocystis ulmi* (Buism.) C. Mor. and in *Larix* to *Dasyscypha willkommii* (Hart.) Rehm. In Europe, *Pseudomonas syringae*, v. Hall, *Melampsora* and *Dothichiza* on poplars are important examples (Richardson 1960). The procedures for resistance breeding include selection of such disease-free individuals and testing their progenies by raising them in a heavily diseased area or through inoculations which will eliminate the escapers and ensure the heritability and selection of the resistance factor. Artificial inoculations where massive spores or mycelium as inocula are used can be so severe that they assess only immunity and not relative resistance. This should, therefore, not be relied upon fully. During selection, due consideration should be given to other desirable characters in the tree necessary for tree improvement. Parents and progenies should be tested against all known biotypes of the pathogen by exposing to natural infection and to inoculation. The rigorous testing through inoculation may not always be necessary with forest trees unlike agricultural crops where genetic uniformity is desired. The improved stock may be outplanted providing means of testing at the same time resistance under different environmental conditions favourable for disease development. Individuals that have not developed disease resistance can be salvaged during improvement cuttings. Further propagation of these trees may be done by establishment of grafts in seed orchards. Intra-specific crosses between the selected individuals may increase the degree of resistance in the progenies. Seed orchards yielding such progenies will thus form valuable basis for further propagation of a resistant stock. Another line of approach may be to develop inter-specific hybrid progenies as is done for resistance against fusiform rust, *Cronartium fusiforme* Hedgc. & Hunt, by crossing short leaf × loblolly pine, short leaf × slash pine and slash pine × loblolly pine where no infection was observed in small plantings of short leaf × loblolly pine although the surounding slash pine plantation was heavily infected (Jewell 1959).

In breeding of white pine for resistance against blister rust at inter-specific level, successful crosses are made between closely related groups of five needle species, namely, *Pinus wallichiana* A.B. Jackson, *P. monticola* Dougl., *P. parviflora* Siev. & Zucc., *P. peuce* Griseb and *P. strobus*. However, in such inter-specific hybridization there is some loss in seed production and in fertility of the resulting F_1's. For this reason, the most promising approach in long-term breeding is the introduction of additional resistant genes from the exotic (donor) species to the native (receptor) species by means of back crossing with the native species as the parent. This will result in strains with fully recovered fertility and adaptation, permitting free gene exchange with the most advanced strains of the native species at any point of an intra-specific resistance breeding programme (Heimburger 1972).

Though genetic resistance has been recognised in a large number of forest tree species, remarkably little is known about the nature of host resistance. Hairiness in leaves can provide complete protection against some virus vectors such as jassids, whiteflies (Robinson 1973). In *Castanea*, high concentration of pyrogallic tannins in the bark of genotypes resistant to *Endothia parasitica* may be responsible for retarding the growth of the fungus (Nienstaedt 1953). Enzymatic activity is responsible for resistance of oaks against *Microsphaera alphitoides* Griff. & Maubl. A close correlation between bark moisture and disease development exists in canker diseases in trees, as in willow due to *Cryptodiaporthe* (Bier 1959 a); in poplar due to *Fusarium* (Bier 1959 b) and in western hemlock due to *Cephalosporium* (Bier 1959 c). In absence of much data on the nature of host resistance, however, breeding forest trees for disease resistance should continue on empirical basis as has been achieved with *Populus* clones. In *Populus*, significant progress has been possible because of the variation in disease resistance between and within the natural species, wide range of genotypes and ease of vegetative propagation for purpose of clonal multiplication (Schreiner 1960).

12. FOREST PLANT QUARANTINE

There are many instances around the world when pathogens inadvertently introduced and established in an area have caused catastrophic damages to susceptible hosts. In natural plant communities, the pathogens and their hosts tend to reach a biological balance through weeding out of susceptible individuals and natural selection of resistant ones. An exotic pathogen, however, may find a susceptible host and may cause economic damage under suitable environmental conditions. In most cases, the pathogens become introduced along with the host plants. *Endothia parasitica* (Murr.) P. J. & H. W. And. in its native home in China does not cause any damage to Asiatic chestnuts which are resistant. The pathogen is believed to have been transferred with the nursery stock from Asia into America and then into Europe where the chestnut is seriously affected. *Phytophthora cambivora* (Petri) Buism. also caused serious damage to chestnuts in Europe where the fungus was inadvertently introduced from its original home in eastern Asia (Krstic 1960). *Ceratocystis ulmi* (Buism.) C. Mor., a fungus of Asiatic origin, does not cause economic damage to Asiatic elms which are resistant. The fungus escaped into Europe during the first world war through baskets made of diseased elms and then to the United States through infected elm timber (Boyce 1961). All American and European elms are highly susceptible to attack by this pathogen which has limited the success of elms as shade and ornamental trees in North America. Yet another instance of a fungus of Asiatic origin is *Cronartium ribicola* Fischer which when transferred to Europe caused damage to eastern white pine. The fungus eventually spread through infected nursery stock to the United States where 5-needle white pines proved very susceptible to the pathogen.

The original home of *Rhabdocline pseudotsugae* Syd., the cause of needle-cast in douglas fir, is North America from where the disease got transferred into Scotland with a shipment of young trees during the first world war. From Great Britain, it spread to Germany, presumably on nursery stock. Its further spread in Europe was along the direction of the prevailing wind (Boyce 1961). *Phaeocryptopus gaumannii* (Rohde) Petr., the cause of decline of *Douglas fir* in young plantations is also supposed to have got transferred from North America into Europe (Krstic 1960).

Cercospora cryptomeriae Shirai, the cause of *Cryptomeria* blight in Japan, is the same as *C. sequoiae* E. & E. which attacks *Sequoia gigantea* (Lindl.) Decne., causing symptoms similar to those in *Cryptomeria.* The fungus got transferred from North America into Japan with seedlings of *Sequoia gigantéa* and found a new host in *Cryptomeria.* The fungus later got transferred to Taiwan with the seedlings of *Cryptomeria* imported from Japan (Imazeki 1960). *Septotinia populiperda* Wat., the cause of leaf blotch of poplars, is innocuous in the United States and Europe. It is believed to have been transferred to Japan with scions of poplars and has become serious under climatic conditions ideal for the development of the disease (Imazeki 1960).

Dothistroma pini Hulbary is a little-known fungus in its indigenous home in the United States. The fungus, however, caused serious damage to *Pinus radiata* D. Don. plantations in East Africa, Australia and New Zealand. In East Africa, further planting of radiata pine was suspended in favour of the more resistant but less productive *Pinus patula* Schlecht. & Cham. In New Zealand, chemical control of the disease was successfully worked out and radiata pine continues to be the most important species being planted there (Anon. 1972). The spread of this pathogen to a distant continent like Africa and to New Zealand possibly occurred through human agency and also through air-borne conidia taken into clouds. Conidia are sufficiently resistant to extremes of temperature, desiccation and irradiation to survive appreciable period of air-borne exposure under drier and warmer conditions (Gibson 1972).

The international spread of dangerous diseases has led countries with economic interest in chestnuts to place embargoes on the import of living chestnuts including wood and bark. In the United States, some states have embargoed elm trees from those states which are not free from the Dutch elm disease. A few countries have placed restrictions on the import of five-needle pines susceptible to stem blister rust. This is the case with India where the rust is present on blue pine and its alternate host, *Ribes*, occurs in the Himalayas from Sikkim to Kashmir. This restriction became necessary for two reasons. Firstly, the susceptible white pines are likely to get severely attacked by the stem rust which may later build up a high inoculum potential and pose danger to the native pine. Secondly, a virulent strain may be introduced against which the native pine may have no resistance.

Imposition of quarantine is the first safeguard against inadvertent entry of diseases and insects. It may, however, be mentioned that quarantine, howsoever faithfully and conscientiously carried out, may not ensure absolute success in excluding diseases and insects indefinitely. This is particularly so if introductions are made through whole plants or cuttings which can be examined only superficially. They cannot be subjected to rigorous testing and speedy examination with the result that latent infections may go undetected. Many harmful microorganisms may be transmitted through soil carried with plants or adhering to plant roots. There is no means to sterilise plants or cuttings without injury or death of the tissues. With rapid rise in international traffic and increasing facilities of transport, it is difficult to keep pace with the growing trade in commodities for quarantine inspection.

In order that quarantine may be effective in preventing movement of important diseases and insects, it is important that regulations have a sound biological base and are administratively feasible. For new outbreaks of diseases or insect pests, information should be available on biology, mode of spread, host range and damage before a decision may be taken to establish a new quarantine regulation. Efficiency of quarantine depends on the knowledge and skill of the inspecting staff. Diseases and insects of forest plants are generally different from those in agricultural crops. Therefore, there should be special staff for doing forest plant quarantine and specialists should be available to deal with diverse groups of diseases caused by fungi, bacteria, viruses and others.

Important diseases and insects attacking forest trees in different parts of the world are known from published literature. These should not, however, form the only basis in compiling list of plants whose entry should be prohibited or restricted in a country. Firstly, it may not always be justifiable to enforce severe restrictions on entry of plants on the basis of diseases and insects which are already well-established in a country. Also, changing pattern of demand in forestry in favour of industrial woods necessitates introduction of fast-growing exotic species. Finally, the performance of an exotic cannot be predicted unless it is raised in test plantings.

All plant quarantine laws aim at preventing inadvertent entry and/or spread of diseases and insects to new areas. Measures adopted to achieve this objective, therefore, have a direct relation to the types of diseases and insects involved. It is important that seeds and other plant material imported into a country are accompanied by a phytosanitary certificate issued from the country of origin certifying that such materials are free from fungi, bacteria, viruses and insects.

Due to dangers inherent in introducing exotics as seedlings, the safest means of importation with the minimum risk of introducing diseases and insects is through seeds. To protect seeds from becoming infected in transit, seeds should be treated with a seed dressing chemical. If, however, importation of living plants or cuttings is desired to be made for a definite purpose, it may be advisable for the exporting country to certify that plants have been inspected in the field and found to be free from diseases and insects in the locality and that cuttings have been obtained from healthy plants. After import, the material can be subjected to post-quarantine inspection when any latent infection eventually becomes evident and timely action can be taken before the pathogen can establish and spread.

For research and experimental purposes, live fungi as spores or mycelium or culture or fungi on living plants may be imported into a country by appropriate authorities if they are to be confined to the laboratory or are not likely to cause infection to any crop in the country. Likewise, plants infested with live insects and intended for purpose of introducing such parasites as predators or for purposes of biological control can be imported if accompanied by a special permit from competent authorities. Dead specimens of fungi and insects may be sent duly certified that they have been killed before despatch.

Introduction of mycorrhiza is effectively and conveniently done by soil inoculum. However, in view of the dangers inherent in the transmittal, with the soil, of nematodes, bacteria, fungi and other potentially dangerous plant pathogens, transferring mycorrhiza through soil from one country to another should be prohibited.

Timber as such and as a packing material is becoming increasingly important in the international trade. Bark on logs or timber provides a suitable substrate for the spread of diseases like oak wilt, chestnut blight, Dutch elm disease, Diplodia die-back and others. Incipient decay in wood not detected from any signs may progress during shipment under ideal conditions for decay, namely, high temperature and humidity and poor ventilation, in the hold of the ship. Spread of dangerous forest insects throughout the world occurs through wood packing material. Timber quarantine needs to be developed with provision for pre-shipment phytosanitary clearance for wood components by the country of origin.

In India, the Directorate of Plant Protection, Quarantine and Storage in the Ministry of Agriculture, Government of India, is the official organisation for quarantine work. The Government passed the Destructive Insects and Pests (D.I.P.) Act in 1914 to restrict or prohibit import of certain plant materials from other countries and also to regulate their movement from one part of the country to another. Various notifications have been issued from time to time under this act. Due to natural barriers with adjacent countries, free movement of plant material from other countries into India is limited. Certain sea and air ports have been declared as ports of entry for plant materials. Imported plants are subject to inspection on arrival at ports of entry where facilities are also provided for fumigation.

Among notable diseases that were introduced in India in the past and have become widespread, mention may be made of the coffee rust, *Hemileia vastatrix* Berk. & Br. in 1876, late blight of potato (*Phytophthora infestans* de Bary) in 1883 and flag smut of wheat (*Urocystis tritici* Koern.). With enactment of D.I.P. Act in 1914 and setting up of quarantine bases, interception of diseases and insects have been made regularly at the ports of entry. In spite of this, many have since gained entry into India. These include San Jose scale of temperate zone fruits, woolly aphid of apple, cottony cushion scale of *Citrus*, bunchy top and mosaic diseases of banana, wart disease and golden nematode of potato (Renjhen, Venketesh and Joshi 1962). Recent records of exotic forest diseases include Dothistroma blight of radiata pine, Diplodia die-back of *Pinus patula* and *P. pseudostrobus*, *Cercospora pini-densiflorae* Hori et Nambu needle cast of radiata and chir pines, Lophodermium needle cast of slash pine and Monochaetia canker of cypress.

Forests require many decades to complete one rotation age. Exotic pathogens and insects may be so damaging to forests that the country's economic and developmental plans may become disturbed. It is, therefore, necessary to strengthen the quarantine base and develop forest quarantine in India.

PART II

DISEASES OF TREES

13. SAL–SHOREA ROBUSTA GAERTN. F.

GENERAL

Sal (*Shorea robusta*), the most important timber species in northern and central India, grows gregariously and tends to form a dense crop. It produces a long, thick tap root and stout laterals from which vertical roots descend. The crown is conical to elongate when young, becoming rounded with a strong branch system in mature trees. The bole is straight and clean. On the basis of maximum height attained at maturity, sal is put in 4 quality classes : I, over 110 ft. ; II, between 90–110 ft.; III, between 70–90 ft. ; and IV, under 70 ft. (Smythies and Howard 1923). The bark of saplings is greyish brown and smooth with a few deep longitudinal cracks. In trees, the bark becomes dark brown, 2–5 cm thick, rough, with deep longitudinal furrows. The sapwood is small, white or pale, with a broad transition region to the heartwood. True heartwood develops when the tree is about 15–20 years old and is accompanied by formation of chemicals which infiltrate into cell-walls and impart natural durability to wood. Tyloses occur in the vessels of the heartwood, but are absent in sapwood and the transition wood between the sapwood and the heartwood. The transition wood which is 2–3 cm in width is not true heartwood and is non-durable like the sapwood though it is similar to heartwood in colour (Choudhury 1942). The growth rings in sal are not seasonal and also false rings may develop, so that the age in sal cannot be determined from ring counts.

Sal occurs between its western limit at about 76°E longitude in Himachal Pradesh and the eastern limit at about 93°E in Assam, and from about 31°N latitude in Una and Dehra Dun southwards to about 18°N in Ganjam in Orissa and Bastar in Madhya Pradesh. Besides its general distribution in the plains, sal extends into the outer Himalayas upto about 1200 m. There is, therefore, a wide range of climatic and edaphic conditions under which the species grows. Based on these conditions, 13 main types, 5 of which are further divided into sub-types, making 21 in all, are differentiated under dry sal, moist sal, coastal sal and wet sal (Champion 1933). None of the types are true climaxes but for the most part are climatic climaxes whose stability is governed by factors like fire, grazing, unregulated felling and lopping. Protection from these favours progression towards evergreen forests and is necessary in dry sal and in relatively dry locations of moist sal. In coastal and wet sal, however, the progression is to be checked and regression from their present condition is essential for regeneration purposes. In moist sal, regression is to be prevented in established crops, while progression has to be checked during regeneration period. Fire is the most important factor influencing ecological processes : its exclusion favours progression, while its operation helps regression. Where a balance is needed, control burning is necessary. Grazing is the second controllable factor operating along similar lines and its exclusion is necessary for the regeneration to establish.

Sal is fire-hardy and withstands fire better than its associates. If fire occurs in the establishment stage, seedlings may be killed back but the roots usually survive to put forth fresh shoots in the next growing season and they may also be burnt if fire recurs and the process may continue. In the meantime, the root stocks may develop sufficiently to throw off vigorous shoots which may escape fire, particularly if this is mild or occurs late in the season. In the whippy and pole stages when the bark is yet thin, fire may cause unsoundness which remains as a permanent defect in the tree. Fires may also expose the wood through which decay fungi may establish in the heartwood and cause progressive heart rot. Considering the necessity of control burning in wet sal and in compratively wet locations in moist sal, fire may be considered as a necessary evil in such forests.

Sal is frost-tender at early age. Frost may injure the cambium, and if extensive, may cause death of the leader or lateral branches or the whole plant, or if localised, may cause cankers through which decay fungi may enter and cause heart rot.

In dry sal, deficient rainfall for 3 or more successive years may adversely affect healthy growth of trees. Heavy grazing results in deterioration of the site. Widespread dying of sal, particularly in dry type in Bihar and Orissa, was due to site deterioration in which the most important factor was lessened soil moisture (Boyce and Bakshi 1959).

Sal coppices well and the system of coppice-with-standards is practised. High sal forests are worked under the Selection system with improvement fellings to remove unsound and overmature trees in the interest of the future crop. This system is now mostly given up in favour of the Uniform system, since it is observed that in sal, regeneration takes place satisfactorily when the canopy is opened out and that sal tends to establish itself naturally in even-aged stands. In the Uniform system, the crop is divided generally into 4 periodic blocks, of 30 years each, in a rotation of 120 years. The final periodic block (PB Old) consists mostly of mature crop between 90-120 years, which is removed in successive fellings to open out the canopy to secure regeneration. At final felling, the forest consists of an even-aged crop, P.B. Young (P. B. IV), which passes into PB III and thence to PB II. In PB II and PB III, together known as PB Inter, thinnings and improvement fellings are carried out. PB II finally passes into PB I (PB Old).

PATHOGENIC DISEASES

POLYPORUS SHOREAE ROOT ROT

Polyporus shoreae Wakef., causing partridge rot, is a serious root parasite on sal. It is not commonly recorded on other hosts except rarely on *Shorea assamica* Dyer (Bagchee 1957) and *Pometia pinnata* Forst. (Sujan Singh, Balwant Singh and Bakshi 1961). Elsewhere, the fungus is recorded on *Hevea brasiliensis* Muell.-Arg. and *Tectona grandis* Linn. f. from Nigeria (Browne 1968).

Occurrence

P. shoreae occurs on sal throughout its range of distribution. It is sporadic in drier types of sal, becoming common in locations with higher rainfall. The fungus is common in Jalpaiguri and Buxa divisions in North Bengal and in Kachugaon, Kamrup, Haltugaon and Goalpara divisions in Assam. In Orissa, the fungus is recorded from Angul, Puri and Ghumsar divisions, while in Bihar, it is recorded from Chotanagpur area in Palamau, Porahat, Saranda and Kolhan divisions. The fungus is common in South Mandla and sporadic in Bilaspur and Raipur divisions in Madhya Pradesh. It is also recorded from sal forests in Uttar Pradesh.

Infection and damage

The fungus is capable of infecting healthy and uninjured roots. The roots are first attacked at the distal end. The infection progresses up the roots to the collar region and does not usually progress up the stem. The infection spreads from the diseased tree to adjoining healthy trees by root contact or root grafting resulting in occurrence of diseased trees in groups which appear scattered throughout the stand.

The disease is reported as damaging in sal plantations of Gonda, Bahraich, Ramnagar and Gorakhpur divisions (Uttar Pradesh) and Buxa division in North Bengal (Glasson 1922, Davis 1933). In Haltugaon, 65 sq km of pure sal forest were affected wholesale (Bagchee 1957). In Goalpara East and West, South Kamrup (Kawasing reserve, PB I), Jarasal reserve (hill sal) and Kulsi reserve in Assam State, number of drying and dead trees was estimated at 10-20 per cent. In Goalpara division, Assam, 2000 ha of sal were affected necessitating the suspension of normal operations in favour of marking dead and wind-thrown trees (Rajkhowa 1958).

Predisposing factors

P. shoreae root rot in sal is economically significant in areas of high rainfall, from 2000 mm and above. In North Bengal and Assam, the high incidence of the disease in mature forests may be correlated with fire protection measures adopted during early age of the crop. Continued fire protection leads to progression to mixed evergreen forests where the disease becomes virulent (Haines 1971, Troup 1921, Bakshi and Boyce 1959). Davis (1933) attributed widespread attack by *P. shoreae* to excess of soil moisture. Fire protection leads to high soil moisture and heavy weed growth under which conditions the roots become susceptible to attack by *P. shoreae*.

Symptoms

Affected trees do not exhibit symptoms till some roots are attacked. The remaining healthy roots are unable to meet the water requirement of the crown, particularly in big trees, which show symptoms of top-dying. Dying of branches gradually extends downwards till trees are dead in a few years or, as may commonly happen, such trees become wind-blown because they loose foothold in soil due to decay of roots. In drier locations, killing of trees is slow, while in the moister types, death is rapid (Bagchee 1957). In some cases, the tree may put forth epicormic branches (Pl. **10**, 1). The fungus causes white pocket rot in the bark and sapwood and the decay is commonly referred to as partridge rot (Pl. **9**, 5). Small, light yellow mycelial felts of the fungus develop on bark of roots. The heartwood usually remains unaffected.

Sporophore

The sporophores of the fungus are usually formed on affected trees at the base (Pl. 10, 2) or on exposed roots of wind-blown trees during the rainy season. Sporophores are sessile, rarely sub-stipitate, dimidiate, sometimes funnel-shaped, soft when fresh becoming brittle when dry, light in weight, single or imbricate; upper surface is brown to black, glabrous ; hymenial surface is brown to dark brown (Bakshi 1971).

Control

In wet sal forests, control burning, a necessary silvicultural measure in such forests, is an effective method of reducing soil moisture and thereby minimising infection by the fungus. Burning also helps in checking weed growth which conserves and increases soil moisture favourable for disease development (Bakshi and Boyce 1959).

In sal forests where incidence of the disease is high, dry marking is recommended. The dead trees and also trees in advanced stages of the disease should be removed and the timber utilised as it remains sound. Working plans should take into consideration the past records of dead and wind-blown trees to arrive at the correct yield.

OTHER ROOT ROTS

Ganoderma lucidum (Leyss.) Karst. and *Fomes lamaoensis* (Murr.) Sacc. and Trott. also cause root rots in sal in India.

HEART ROTS IN SAL

Sal has the reputation of being one of the most naturally durable timbers in service and is not easily subjected to decay or attack by termites and borers. However, the living tree is liable to attack by many wood-destroying fungi which cause decay of economic significance in the heart-wood. Bagchee (1954, 1959, 1961) did pioneering work on heart rots of sal in India by his extensive surveys of sal forests and study of the biology of the pathogens, analysis of factors predisposing forests to attack by fungi and outlining measures to minimise decay. Later, an intensive sample survey was conducted in a sal forest by Bakshi, Rehill and Choudhury (1963) for estimation of decay.

Infection

In high sal forests, heart rots establish in trees from trunk as well as butt infections, while in coppice forests, they establish mainly on the stool from where they progress up the tree when the heartwood develops in coppice shoots.

High forests. In sal, the thick bark and the sapwood afford protection to the heartwood from infection by decay fungi. However, with age, the heartwood may open up due to various causes when decay fungi may establish in it. Wounds constitute the main avenues for infection. Though sal is fire-hardy, fire is the most important agency causing wounds, particularly at the bases of trees (Pl. 7, 4). Injury due to fire results in fire-scars which, if small, may eventually become occluded in the tree. If decay does not establish on fire wounds, fire unsoundness remains as a defect at the base of the tree. If decay fungi enter through fire wounds, they establish and cause heart rots.

Wounds on trees are also caused by breaking of branches, lightning, browsing by animals and during felling and logging operations. Infection depends on the size of the wound, the bigger its size, the greater is the chance of infection. Tree wounds increase in number with tree age when healing of wounds slows down and, therefore, the incidence of wound parasites, like *Hymenochaete rubiginosa* (Schrad.) Lev., becomes greater with age (Bakshi, Rehill and Choudhury 1963).

Dead branches, branch stubs and knots constitute important infection courts. 70-80 per cent infections in sal may occur through branch stubs (Bakshi, Rehill and Choudhury 1963).

In frosty locations, the young sal crop may be affected by frost when leaders and lateral branches are killed back. Buds may be killed after sprouting resulting in a large number of knots (Pl. 7, 1). Frost may also cause splitting of the bark through which *F. caryophylli* (Rac.) Bres. may infect the tree (Bagchee 1961). The fungus also infects the tree through killed-back leaders and dead lateral branches and knots. Decay due to *F. caryophylli* may establish in the tree even before true heartwood develops. Eventually, the fungus attacks the heartwood.

Coppice forests. Coppice sal forests are managed on a short rotation of about 10 years for fuelwood or on a comparatively longer rotation, about 30-40 years, for transmission poles. Coppicing results in wounds on the stool. *Hymenochaete rubiginosa, Fomes caryophylli* and *F. fastuosus* (Lev.) Cke. are the principal fungi which infect stools

PLATE 7

HEART ROT IN SAL—INFECTION COURTS

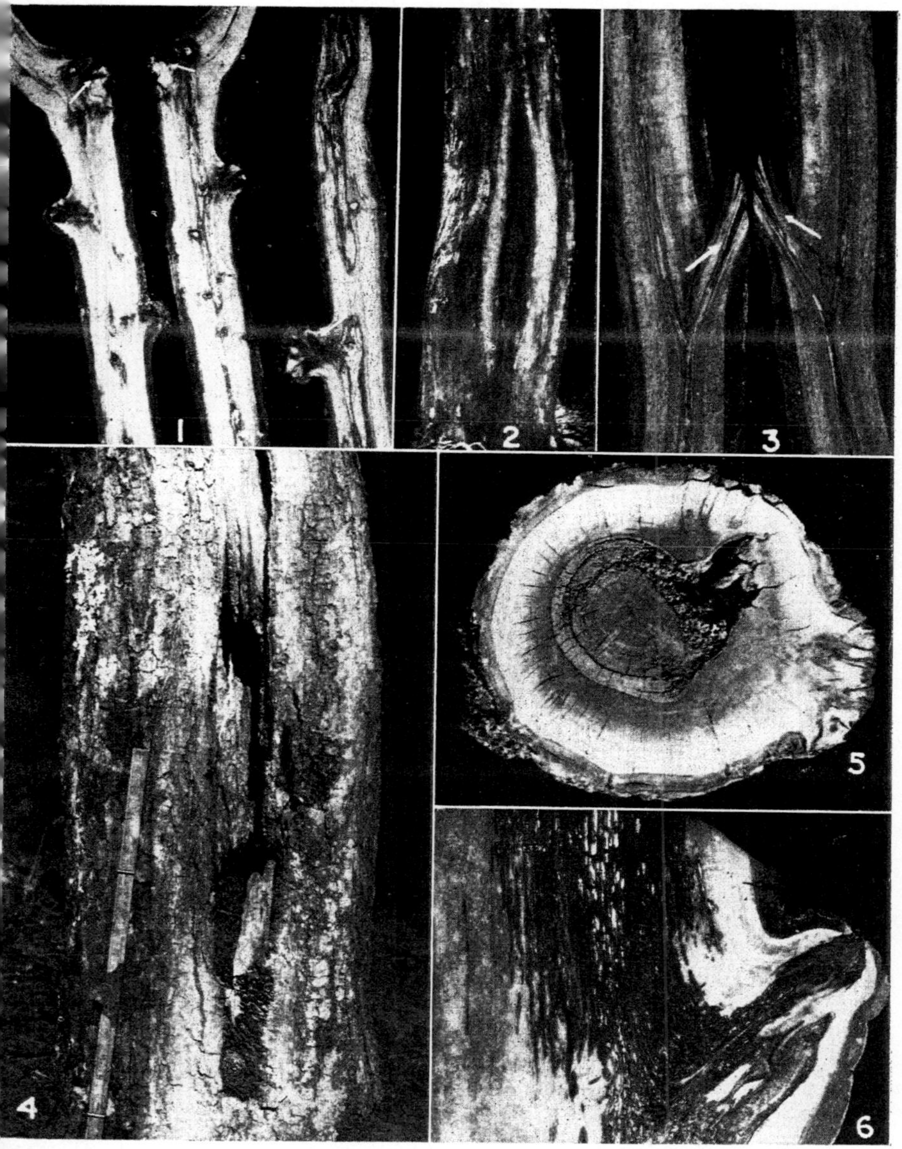

1–3. Infection courts of *Fomes caryophylli* : 1. Decay entering through branch stubs. 2. Lightning scar. 3. Decay entering through dead branch. 4, 6. Infection courts of *Hymenochaete rubiginosa* : 4. Fire scar and sporophores. 6. Decay entering through branch stub. 5. Decay entering through branch stub and forming a column in wood. (1, 3 Photo—FRI)

PLATE 8

HEART ROT IN SAL—INDICATORS

1–5. Indicators due to *Fomes caryophylli* : 1, 2. Punk knots on stem. 3. A punk knot isolated from decayed wood. 4. Decay pattern and punk knots. 5. Sporophore. 6. Swollen or bottle-shaped stem.

through such injuries and cause decay in the heartwood of the stools (Pl. **9,** 2). In a coppice shoot when heartwood develops and bridges with the heartwood of the stool, the decay from the stool migrates into the coppice (Pl. **8,** 1, 2). The incidence of decay in coppice forests is high even though such forests appear outwardly healthy (Pl. **9,** 1) (Bakshi 1957 c).

Indicators of decay

External indications of decay are valuable in identifying decayed trees, estimating the volume of sound timber vis-a-vis decayed wood in the tree and categorising the tree in the correct fitness class during marking prior to felling. The usual indicators for decay in sal are sporophores (Pl. **8,** 5), punk knots (Pl. **8,** 1, 2), wounds (Pl. **7,** 4) swollen bole (Pl. **8,** 6), knots, branch stubs (Pl. **7,** 1), dead branches (Pl. **7,** 3) and forks. The relative importance of each of these indicators in indicating decay was worked out in sal by Bakshi, Rehill and Choudhury (1963). In advanced stages of decay, sporophores develop on the host. Therefore, sporophores indicate decay in cent per cent cases.

Punk knots are a characteristic symptom of attack by *Fomes caryophylli* on sal, though they are not always formed in all locations. Punk knots consist of decayed wood and are soft and punky (Pl. **8,** 3). These cylindrical cores, upto 2 cm in diameter and upto about 6 cm in length originate from different depths of sapwood and heartwood (Pl. **8,** 4) and emerge horizontally through the bark and open up at the surface as 'eyes' or pox-marks (Pl. **8,** 1, 2). The surface becomes hard, woody and black (Pl. **8,** 1, 2). Sporophores may develop on punk knots in jaman (*Syzygium cumini* (Linn.) Skeels) but rarely in sal. Punk knots indicate decay in cent per cent cases.

Wounds, knots, branch stubs and dead branches which constitute infection courts may also indicate decay, though not invariably so and, therefore, each of these indicators has to be assessed on percentage basis.

The portion of the stem containing decay in advanced stages exhibits perceptible swellings giving bottle-shaped appearance to stems (Pl. **8,** 6). This region may emit a hollow sound when struck with a blunt implement, such as the back of an axe. Swollen bole indicates decay in most cases.

Marking rules

Green sal marked for felling is classified according to the soundness of growth form. In Uttar Pradesh, three classes are recognised : (1) sound, expected out-turn over 67 per cent of average saw out-turn ; (2) fit, expected out-turn 33–67 per cent, and (3) unfit, expected out-turn below 33 per cent. In some states, only two classes, sound and unsound, are recognised. The marked trees are grouped into different diameter classes, above 20 cm d.b.h. at intervals of 10 cm, that is, 20–30 cm, 30–40 cm, 40–50 cm, 50–60 cm, 60 cm and above. The number of trees falling in different diameter classes and soundness classes is indicated in the sale list when a forest is auctioned.

It is a common experience that green sal marked for felling may not always produce the expected out-turn of timber after conversion. This is because decayed trees may not always present reliable symptoms of heart rot thereby resulting in faulty marking. The indicators of decay in sal and the associated decay may help in evolving pathological marking rules which together with the observations on the physical form of the tree, such as tree diameter and available clean bole should guide the forester in marking trees into the correct fitness classes. However, much work is necessary in sal forests of different locations and quality classes to assess the indicators of decay and the associated decay volume. For example, while sporophores invariably indicate decay, it is necessary to distinguish between sporophores of different species to evaluate the amount of decay in the tree associated with each fungus.

Causal fungi

Though considerable loss in wood occurs through decay in sal trees, it is not quantified for various quality classes in different locations. In a sample survey of over 700 trees in quality class III in Timli (Dehra Dun), Bakshi, Rehill and Choudhury (1963) observed that decay was present in over 70 per cent trees entailing a loss of about 10 per cent timber. Short accounts of principal fungi causing heart rots of economic significance in sal and damage caused by them individually are given below :

1. ***Hymenochaete rubiginosa*** (Schrad.) Lev.

Occurrence. The fungus is widely distributed in India in both tropical and temperate regions and attacks a wide range of hosts in broadleaved species. The fungus occurs on sal and its associates, namely, laurel (*Terminalia alata* Heyne ex Roth), bahera (*Terminalia bellirica* Roxb.), lendi (*Lagerstroemia parviflora* Roxb.), benteak (*Lagerstroemia lanceolata* Wall.), bijasal (*Pterocarpus marsupium* Roxb.) and axlewood (*Anogeissus latifolia* (Roxb.) Bedd.). The fungus also attacks oaks in the Himalayas. When affected trees are felled, the fungus continues its activity on the stumps on which sporophores develop freely. The fungus may also remain active on logs and converted wood.

Damage. In sal, *H. rubiginosa* commonly enters through fire injuries and may account for nearly 50 per cent of the total decay due to all causes. Incipient decay is charcterised by a dark brown discolouration of the heartwood. Later, typical decay as white pocket rot develops (Pl. **7**, 6; **9**, 2).

Sporophore. Sporophores are usually sessile, sometimes resupinate, corky becoming rigid on drying ; upper surface is dark brown, concentrically zoned, velvety when young soon becoming smooth ; hymenial surface is brown, smooth, setae present.

2. *Fomes caryophylli* (Rac.) Bres.

Decay due to *Fomes caryophylli* is commonly known as 'gauj' or 'ankhya-rot', because of the characteristic appearance of punk knots which look like pox-marks or eyes ('ankhya') of sambhar deer (Pl. **8**, 1, 2). However, decay due to *F. caryophylli* is not always associated with punk knots, possibly due to existence of physiologic forms in the species (Bagchee 1961).

Occurrence. The fungus is widely distributed in all sal forests attacking the species and also its associates, namely axlewood (*Anogeissus latifolia* (Roxb.) Bedd.), sandan (*Ougeinia oojeinensis* (Roxb.) Hochreut) and kasi (*Bridelia retusa* Hook. f. non. Spreng.) in North India, karada (*Cleistanthus collinus* (Roxb.) Benth. & Hook. f.) in Peninsular India, jaman in sal forests with evergreen mixture, bahera, myrabolan (*Terminalia chebula* Retz.) and mahua (*Madhuca longifolia* (Koenig) MacBride var. *latifolia* (Roxb.) Chev.) in sal forests with dry deciduous admixture and *Mallotus philippensis* Muell.-Arg. and *Cordia dichotoma* Forst. f. in scrub forests (Bagchee 1961).

Damage. In sal, *F. caryophylli* may account for over 25 per cent of the total decay due to all causes. In incipient stage of decay, the wood is stained dark brown. In advanced stage, white rot is produced and wood becomes spongy (Pl. **7**, 1, 3). Yellow mycelial mat may develop in decayed wood. Decay may also occur as circular rings associated with punk knots which arise at different depths of the wood and open up in the bark as 'eyes' or pox-marks (Pl. **8**, 4).

Sporophore. Sporophores are perennial or annual, usually resupinate (Pl .**8**, 5), woody, small; upper surface is brown to black, sulcate, crusty, faintly hairy when young, soon becoming glabrous ; hymenial surface is dull woody brown, pore tubes stratified, oblique (Bakshi 1971).

3. ***Fomes fastuosus*** (Lev.) Cooke

Occurrence. *F. fastuosus* possesses a wide host range among broadleaved species all over India. It occurs in all sal forests and is common in those of coppice origin. It also attacks the associate species of sal, namely, laurel, bahera, myrabolan, lendi, benteak and axlewood. The fungus is also recorded on mahogany (*Swietenia mahagoni* Jacq.) and irul (*Xylia xylocarpa* Taub.) among other hosts listed by Bagchee (1960). After the affected trees which are usually attacked at their bases are felled, the fungus continues to grow on stumps on which sporophores develop.

Damage. *F. fastuosus* is a wound parasite, infecting trees through injury principally caused by fire. The fungus primarily causes butt rot and decay may extend into the main roots. In coppice sal forests, the fungus is usually located in the heartwood of stools to subsequently progress into coppice growth when heartwood is formed in the latter. Decay in early stages causes dark brown discolouration. White pocket rot develops in late stages of decay. The tissues in the pockets get disintegrated early leaving the pockets empty in advanced decay.

Sporophore. Sporophores are perennial, sessile, hard, woody, large; upper surface is dull brown turning black when old, zonate, slightly cracking ; hymenial surface is dark brown, pores regular, pore tubes indistinctly stratified (Bakshi 1971).

4. ***Trametes cubensis*** (Mont.) Sacc.

Occurrence. *T. cubensis* is recorded from all sal forests though it is not as common as the fungi listed earlier. The fungus also attacks laurel, axlewood and benteak—the common associates of sal and also *Eucalyptus citriodora* Hook.

Damage. Total volume of decay in sal and also infections due to *T. cubensis* are much lower than those caused by *H. rubiginosa* or *F. caryophylli.* However, average decay volume per infection by *T. cubensis* is much greater than that caused by other fungi (Bakshi, Rehill and Choudhury 1963). The fungus causes brown cuboidal rot. Thin white to pale mycelial mat is present in crevices of decayed wood.

Sporophore. Sporophores are usually sessile, applanate, corky becoming rigid later ; upper surface is white to light brown turning black at base, soft tomentose ; hymenial surface is white to pale brown, pores regular (Bakshi 1971).

5. *Trametes incerta* (Curry) Cooke

Occurrence. *T. incerta* is recorded on sal and also other hosts like laurel, lendi among others (Bagchee 1959).

Damage. The fungus causes dark discolouration in the heartwood followed by development of white pocket rot.

Sporophore. Sporophores are sessile, applanate, hard, woody; upper surface is dirty white, grayish turning black with age, crusty, rimose: hymenial surface is woody brown, pores irregular becoming lamellar when old (Bakshi 1971).

Other heart rot fungi in respect of which the extent of damage on sal is not known are *Fomes senex* (Nees. & Mont.) Cke., *Ganoderma applanatum* (Pers. ex Wallr.) Pat. and *Hexagonia sulcata* Berk.

Control of heart rot

Losses due to heart rot in sal may be prevented or reduced to an acceptable level through management practices and silvicultural measures which are mostly aimed towards minimising development of infection courts, particularly wounds, on trees. Control measures are outlined separately for high forests and for coppice forests.

High forests

Frost. In areas where frost is common, the injury to the young crop may be prevented by adjustment of the canopy. In the Uniform system, this is possible by adjusting the successive fellings in the final periodic block. In the Selection system, canopy should not be too much opened out to obtain regeneration.

Suppression. Suppression may not occur in an even-aged crop as in the Uniform system. However, in Selection system, trees in the lower storey may suffer from suppression due to a thick overhead canopy resulting in development of dead branches, branch stubs and knots, all of which constitute infection courts for heart rot fungi. Canopy adjustment is, therefore, important in preventing suppression and also damage by frost.

Fire. Fire protection measures in dry and in moist sal, particularly after the establishment stage, are important to minimise fire damage. Where accidental fires are common, it is advisable to practise early control burning in order to minimise the risk of severe fire injury. In addition, other fire protection measures may be adopted. During control burning, adequate precautions may be taken to keep down fire damage to the minimum.

Stand management. Thinnings and improvement fellings should always be carried out with a view to stand improvement when deformed, crooked, injured, diseased and dead trees should also be removed from the stand. During cultural operations, all trees showing injury due to fire, frost and also those exhibiting the important indicators of decay, like sporophores and punk knots, should be removed. This will not only remove the infection source but also aid in stand improvement.

During felling operations, falling trees may cause injury to residual trees. The resultant wounds become infection courts for decay fungi. Injury to the standing crop may also occur during logging operations. Such injuries can be prevented by lopping thick lateral branches prior to felling and through mechanised logging. Residual trees badly damaged during felling may also be marked for removal.

In companion trees exhibiting forking up to a height of about 1 m from the ground level, it is usual to mark one of them for felling leaving the other for future working to avoid drastic opening of the canopy. Basal rot, if present, is likely to extend as trunk rot in the two companion trees whose heartwood is bridged at the region of the fork. If after felling one of them, decay is noticed at the cut end, it may be advisable to fell the other. If, however, decay occurs as trunk rot in only one of the companion trees, the other may be left standing if decay is absent in the region of the fork.

Pathological rotation age. In stands where decay hazards are high, particularly from the early age, decay may become significant with age. In advanced age, destruction of wood due to decay may proceed at about the same rate or may even exceed the rate of production of new wood before the stand attains normal rotation age. For such stands, pathological rotation age should be worked out and the forests felled at that age to obtain maximum return of sound wood.

Coppice forests

The incidence of decay in stools in coppice sal forests is high in most cases. Forests which are worked on a short rotation age of about 10 years mainly for fuelwood may continue to be managed thus because the decay in the stool will not migrate into coppice at that age when no heartwood develops in the latter. However, the high incidence of decay in stools will limit any attempt to increase the rotation age of such forests to obtain healthy timber free from decay. In the management of such derelict sal forests, it is necessary to change over gradually from the coppice to the seedling crop in 10–15 years. For this, the area can be naturally regenerated for which silvicultural measures like fire protection, closure to grazing and contour trenching to withhold soil moisture are necessary to ensure regeneration and its establishment. The area may also be planted up as an alternative. When the seedling crop is established, the coppice crop may be felled and the stools prevented from further coppicing by girdling the stool to a depth of about 5 cm and killing the effective tissue which develops coppice with poisons like diesel oil, 2, 4–D or 2, 4, 5–T. Such decayed stools if left behind, will disintegrate in a few years and will not pose any danger to the seedling crop (Bakshi 1957).

STEM CANKER AND WITCHES'-BROOM

The disease has been sporadically recorded in South Kheri, Bahraich and Dehra Dun Forest Divisions, Uttar Pradesh and Mandla and Bilaspur Forest Divisions, Madhya Pradesh. The symptoms of the disease include knotted twigs, prominent cankers with terraced callus and formation of witches'-brooms. The causal organism is identified as *Macrophoma shoreae* Bagchee (Bagchee 1953).

The fungus probably infects smaller branches through hailstone injury. Pycnidia embedded in the cracked bark are dark brown to black, spherical to oblong and produce hyaline one-celled spores. The fungus produces prominent haustoria in the parenchyma of the wood and in the adjoining callus tissues. The disease was successfully reproduced in potted plants by artificial inoculations (Bagchee 1953).

SEBAÇINA DISEASE

Sebacina alutacea Wakef., is reported to attack bases of young sal saplings and is reported from Ganjam in Orissa (Wakefield 1922). The fungus produces grayish yellow, widely effused incrustations with sterile, white, fimbriate margin.

LEAF SPOTS

A *Cercospora* sp. causes diffused gray-coloured necrotic spots on leaves of sal in submontane region of Uttar Pradesh and Chotanagpur hills of Bihar. The necrotic regions fall off after the rains producing a shot hole symptom (Bagchee 1953).

Another leaf spot in sal due to an unidentified species of Sphaeropsidales occurs in overcrowded, unthinned sal forests of tarai region (Uttar Pradesh). The spots are at first pinkish, later turn grayish-brown with sunken pycnidia (Bagchee 1953).

SOOTY MOULDS

Three sooty moulds are recorded on sal leaves. *Meliola* sp. is recorded in tarai sal forests from North Kheri to Dehra Dun, Uttar Pradesh. The perithecia are stalked and elongate with crustose subicle. The asci have dark, elongate, muriform ascospores (Bagchee 1953).

Asterina sp. is recorded in the moist zones on the leaves of young sal. Hyphae are knotted to barrel-shaped, dark brown and produce sori consisting of radiating cells. Perithecia are dark and produce two-celled brown spores. The fungus also produces conidia (Bagchee 1953).

Capnodium sp. is recorded on sal leaves during spring and summer covering the upper surface as a black crust. Dark hyphae are superficial and do not penetrate the epidermis (Bagchee 1953).

The sooty moulds usually grow on a sugary 'honey dew' produced on leaves by certain aphids and scale insects. The hyphae of the fungi do not penetrate the living cells under the cuticle. The fungi, however, develop thick black coating on leaf surface resulting in some interference with the physiological functions of the host (Boyce 1961).

THREAD BLIGHTS

Marasmius gordipes Sacc., causing horse hair blight, and *Polyporus* sp. causing thread blight are recorded by Bagchee (1953) on young sal regeneration. *Corticium repens* Berk. causes white thread blight of young seedlings and saplings resulting in defoliation, death of young terminal shoots and consequent die-back of affected seedlings. Bagchee and Ujagar Singh (1954) recorded *Marasmius cupressiformis* Berk. causing black thread blight in sal. Huque (1964) recorded *Marasmius sarmentosus* Berk. causing thread blight in sal in Pakistan.

PHANEROGAMIC PARASITES

Four species of mistletoes, namely, *Dendrophthoe falcata* (L.f.) Ettingsh., *Helixanthera parasitica* Lour., *Loranthus scurrula* L. and *Macrosolen cochinchinensis* (Lour.) v. Tiegh. and *Cuscuta reflexa* Roxb. are recorded on sal. *D. falcata* and *H. parasitica* are serious parasites of sal in India and Pakistan (Browne 1968). *L. scurrula* and *M. cochinchinensis* are serious on plantation sal (Davidson 1945).

PHYSIOLOGICAL DISORDER

DYING OF SAL

Dying of sal occurs from time to time throughout its range of distribution. Some mortality may be attributed to parasitic causes, like *Polyporus shoreae*, while insects may be responsible for partial or severe defoliation, which, if continued for a few successive years, may result in trees becoming moribund or dead. However, most severe and extensive top-dying and mortality occurring in different parts of the country is not connected with any parasitic cause and appears to be due to physiological disorders (Boyce and Bakshi 1959).

Damage

In Ramnagar Division (Chilkia and Phanto Ranges), U.P., the dying of sal in high forests in quality class III is believed to be due to lessened soil moisture. A layer of sand, about 30-60 cm deep, mixed with gravel occurs at a depth of 180-210 cm below the surface and may represent the former bed of river which changed its course and now lies buried several metres below the surface owing to super-imposition of clayey belts (Troup 1921). This sandy layer is unable to retain moisture and also does not allow moisture to pass from the water-table located at great depths to the upper layers where roots mostly occur. Dying was also noticed on poor sites in other places in Ramnagar and Gorakhpur Divisions, Uttar Pradesh, where the soil is dry and compact due to heavy grazing and frequent fires and the undergrowth is scanty.

In Daltonganj Division, Bihar, dying of sal which occurred on a large scale around 1950 but which began sporadically some years earlier became catastrophic during the 1960's (Pl. 9, 3). Over an area of thousands of hectares, the dying was so widespread that sal which used to be a dominant species in the area became a minor component of the forest. The associates of sal also sufferred from mortality though to a lesser extent. The mortality was high in the western parts of the Division but decreased in intensity towards the eastern parts with moister and deeper soil and moderate ground vegetation. Further east, in Kolhan and Porahat Divisions, where rainfall is high, mortality was low and probably within the limit of normal dying of trees in the forest. In Orissa, large scale dying of sal in quality classes III-IV occurred in Sambalpur and Deogarh Divisions during the 1950's.

Symptoms

In general, the dying of sal is characterised by die-back of topmost branches in the crown. In course of time, dying extends downwards to the lower branches and the stem. Epicormic branches may develop profusely on the main stem in an effort on the part of the tree to overcome the adverse growing conditions. Finally, the tree dies in about 1-2 years. The root system remains healthy full of sap even in advanced stages of dying and is the last to die.

Cause

From an extensive survey of mortality areas, Boyce and Bakshi (1959) opined that dying of sal, particularly on poor sites, is due to adverse factors like grazing and burning which result in deterioration of site as typically depicted in large scale dying of sal in Daltonganj (Bihar). Over the years, the soil became compact and hard due to grazing. Much top soil is lost and erosion is severe due to lack of undergrowth or litter. Sal regeneration is absent. Rain does not percolate into the soil but flows away in the terrain. Repeated fires desiccate the soil. These two factors are responsible for lessened amount of soil moisture. Drought is likely to affect sal, particularly in dry type, where lowering of sub-soil

DISEASES OF SAL

I. Coppice sal forest. 2. Decay in stool bearing two old and one current year's growth. 3. Dying of sal. 4. *Hypoxylon mediterraneum*, stroma on dying stem. 5. Decay in sapwood due to *Polyporus shoreae*.

PLATE 10

ROOT ROT IN SAL

1. Mortality due to *Polyporus shoreae*, living trees exhibit epicormic branches o
stem. 2. Sporophore of *Polyporus shoreae*. (Photo—FRI).

moisture to a slight degree continuously for 2-3 years, may become a critical factor in the healthy growth of sal (Troup 1921). Mortality in sal in Bihar was also ascribed to drought by Seth, Khan and Yadav (1960).

Hypoxylon mediterraneum de Not. var. *microspora* Miller may be associated with dying or dead sal. The fruit bodies of the fungus appear on the bark as black stroma (Pl. **9**, 4) giving a charred appearance to the bark. In the dying trees, the stroma normally occurs on the dead bark immediately adjoining the living, on branches or main stem. As the dying continues downwards, the stroma appears on the dead bark though in rare instances, such stroma may also occur on the living part of the tree. The fungus attacks trees weakened due to other causes, starts out as a parasite and ends up with stroma developing on the dead wood (Bakshi 1963). Elsewhere, in the Mediterranean region, *H. mediterraneum* is reported to occur in epidemic form on different species of *Quercus*, *Eucalyptus* and other genera (Malencon and Marion 1951, 1952). The authors remarked that the fungus is not highly pathogenic but gradually kills the trees.

Control

As large-scale dying of sal is generally due to site deterioration, the most important single factor being the lessened amount of soil moisture, it is necessary to improve upon the site. This may be possible through various silvicultural measures. The site may be improved through measures like fire protection, closure to grazing and contour trenching, to hold water run-off and thus increase soil moisture. The deterioration of site occurred over many decades and it will take a long time to restore the site to a state where sal may regain its status as a normal component of a healthy forest. Sal regeneration will come up when site conditions improve. Maintenance of stand vigour will ward off the attack by *H. mediterraneum*.

14. SISSOO–DALBERGIA SISSOO ROXB.

GENERAL

Sissoo (*Dalbergia sissoo*) is a large deciduous tree with a loose spreading crown and is a strong light demander. The species occurs throughout the plains in North India, from Punjab to Assam, ascending upto about 100 m in the Himalayas. It grows gregariously on alluvial ground on river banks, as one of the first colonisers along with khair (*Acacia catechu* Willd.) and also comes up naturally on freshly exposed soils. The heartwood is strong and durable, brown with dark figures, for which it is prized for furniture and general constructional work. It is an important timber for gun carriage parts. It also makes good veneers. Due to the value of the timber, sissoo is raised in plantations. The species is also planted in avenues and along canal banks.

Given irrigation, it is a successful species for reclaiming sandy wastes and converting them into valuable plantation areas. The soil moisture is different at different depths (Deogun 1943). Above the water-table lies the capillary zone where water from water-table moves by capillary action. Above this is the field capacity zone which represents the maximum moisture content to which the soil can be wetted under the action of gravity. The top most layer, termed pellicular or dry zone, does not normally part with any moisture for use by sissoo roots. Under irrigated conditions, the surface soil is moistened to field capacity, the depth depending on the time, frequency and depth of irrigation channel. Flood and shallow (15 cm) irrigation tend to leave a dry zone below the growing roots and trees tend to remain shallow-rooted. Deep (30 cm) irrigation brings a broader profile under field capacity level for trees to send deeper roots into the capillary zone when further irrigation may not appear necessary.

Sissoo does not grow healthy in soils of heavy texture. The species can withstand high soil moisture under good drainage conditions. It cannot withstand water-logging even for a short period (Streets 1962) but can sustain long periods of drought.

PATHOGENIC DISEASES

WILT DISEASE

Occurrence

Wilt disease is prevalent in the states of Uttar Pradesh and Punjab in plantations raised on unsuitable sites. The disease is absent in natural forests and in plantations on suitable sites.

Predisposing factors

The incidence of wilt is correlated with soil texture and soil moisture. The disease is absent in soils containing a high proportion of sand and a correspondingly low silt and clay, provided the site is not water-logged and drainage is good. With decrease in sand and proportionate increase in silt and clay, the disease begins to manifest, the incidence becoming greater as the proportion of sand decreases and that of silt and clay increases (Bakshi, Arora and Sujan Singh 1957). Absence of wilt in natural sissoo forests where soil is loose and sandy supports these observations. In stiff and clayey soils where aeration is poor, roots do not develop healthy and may also die from asphyxiation. *Fusarium solani* enters through dead and weakened roots from where it progresses into healthy tissues.

Wilt in sissoo, like many vascular wilts, is favoured by high soil moisture in sites with poor drainage. In bhabar tract immediately below the foothills of the Himalayas, the natural water flows at a great depth. With good rainfall and adequate drainage, sissoo generally grows healthy in the bhabar tract. However, in the tarai, lower down the bhabar tract, the natural water is brought to the surface by the underlying clayey deposits. The tarai is thus swampy with high water-table which may come up to the surface during the rains. Sissoo does not grow healthy in the tarai and may suffer from wilt.

Symptoms

Wilt disease is systemic and manifests itself during rains between July and September. Trees of advanced age are usually susceptible to the disease. Symptoms of the disease include yellowing and death of leaves in acropetal succession up the tree. Eventually the entire tree presents a yellow appearance. In later stages, leaves drop off rendering the branches increasingly bare (Pl. **11**, 1). Affected trees die within a few months. The pathogen is mostly restricted to roots and in some cases may extend to stem bases. The outer sapwood exhibits a characteristic pink to reddish pink stain (Pl. **11**, 3). In diseased roots, vessels are plugged with hyphae and infiltrated with jelly-like substances which hinder sap flow to the crown resulting in wilt symptoms. Unidentified toxins are produced by *F. solani* in culture and these toxins may also be responsible for causing wilt.

Cause

Sissoo wilt is caused by *Fusarium solani* (Mart.) Appel & Wr. f. *dalbergiae* Gordon (Bakshi 1954, Bakshi and Sujan Singh 1954). The latter authors (1959) proved the pathogenicity of the fungus. *F. solani* is a normal soil inhabitant and possesses wide powers of saprophytic colonisation (Bakshi 1955). The fungus exhibits tolerance to a wide range of pH in culture with optimum growth around pH 4.6. pH of healthy and diseased roots lies between 4.6 to 5.2 (Bakshi, Arora and Sujan Singh 1957) and is, therefore, ideal for infection by the wilt pathogen.

Control

Wilt in sissoo is common in plantations raised on stiff soils with inadequate drainage. If sissoo is raised on a site having light textured soil with adequate soil moisture and good drainage, trees grow healthy free from wilt disease.

GANODERMA ROOT ROT

Root disease in sissoo due to *Ganoderma lucidum* (Leyss.) Karst. is common both in natural forests and in plantations. Trees of advanced age are normally attacked by the pathogen. The disease is prevalent in both light and heavy textured soils. In light soils, the pathogen spreads rapidly on roots and trees are killed in a short period. In stiff soils, however, the fungus spreads slowly on roots. The fungus spreads through root contact and, therefore, in a pure plantation, the spread of the disease is rapid resulting in large gaps in the forest. The affected trees exhibit a stagheaded appearance (Pl. **11**, 2) in which condition they may continue for a few years before being killed. A detailed account of the pathogen, its mode of infection and spread and control are given in Chapter 15.

POLYPORUS GILVUS ROOT AND BUTT ROT

Polyporus gilvus (Schw.) Fries possesses a wide host range among important broadleaved species like sissoo, khair, sal (*Shorea robusta* Gaertn. f.), rosewood (*Dalbergia latifolia* Roxb.), babul (*Acacia nilotica* (Linn.) Willd. ex Del. ssp. *indica* (Benth.) Brenan), bijasal (*Pterocarpus marsupium* Roxb.), siris, oaks among others. The fungus is primarily a wound parasite and mostly attacks sissoo in plantations and may be found in association with *G. lucidum* or the wilt pathogen, *F. solani.* Trees of advanced age are attacked. Affected trees exhibit stagheaded condition. Sporophores which develop readily on root and stem bases are sessile, brown and corky. The fungus causes restricted decay as white rot in the sapwood and to a limited extent in the heartwood (Bakshi 1971). Prevention of injury to trees will minimise attack by *P. gilvus*.

Serious wilt of sissoo seedlings in nurseries due to *Fusarium oxysporum* Fr. which also causes seedling wilt in many other forest species and *Rosellinia aquila* (Fr.) de Not. are reported from Pakistan. Web blight caused by *Pellicularia filamentosa* (Pat.) Rogers causing considerable damage to sissoo seedlings and other tree species is reported from India and Pakistan (Browne 1968).

LEAF SPOTS

Cercospora sissoo Syd. attacks leaves of sissoo, mostly on the lower surface, producing yellowish to grayish-green discolourations. Pustules are mostly intra-epidermal. Stromata are brown with simple or forked conidiophores. The disease is recorded from Uttar Pradesh and Delhi (Sydow and Mitter 1933).

Colletogloeum sissoo (Syd.) Sutton, causing imperceptible spots and recorded from Varanasi (U.P.) on sissoo leaves, is similar to *Cercospora sissoo* from which it differs only in size and number of septation of conidia (Pavgi and Singh 1971).

Phyllachora dalbergiae Niessl attacks upper surface of leaves and produces shining black, cushion-like stromata which occur scattered or in clusters. The fungus is recorded from Calcutta (West Bengal) and Pusa (Bihar) (Saccardo 1883).

Phyllachora spissa Syd. forms densely aggregated, dot-like dark stromata on irregular brownish infection spots in sissoo. The parasite is recorded from Wynaad (South India) and Khasi and Jaintia Hills (Assam) (Saccardo 1926).

Phyllosticta sissoo Died. produces round to irregular, grayish-brown infection spots which sometimes cover the entire leaf surface in sissoo. Dark brown pycnidia are produced on lower surface of leaves in densely aggregated groups. The fungus is recorded from Pusa (Bihar) (Saccardo 1931, Da Costa and Mundkur 1948).

Mycosphaerella dalbergiae Muller & Ahmed causing leaf spot in sissoo is reported from Pakistan (Browne 1968).

POWDERY MILDEW

Phyllactinia dalbergiae Pirozynski, the powdery mildew, produces yellowish, persistent dense mycelium on the lower surface of sissoo leaves. The fungus is recorded from Dehra Dun, Allahabad (U.P.), Pusa (Bihar), Poona, Bombay and Nagpur (Maharashtra) (Pirozynski 1965).

LEAF RUSTS

Maravalia achroa (Syd.) Arthur & Cumm. attacks leaves and juvenile twigs of sissoo and rosewood seedlings in nurseries. The affected parts are killed resulting in die-back and subsequent death of affected plants. Uredinial sori are yellowish and formed on the lower surface of leaves. Telia are colourless and pulvinate. The rust on seedlings in nurseries may be controlled by sulphur fungicides (Bakshi and Sujan Singh 1967). The disease is recorded from Bihar and U.P.

Uredo sissoo Syd. & Butler attacks sissoo of all ages in nurseries, plantations and natural forests. Rust sori mature about January when infected leaves fall off alongwith uninfected leaves during normal leaf-fall. Uredia appear about November on the lower surface of leaves as minute, powdery, densely scattered pustules surrounded by numerous paraphyses. The rust occurs all over sissoo forests in North India (Bakshi and Sujan Singh 1967).

PHANEROGAMIC PARASITES

Tolypanthus involucratus (Roxb.) v. Tiegh. (= *Loranthus involucratus* Roxb.) attacks branches and twigs. The mistletoe which develops into bushes of different sizes occurs in sissoo forests in North India. *Cuscuta reflexa* Roxb., *Dendrophthoe falcata* (L.f.) Ettingsh. and *Loranthus pulverulentus* Wall. are also recorded on sissoo in India. Other records include

Tapinanthus dodoneifolius (DC.) Dans. from Nigeria and *Tapinanthus* sp. from Sudan (Browne 1968).

The mistletoe on sissoo may be controlled with copper sulphate and fernoxone (Kadambi 1954) at dosages of 1/4 oz of the former and 1/32 oz of the latter per tree, injected into holes made into the sapwood and directed downwards at 45 degrees. The mistletoe bush begins to dry up within a few days and dies in 2 months. The treatment becomes more effective when carried out in spring, when new flush appears, than in autumn. At higher dosages, the chemicals are toxic to the host plant.

POD FUNGI

Glomerella cingulata (Stonem.) Spaulding & Shrenk and *Septothyrella dalbergiae* Pavgi & Singh are recorded on pods from Varanasi (U.P.) (Pavgi and Singh 1971). A species of *Phoma* is common on pods of sissoo forests in western U.P.

OTHER DISEASES

Other disease records on sissoo include heart rots due to *Fomes robiniae* (Murr.) Sacc. & D. Sacc. and *Ganoderma applanatum* (Pers. ex Wallr.) Pat. from India, branch canker due to *Hypoxylon hypomiltum* Mont. from Pakistan, horse hair blight caused by *Marasmius equicrinis* F. Muell. from India and sooty mould due to *Meliola bicornis* Wirt. from Ghana (Browne 1968).

PHYSIOLOGICAL DISORDER

In non-irrigated plantations, widespread mortality in sissoo occurs in different plantations in U.P. State. During 1971, 20–30 per cent mortality was reported in a pure sissoo plantation in Piplee Block, Bareilly Forest Division and Sahanpur, Bijnor Forest Division and in a mixed sissoo-khair plantation in Gangapur-Patia, Tarai and Bhabar Forest Division. In general, the dying trees exhibit pale yellow foliage, die-back, stagheadedness and finally death of affected trees. Trees thus weakened may be attacked by the wilt pathogen, *F. solani* or the root rot fungus, *G. lucidum*. In the latter case, affected trees may be wind-blown.

In the tarai, the water-table is high and may come upto the surface during rains creating water-logged conditions. As a result, roots die of asphyxiation affecting the health of trees which become stagheaded and finally die. Sissoo can withstand high soil moisture only for a short period under conditions of good drainage. The species cannot withstand water-logged condition even for a short duration (Streets 1962) but can tolerate fairly long dry periods. Selection of suitable site with loose texture, adequate soil moisture with good drainage is, therefore, important in raising sissoo healthy.

In irrigated plantations, mortality in sissoo occurs under shallow irrigation when irrigation is discontinued as happened in Khandakheri, Karnal Forest division, Haryana State. As explained earlier, shallow irrigation encourages shallow root system which is unable to draw water from the water-table and trees dry out under low rainfall conditions. In dry locations, therefore, deep irrigation is recommended when trees develop deep root system which can draw water from the water-table when irrigation is stopped (Deogun 1943).

PLATE 11

DISEASES OF SISSOO

1. A wilted tree in a natural sissoo forest. 2. Stagheaded trees in a plantation.
3. Discolouration in outer living wood in sissoo root due to *Fusarium solani.*
4. *Ganoderma lucidum,* fruit bodies at tree bases. (4, Photo—FRI).

PLATE 12

DISEASES OF KHAIR

1. *Ganoderma lucidum* sporophores at tree bases. 2–4. Heart rot due t *Fomes badius*: 2. Sporophore. 3–4. Decay in logs. (2, Photo—FRI).

15. KHAIR–ACACIA CATECHU WILLD.

GENERAL

Khair (*Acacia catechu*) is a moderate-sized deciduous tree with a light feathery crown. The heartwood is light red to dark brown. The wood is largely used for extraction of cutch and *katha* which are obtained by boiling heartwood chips. Cutch is largely used for dyeing and tanning, while *katha* is used mainly as an ingredient for chewing betel leaves (*pan*). The wood is also used for posts and agricultural implements. It is also excellent for fuel and charcoal. The species grows gregariously on alluvial beds along river courses as one of the first colonisers of such soils, along-with sissoo (*Dalbergia sissoo* Roxb.). It also occurs in drier locations away from water courses mixed with other species characteristic of dry regions. The species is widely distributed in India. Khair is frost-resistant and fire-hardy. It can be severely damaged by porcupines which gnaw the bark and in extreme cases, may girdle the trees. Characteristic deep cankers of unknown origin occur commonly on trees and they are characterised by the presence of narrow zones of callus tissue. Due to value of the timber and its importance as a cash crop tree, khair is one of the preferred species raised in plantations.

PATHOGENIC DISEASES

GANODERMA ROOT ROT

Ganoderma lucidum (Leyss.) Karst. causes serious root rot mortality in reforested stands not cleared of residual roots and stumps. Bakshi, Reddy and Sujan Singh (1975) studied the problem in detail, a brief account of which is given below.

Site preparation

In preparing site for plantations, degraded natural forests are clear-felled. Large stumps and roots are removed and debris is burnt. The soil is worked and two agricultural crops are raised during the first year. In the second year, contour trenches 0.3 m wide, 0.5 m deep and 5 m apart are made. Stumps and roots inside the trenches are removed upto the depth of the trenches. All the same, old roots may remain on the sides and below the base of trenches. Also, a large number of stumps and roots remain in between the lines. Before the rains, the trenches are filled and sown with khair, mostly pure or sometimes in mixture with sissoo, semul (*Bombax ceiba* Linn.), ailanthus (*Ailanthus excelsa* Roxb.) and other species when such mixture generally is not intimate but in pure blocks. Normal operations include cleaning and weeding. In khair, thinning commences in the third year and is continued in subsequent years at regular intervals.

Infection

Infection due to *G. lucidum* may be traced back to residual stumps and roots of the previous forest cover. The pathogen possesses a wide host range among broadleaved species which go into the composition of natural forests. The common genera including species attacked by *G. lucidum* are *Acacia* (*A. catechu*, *A. auriculaeformis* A. Cunn., *A. melanoxylon* R. Br. and *A. nilotica* (Linn.) Willd. ex Del. ssp. *indica* (Benth.) Brenan), *Acrocarpus* (*A. fraxinifolius* Wt.), *Albizia* (*A. chinensis* (Osbeck) Merr., *A. lebbeck* Benth. and *A. procera* Benth.), *Areca* (*A. catechu* Linn.), *Cassia* (*C. javanica* Linn., *C. nodosa* Ham. and *C. siamea* Lam.), *Casuarina* (*C. equisetifolia* Forst.), *Cocos* (*C. nucifera* Linn.), *Dalbergia* (*D. sissoo* Roxb. and *D. latifolia* Roxb.), *Delonix* (*D. regia* (Bog. ex Hook.) Raf.), *Eucalyptus* (*E. tereticornis* Sm. so called Mysore hybrid), *Hevea*, *Jacaranda* (*J. acutifolia* Humb. & Bonpl.), *Lannea* (*L. coromandelica* (Houtt.) Merr.), *Mangifera* (*M. indica* Linn.), *Melia* (*M. azedarach* Linn.), *Morus* (*M. alba* Linn.), *Pinus* (*P. roxburghii* Sargent), *Pleiogynium* (*P. cerasiferum* (F.v.M.) Parker), *Pongamia* (*P. pinnata* (Linn.) Pierre), *Pterocarpus* (*P. marsupium* Roxb.), *Sterculia* (*S. villosa* Roxb.) and *Terminalia* (*T. alata* Heyne ex Roth) among others. Species of *Eucalyptus* and *Populus* are also among the susceptible hosts. The fungus is normally endemic in natural forests where it does not cause any serious epidemic. This is due to the fact that in a natural stand, some species may be resistant to the pathogen, while others may exhibit varying degrees of susceptibility. Very susceptible hosts may be killed outright. In less susceptible species, the fungus colonises a small mass of roots in relation to the large mass of healthy roots reacting strongly against infection. Such trees continue to live though they harbour the fungus in some roots. When the forest is clear-felled, the host resistance is lost and *G. lucidum* quickly spreads on residual roots and stumps to build up a high inoculum potential. If such a site is not cleared of infected stumps and roots, the susceptible species subsequently planted in the stand become infected.

Seedlings of khair develop deep tap roots which may come in contact with old roots infected by *G. lucidum* lying deeply buried in the lines. The lateral roots which are poorly developed initially, become long spreading when seedlings are 3–4 years old. Such lateral roots may come in contact with old infected roots and stumps in between the lines and contract the disease. Both types of infections are responsible for initiating new infection centres in the lines. These new centres are responsible for subsequent lateral spread of the disease in the line by root contact between diseased and adjoining healthy plants.

The development of new infection centres in lines declines with increase in plantation age because the infected roots of the previous stand disintegrate in course of time. On the other hand, the lateral spread of the disease which is nearly absent in the early age increases with plantation age. This is explained by the facts that in the early age, lateral roots are poorly developed and, therefore, the lateral spread of the disease through root contact is minimum. Also, in the early age, the infected plants are killed and their small mass of roots disintegrates in the soil quickly to be of any consequence in the lateral spread of the disease. However, with increasing age, the chances of lateral root contact become greater and the incidence of lateral spread of the disease increases, resulting in death of adjoining plants in a row.

Damage

G. lucidum infects roots through intact as well as injured surfaces. The pathogen attacks and kills the bark and causes white spongy rot in the sapwood. Khair is susceptible to attack by *G. lucidum* at all ages. Affected plants exhibit pale foliage which eventually dries up. Young plants are killed soon after infection. Mature trees also are killed when most of the roots become affected. The fungus produces thin white mycelial mat between the bark and the wood. It possesses ectotrophic growth in the form of small, thin mycelial aggregates on bark of roots and also on solid objects, like stones and soil lumps which are in direct contact with infected roots. It does not possess any free mycelial spread in the soil. Its spread from a diseased to a healthy plant occurs by direct root contact. Therefore, in line plantings, the spread is rapid from plant to plant once the infection establishes in the line. Mortality starts from the third year and increases with age of the plantation. At 7 years' age, mortality as high as 40 per cent has been recorded.

Host susceptibility

In plantations where khair and other preferred species like sissoo, semul, ailanthus, kanju (*Holoptelea integrifolia* Planch.) and teak (*Tectona grandis* Linn. f.) are raised on a reforested stand not cleared of residual stumps and roots, the risk of *G. lucidum* root infections is high. Khair and sissoo are the most susceptible hosts, semul is attacked only when vigour is poor, while others are resistant. Pure culture inoculations on these hosts confirm this observation.

Sporophore

Fruit bodies of *G. lucidum* commonly develop on affected plants, usually at the base (Pl. **12**, 1) or apparently on the ground but in the latter case, the fruit bodies are invariably attached to decayed roots. Due to presence of sporophores, it is easy to locate the infected plants.

Sporophores are usually stalked, rarely sessile, corky, later woody. Stalk and upper surface are shiny laccate, lightly zoned. Hymenial surface is white when fresh turning light brown on drying, pores minute (Bakshi 1971).

Control

Mortality in khair due to *G. lucidum* may be minimised or kept under control by adopting one or more of the following procedures :

1. **Mechanised plantations.** *G. lucidum* is endemic in the natural forests but becomes epidemic if a plantation is raised after clear-felling a natural broadleaved forest without clearing residual roots and stumps. The disease risk is minimised to a large measure by raising mechanised plantations where stumps and roots are mostly removed. However, in this operation also, small infected roots may be left behind which may serve as sources of infection. If any mortality occurs because of infection from such sources, the dead trees should be removed along with decayed roots. It may be mentioned that mechanised plantations possess the advantage of better soil working thus leading to increased plant vigour which, in species like semul, may impart increased resistance against the disease.

2. **Extraction of stumps.** If it is not possible to raise mechanised plantations, the incidence of the disease can be minimised if the

stumps and roots present in the lines are removed completely, unlike the current practice where they are removed only upto the depth of the trench.

3. **Isolation trenches.** Once the disease establishes in the line, its further spread is rapid from plant to plant through root contact. This lateral spread can be checked by isolating the row of diseased and dead plants from the adjacent healthy plants by digging trenches about 0.3 m wide, 0.7 m deep and 1.5 m long across the line. It may be advisable to dig an additional trench which should include one apparently healthy plant immediately after the first trench. Such plants may harbour infection but the infection may not be in a sufficiently advanced stage to produce disease symptoms. This will provide an additional safeguard in containing the disease within the trenched area.

4. **Planting resistant species.** In a reforested stand where the risk of *G. lucidum* root rot mortality is apprehended, it may be advisable not to plant susceptible species, like khair and sissoo, in the first rotation but to raise resistant species, like ailanthus and semul. Ailanthus has a rotation age of about 10 years and semul about 40 years. At the end of the rotations, susceptible species may be planted as, by then, the decayed roots are likely to disintegrate and source of infection will thus no longer be present.

5. **Mixed stands.** If, however, susceptible species, like khair and sissoo, are raised in a reforested stand with high risk of root disease, it may be necessary to raise them in intimate mixture with resistant species, like semul and ailanthus, which will break the continuity of a pure line of khair and other susceptible species and thereby prevent rapid spread of the disease in the plantation.

POLYPORUS GILVUS ROOT ROT

Polyporus gilvus Fr. is a wound parasite attacking roots and tree bases causing white spongy rot with zone lines in sapwood. The fungus is a weak parasite and does not cause any serious damage to forests.

FOMES BADIUS HEART ROT

Occurrence

Fomes badius Berk. causes heart rot in khair and is common in all khair forests, both natural and planted, throughout its range of distribution.

The fungus is mostly restricted to khair and occasionally attacks babul (*Acacia nilotica* ssp. *indica*) and lal-khair (*A. chundra* (Roxb.) Willd.).

Infection and damage

Fomes badius is a wound parasite. The fungus infects khair through injuries due to mechanical causes and damage caused by animals.

The fungus causes decay in the heartwood only (Pl. **12**, 3). Sapwood remains healthy and, therefore, infected trees continue to grow and may exhibit all outward appearances of healthy trees. However, the heartwood is decayed progressively with age and becomes useless for obtaining cutch and *katha*. Furthermore, multiple infections are common as evident from the development of sporophores at different heights on the tree. This results in development of decay at different regions of the tree resulting in almost complete loss of the tree.

In initial stages of decay, the colour of the heartwood changes to a deeper shade. In advanced decay, the hard, durable and heavy heartwood becomes yellow, spongy and light and presents a mottled appearance due to formation of long, black, indistinct streaks and zone lines (Pl. **12**, 4).

Sporophore

Sporophores develop freely on affected trees and, therefore, their presence is a ready means of identifying diseased trees. Sporophores are perennial, sessile, hard, woody and easily detachable from the host. The upper surface is brown or black, cracking with age and the hymenial surface dull brown with minute pores (Bakshi 1971).

Control

1. Since *Fomes badius* is a wound parasite, control of avoidable injuries inflicted to trees will reduce the incidence of the disease. Porcupine-proof fence in plantations will ensure freedom from injury to trees due to the animal.

2. Sporophores of the fungus commonly develop on diseased trees. Being perennial, they remain infective for many years through basidiospores that develop in them. It is, therefore, necessary to systematically remove and destroy the sporophores. These sporophores can be easily knocked off by a pole, collected and burnt.

3. In khair, trees above 20 cm d.b.h. are marked for felling. Green but diseased trees below this exploitable diameter class are normally not removed. No useful purpose is served by retaining the diseased tree in the stand because the heartwood becomes progressively decayed and their early removal may enable salvaging of greater volume of sound wood. Also, such trees develop sporophores and thereby increase the risk of infection to healthy trees. Therefore, during cultural and particularly felling operations, all decayed trees, as identified by the presence of sporophores, should be removed irrespective of their diameter class.

Other fungi causing heart rot in khair are *Fomes fastuosus* (Lev.) Cke., *F. senex* (Nees & Mont.) Cke. and *Ganoderma applanatum* (Wallr.) Pat.

POWDERY MILDEWS

Erysiphe acaciae Blumer is recorded on foliage of khair from Poona. The disease is uncommon and unimportant (Browne 1968). *Phyllactinia acaciae* Syd. is recorded on foliage of khair from Sagar, Madhya Pradesh (Tandon and Chandra 1963–64).

POWDERY LEAF SPOT

The disease is caused by *Microstroma acaciae* Syd. which produces snowy-white tufts on lower surface of leaves of khair. Stromata are sub-epidermal. Basidia burst out from the stromata. The disease is reported from Majhgawan (U. P.) (Sydow, Mitter and Tandon 1937).

LEAF RUST

Ravenelia tandonii Syd., the leaf rust, is common on khair in North India. The infected pinnules are shed prematurely. Minute uredinia and telia develop below the cuticle on the upper surface of pinnules. Spore heads are chestnut brown. Cysts are round to elliptical. Pedicels are hyaline, short and deciduous (Bakshi and Sujan Singh 1967).

PHANEROGAMIC PARASITE

Macrosolen cochinchinensis (Laur.) v. Tiegh. is the only mistletoe recorded on khair (Browne 1968).

16. TEAK–TECTONA GRANDIS LINN. F.

GENERAL

Teak (*Tectona grandis*) is indigenous to India, Burma, Thailand, Indonesia and other islands in the Far East. The species grows naturally and is also planted in central and southern India, while in the North, it is widely planted but the quality is average. However, even the low quality teak is considered valuable. It is a large deciduous tree with a tall and cylindrical bole and a rounded crown. The species is not naturally gregarious. The majority of teak forests occur naturally on hilly or undulating terrain and flat alluvial ground with good sub-soil drainage combined with fair rainfall. Teak is normally a component of mixed deciduous forests. The species is a strong light demander, is sensitive to frost and drought and possesses a great power of resisting the effects of fire than most of its associates though fire may cause hollowness in lower part of the bole.

In the moister and better type of teak forests, the species is worked under the Selection system. In plantations, the normal rotation age is about 80 to 100 years. In dry parts of India where teak does not reach any great size, the system of coppice-with-standard is largely followed with a rotation age of 30 or 48 years.

PATHOGENIC DISEASES

BACTERIAL WILT

Occurrence

The disease has a wide distribution in the tropics and warmer parts of the temperate zone. The wilt pathogen, *Pseudomonas solanacearum* (E.F. Smith) E.F. Smith has a wide host range in plants belonging to 17 families of which Solanaceous plants are most susceptible (Spaulding 1961). Wilt of teak seedlings has been recorded from the Philippines (Rolden and Andres 1953), Malaysia (Mitchell 1962) and Burma (Doo 1968).

Infection

The bacterium occurs in the soil and infects plants mainly through injury. Inside the host, the pathogen is located in the vessels which show signs of discolouration and necrosis.

Symptoms

Small patches of brown tissue first appear between the veins in the leaf, sometimes spreading to give it a scorched appearance. This is followed by wilting of leaves and the plant. The brown spotting may not always occur and younger leaves may completely wilt followed by yellowing of the whole plant. The dark discolouration of vascular tissues occurs in the roots. The main root and the lower part of the stem become soft and moist, later turning black and get decayed. The effect is usually more rapid on young seedlings than on old transplants but may be fatal to either. The disease may take a heavy toll of seedlings and may threaten to disrupt the raising of the nursery stock.

Causal organism

The bacterium is rod-shaped with rounded ends, motile with a single polar flagellum, non-spore forming, aerobic, gram-negative and non-acid fast. Thermal death point is 52°–53° C, optimum growth occurs between 25°–30° C. *Pseudomonas tectonae* Rolden & Andres is probably the same species (Kelman 1953 quoted from Mitchell 1962).

Control

1. High soil moisture and, in some cases, water-logging in the nursery favour the bacterial wilt in teak. Suitable site should be selected for nurseries with good drainage. Low-land areas should be avoided.

2. The affected plants should be dug out and burnt.

3. Root injuries during weeding or transplanting constitute the chief avenues for the entry of the parasite. Weeding may be minimised by using chemical fertilizers instead of farmyard manure as the latter encourages weed growth.

4. In "sick" nursery beds, the soil should be sterilised by treatment with formalin, about 2 weeks prior to planting.

5. Nurseries where the disease occurs for successive years should be abandoned for sometime and a new nursery site for teak selected on

unbroken forest soil in which the possible introduction of contaminated soil should be avoided.

ARMILLARIA ROOT ROT

Armillaria mellea (Vahl.) Quel. causes root rot in teak in south Celebes, Indonesia (Van Hall 1923), Sudan (Hocking 1966) and in Nyasaland (Gibson and Corbett 1964) where the pathogen occurs on a wide range of hosts. In the last instance, rhizomorphs of the fungus do not develop in majority of cases and the mycelium under the bark of the host frequently possesses small perforations behind the growing edge. Though both these field characters suggest that the disease is caused by *Armillaria tabescens* (Scop. ex Fr.) Emel., the pathogen produces vigorous rhizomorphs in culture on which basis, it may be identified as *A. mellea*, which varies between strains and with environment.

FOMES LAMAOENSIS ROOT ROT

F. lamaoensis (Murr.) Sacc. and Trott., the cause of brown root rot is recorded on teak in Indonesia (Spaulding 1961).

FOMES NOXIUS BASAL ROT

Fomes noxius Corner causes basal rot in teak in plantations in Indonesia (Van der Goot 1935) and Tanzania (Browne 1968).

HELICOBASIDIUM COMPACTUM ROOT ROT

Occurrence

H. compactum Boed. causes root rot in teak and also other plants like coffee, tea, rubber and a variety of different hosts (Boedijn and Steinmann 1930). The fungus is also recorded on *Albizia falcataria* (L.) Fosberg from Indonesia, chir from Union of South Africa (Spaulding 1961) and on teak in East Africa (Hocking and Jaffer, 1966, 1966 a, 1967). Serious damage to teak, possibly due to *Helicobasidium* sp., is reported from Sudan (Vidal Hall and Williams 1956).

Symptoms and damage

The disease occurs both in the nursery and also in newly established plantations in East Africa, particularly on sites with poor drainage. In Lunguza teak nursery, the disease was responsible for the loss of 10,000 seedlings (Hocking and Jaffer 1966 a). Although the disease spread rapidly in young plantations at the rate of 2.5-5.5 m in the year (Hocking and Jaffer 1967), its importance declined sharply as the trees matured (Hocking 1966).

The fungus forms an extensive mycelium in the soil, hyphae from which run up to the roots of plants, penetrate the tissues and form a cushion-like covering, brownish-violet when dry and dark purple-brown to black when wet, on the base of the stems, often girdling them. The symptoms include decline of the foliage, production of sprouts followed by death of the plant.

Causal organism

Hyphae are brown, branched, septate and thin-walled. Basidia, which are borne directly on hyphae, are at first straight, later curved.

Control

In East Africa, the disease is absent in nurseries established on well-drained sites. In plantations where disease may be severe mostly on level, low-lying and poorly-drained sites, trenching to isolate diseased patches is advised (Hocking and Jaffer 1966 a). The spread of the disease may be arrested when dead and dying trees are removed, possibly as a result of admission of sunlight (Vidal Hall and Williams 1956).

PENIOPHORA ROOT ROT

Occurrence

P. rhizomorpho-sulphurea Bakshi & Sujan Singh has so far been recorded only on teak in plantations in Dehra Dun where this pathogen, in association with *Polyporus zonalis* Berk., causes root rot and wind-throw of trees (Bakshi, Sujan Singh and Ujagar Singh 1966).

Infection and damage

P. rhizomorpho-sulphurea generally colonises teak stumps after coppicing or thinning. The fungus eventually migrates from decayed stumps into the coppice shoots and also progresses into roots upto the distal ends from where healthy roots of adjoining trees coming in contact with them become infected. The fungus also develops rhizomorphs which spread from infected trees into the soil upto a depth of 1 m and laterally upto 2–3 m or more. In the soil, rhizomorphs are thick and rarely branched. However, after coming in contact with roots, they become thinner, branch profusely and their tips penetrate the host. The rhizomorphs are initially white and develop deep yellow shade later.

The fungus attacks only the sapwood in which it causes yellow fibrous rot. Yellow mycelial mat of the fungus develops in the decayed wood. In late stages of decay, the wood tends to separate along annual rings giving the appearance of laminated decay. The fungus is mostly restricted to roots and affected trees become wind-blown due to decayed roots losing foothold in the soil.

Teak is a pronounced light demander and requires adequate space for root development. Thinning is, therefore, important in stand management (Troup 1921). Wind-throw of trees as a result of root rot exercises a beneficial effect in an overstocked stand by providing spacing, if large openings are not created and the spatial distribution of stems in the remaining crop is fairly regular (Bakshi, Reddy, Puri and Sujan Singh 1972). Such a beneficial effect was also shown in a Douglas fir plantation affected by root rot (Foster and Johnson 1963). However, if the root disease spreads and results in further wind-throw to reduce the stocking to a level lower than that prescribed, it will be detrimental to the stand and will adversely affect productivity.

Sporophore

Sporophores are thin, appressed, and sulphur-yellow in colour (Bakshi, Sujan Singh and Ujagar Singh 1966).

Control

It is difficult to control a pathogen like *P. rhizomorpho-sulphurea* which has the ability of saprophytic colonisation and spreads freely in the soil by rhizomorphs. The disease is serious in wet locations where coppicing of teak may be avoided.

POLYPORUS ZONALIS ROOT ROT

Occurrence

P. zonalis has been recorded in plantations of teak from Dehra Dun (Bakshi, Sujan Singh and Ujagar Singh 1965) and Cooch Behar, West Bengal (Bakshi, Reddy, Puri and Sujan Singh 1972). The fungus is also recorded on living *Bauhinia purpurea* Linn., jaman (*Syzygium cumini* (Linn.) Skeels), *Ficus benghalensis* L. and also on stumps and dead wood of various timber species (Bakshi, Sujan Singh and Balwant Singh 1963).

Infection and damage

P. zonalis causes root and butt rot. It attacks heartwood though sometimes sapwood also is attacked. In the early stages, the fungus causes white mottled rot with orange-yellow lines. Eventually, pockets develop in the bleached areas. Decay of roots and butt results in wind-throw of trees.

Sporophore

P. zonalis commonly develops sporophores at the bases of trees in imbricate clusters. Sporophores are usually sessile, leathery when fresh, hard and rigid when dry. Upper surface is pinkish-buff when young, becomes reddish-brown with age and is concentrically zonate. Lower surface is white to pinkish with a silky sheen.

Control

P. zonalis is a wound parasite. Freedom from injury to trees is, therefore, important in preventing attack by *P. zonalis*. Infection may originate from the nursery stock and, therefore, only healthy stumps should be selected for planting.

OTHER ROOT ROTS

Xylaria thwaitesii Cke. from Indonesia, *Polyporus shoreae* Wakef. from Nigeria and *Ustulina deusta* (Fr.) Petrak are other root rot fungi recorded on teak (Browne 1968).

PINK DISEASE

Pink disease caused by *Corticium salmonicolor* B. & Br. occurs commonly in teak in high rainfall areas of Kerala and Karnataka States. However, the disease is not so far damaging to teak in India. Elsewhere, the disease is recorded from Indonesia (Schwarz 1926). The pathogen attacks trees of all ages, infection occurring through lenticels and wounds. Infected young branches may be killed outright, while infection in older branches results in the formation of cankers and longitudinal bark-cracking. Cankers resulting from *C. salmonicolor* probably serve as infection courts for heart-rot fungi which subsequently decay the heartwood and cause hollowness when decayed wood gets disintegrated. The incidence of hollowness was found to be as high as 73 per cent (Altona 1926).

The pathogen produces four growth forms out of which only three, namely, cobwebby, pustule and pink incrustation forms are recorded on teak. The disease is described fully under *Eucalyptus* (Chapter 25).

NECTRIA STEM CANKER

Nectria canker on teak is caused by *Nectria haematococca* Berk. & Br. which is known in its imperfect state as *Fusarium solani* (Mart.) Appel & Wr. In the latter form, the fungus is recorded from Tanzania and East Africa, where it causes cankers on scattered groups of middle-aged planted teak established on a low-lying ground with high water-table which is, therefore, unsuitable for teak. The cankers appear as slightly sunken bark fissures upto 40 cm long, mostly located at or near the ground level, accompanied by a slight flow of reddish sap. Removal of surrounding bark exhibits pink stain in the underlying wood. When teak seedlings are inoculated with the pathogen through collar wounds, lesions are produced which later extend, girdle and kill the plants (Hocking 1968).

In its perfect state, *Nectria haematococca* is recorded on teak in plantations in northern India from Dehra Dun and Gorakhpur (U.P.). Trees die from top downwards, leaves turn yellow and are shed off prematurely. Water-soaked lesions develop under the bark, followed by the appearance of conidiophores bearing conidia. Later, pink to red perithecia develop in the crevices of the bark. The pathogenicity of the fungus has been proved (Bagchee 1947).

DECAY IN DRY COPPICE TEAK

In dry locations in western part of central India, teak is worked on a system of coppice-with-standard on a rotation of 30 or 48 years. At the end of the rotation, trees commonly exhibit unsoundness at the base and this often extends upto 2–4 m in the stem. The incidence of unsound trees may be as high as 50 per cent. The unsoundness in the heartwood is due to decay caused by *Fomes lividus* (Kalchbr.) Sacc. and *Polyporus zonalis* causing white spongy and white mottled rot respectively. A species of *Phialophora*, a wood-inhabiting non-decay fungus, is also associated with decay columns in teak. Prior colonisation by *Phialophora* conditions the wood to make it more susceptible to attack by decay fungi in culture (Sujan Singh and Tewari 1970) and this may hold true in the tree.

In view of the fact that incidence of decay in dry coppice teak is high, factors predisposing trees to decay were analysed and also measures whereby such decay could be prevented were worked out by Sujan Singh, Puri and Bakshi (1972), a brief account of which is given below :

Decay in relation to type of shoots (Pl. 13, 1–5)

In the system of coppicing as laid down in Working Plan prescription, trees are cut flush with the ground for obtaining coppice growth. In practice, for the ease of wielding the axe or saw, the trees are felled above the ground and the cut surface is made sloping at the edges to make it flush with the ground. Two types of shoots may originate from a stool, namely, callus shoots which arise out of callus growth resulting from the cambial activity at the cut end of the stool between the bark and the wood, and side shoots arising due to activation of dormant buds on the side of the stool. Callus shoots which tend to occlude the parent stool during their growth, get organically connected with the stool. Decay when present in the stool migrates into callus shoots when heartwood develops in them and establishes a link with the decaying heartwood of the stool. On the other hand, side shoots, originating at or below ground level, develop their own root system and in course of time generally become independent of the parent stool. Such shoots, therefore, remain free from decay as their heartwood develops no organic connection with that of the stool. In the system of coppicing at ground level, incidence of decayed trees is high because the system encourages development of callus shoots, while side shoots are fewer due to cutting away of dormant buds.

Control of decay in relation to stand management

High coppicing discourages the development of callus shoots owing to the rapid drying and shrinkage of the wood at the cut surface causing its separation from the bark (Troup 1921). In high coppicing in a dry locality, the base of the stool will still remain alive even if the upper part dries up and this is likely to encourage development of low side shoots. It is thus evident that high coppicing is likely to discourage development of callus shoots and encourage growth of low side shoots. This was confirmed from a detailed field study in dry coppice teak forests of Gujarat State where coppicing heights were adjusted from 0 to 25 cm at interval of 5 cm. Callus and high side shoots developed less frequently with increasing coppicing heights while the trend was reverse for low side shoots. A coppicing height between 10–15 cm was found ideal for encouraging development of low side shoots with corresponding low development of callus and high side shoots. At these heights adequate number of low side shoots develop which are recommended to be retained for the future crop, as low side shoots develop healthy free from any decay even if decay may be present in the stool. Elsewhere, working with oak stump sprouts, Roth and Sleath (1939) and Roth and Hepting (1943) remarked that sprouts arising above the ground level are likely to develop decay and those arising at or below ground level are likely to remain free from decay because the chance of the heartwood bridge between the sprout and the stool is more in the former than in the latter. In case of teak, field observations show that low sprouts normally remain healthy free from decay.

Heart rot in teak may also be caused by *Ganoderma applanatum* (Wallr.) Pat. and *Polyporus rubidus* Berk.

LEAF SPOTS

Cercospora tectonae Stevens causes dull-brown to grayish-brown or white angular spots on teak leaves. Spots coalesce into large blotches. The pathogen is recorded from Hawaii, Trinidad and India (Spaulding 1961).

Phyllosticta tectonae Syd. & Butl. is reported on teak leaves from Assam (Da Costa and Mundkur 1948).

Sphaceloma tectonae Bitanc causes white leaf spots on teak leaves in Poona, Maharashtra (Sarbhoy, Girdharilal and Varshney 1975).

Xanthomonas melhusi Patel, Kulkarni & Dhande, the cause of bacterial leaf spot in teak, is reported from Maharashtra State. Infection spots are angular, several coalescing into large lesions which are water-soaked, translucent, turning brown to deep brown when dry and raised on the upper surface due to accumulation of bacterial ooze (Patel, Kulkarni and Dhande 1952).

Calderiomyces sp. causes severe infection on lower surface of teak leaves in Dehra Dun (U.P.). However, the damage is not significant (Bakshi, Reddy, Puri and Sujan Singh 1972).

POWDERY MILDEWS

Phyllactinia corylea (Pers.) Karst. attacks teak leaves (Bagchee 1952).

Phyllactinia guttata (Fr.) Lev. forms white, powdery coating on teak leaves forming minute, globular dark coloured cleistothecia over the white fungus weft. It attacks the lower surface of leaves of teak and many broad-leaved trees listed by Spaulding (1961) resulting in defoliation. The fungus is recorded from North America, Europe and Asia.

Uncinula tectonae Salm. is common on teak in central and southern India. It also attacks leaves of *Cordia macleodii* (Chatterji 1912). The mildew is widespread in teak nurseries and forests. However, the mildew may not appear to be seriously damaging as the infected leaves are shed along with healthy leaves during normal leaf-fall. Due to heavy coating of leaves by fungus growth, photosynthetic activity is retarded which may affect growth.

The mycelium of *U. tectonae* is restricted to the upper leaf surface and adheres to the host by serrated appressoria. The infected leaves appear to be coated with a dull white mycelium dotted with black fruiting bodies presenting a bluish-gray mealy appearance. In the teak forest, nearly all trees may become infected. The mycelium sends in haustoria which are restricted in the upper epidermal cells only. Conidia which develop abundantly on the leaf surface are wind-dispersed to cause fresh infection. Hot days and cool nights favour conidial germination and growth of the fungus (Damle 1960). The mildew begins to appear soon after the rains. The mycelium bears conidiophores bearing conidia in chains. In November, conidial production declines and perfect state in the form of cleistothecia develops. Appendages develop profusely all over the cleistothecium except on its basal side and are unbranched with the tips uncinate or helicoid. Sclerotial bodies with few or no appendages and resembling cleistothecia in appearance develop on the host. However, no asci develop in them. These sclerotia are meant for over-summering and germinate under favourable conditions (Damle 1960).

LEAF RUST

Olivea tectonae (Racib.) Thirum., the teak leaf rust, occurs throughout the range of distribution of the host in India. It also occurs in Bangla Desh (Khan 1951). The disease is especially common in the nurseries.

Leaves are commonly attacked usually from October to February. At first uredia develop followed later by telial sori which may be intermixed with uredia or produced in separate sori. Uredial and telial sori are orange-yellow and occur on the lower surface of the leaf and almost plaster the undersurface. The upper surface of the leaves may present a gray appearance due to the formation of flecks which correspond to the position of the sori on the lower surface.

The rust was earlier known only in the uredial stage and described as *Uredo tectonae* by Raciborski (1900). Based on telial stage, the rust was placed under *Chaconia* by Ramakrishnan and Ramakrishnan (1949) and *Olivea* by Thirumalachar (1949). The rust belongs to *Olivea* which is characterised by telia developing on thick cellular base, a character which is absent in *Chaconia.*

Infection due to teak leaf rust causes premature defoliation. The damage is especially severe in nurseries due to density of the crop. Early defoliation results in retardation of plant growth.

The teak rust can be controlled by application of sulphur-based fungicidal spray particularly in the nurseries where such a spray is economically justifiable due to the density of the crop. In the forest, the rust may be kept under control by opening the canopy (Khan 1951).

THREAD BLIGHT

Marasmius sp. causing thread blight in teak is recorded from Indonesia and Burma. Affected leaves develop, on the undersurface, a white mycelium originating from a thick light brown cord following the main vein and petiole. Mycelial cord is somewhat inconspicuous due to the presence of leaf hairs and a lustre due to higher air content of the infected cell layers. Branches and leaves are penetrated, the latter dying and turning brown, thin and papery (Schwarz 1926).

PHANEROGAMIC PARASITES

The mistletoe, *Dendrophthoe falcata* (L.f.) Ettingsh. (*Loranthus longiflorus* Dest.) is widespread on teak in India. The mistletoe is also common on teak in Indonesia and Trinidad. The infection results in mutilation and death of branches or tops and finally in death of trees in case of heavy infections. Sanitation cutting of affected branches, 0.3 metre below the point of infection and, if necessary, of trees, particularly when trees are leafless, keeps the parasite under control. The method, however, is laborious, time-consuming and uneconomical. Chemical herbicidal spray with 1 : 1-dimethyl 4 : 4-bipyridylium during early summer when teak is leafless is efficaceous. The chemical is not absorbed through the non-chlorophyllous part of the tree and is washed off into the soil where it gets inactivated (George 1966).

Phthirusa adunca (G.F.W. Mey) Maguire attacks young teak plants in West Indies. Promising results on control were obtained by spraying the affected plants with 5 per cent paraquat and 0.2 per cent 2, 4-D (Anon. 1965).

Other mistletoes recorded on teak are *Macrosolen cochinchinensis* (Laur.) v. Tiegh., *Phoradendron piperoides* (H.B.K.) Trelease from West Indies and *Tapinanthus* sp. (Browne 1968).

ALGAL PARASITES

Stomatochroon sp. from Indonesia (Bakshi 1964) and *Cephaleuros* sp. from Nigeria and Tanzania (Gibson 1964) are the algal parasites recorded on teak.

PHYSIOLOGICAL DISORDER

WATER BLISTER

The phenomenon of exudation of sap in certain teak plantations in Nilambur, Kerala State, has been known for many years (Lushington 1896). This phenomenon is seen in teak plantations located along river banks where water-table is high and, therefore, abundant soil moisture is available to the trees. In such locations, trees may exude sap from the stem and such exudation occurs mostly at the basal part of the tree and sometimes upto a height of about 3.5 m from the ground. The sap drips down bark and eventually dries up leaving a brownish to blackish discolouration (Pl. **13**, 6). In an active blister, if the point of exudation of sap is opened into the wood to a depth of a few millimeters, the fluid comes out in the form of a spout (Pl. **13**, 7) and may be collected by litres. The sap flow eventually stops from the blister when it will not bleed further if injured in the manner stated above. Water blister is not observed in plantations raised on high ground or on sites with low water-table.

The exuding sap is light yellow and becomes slimy on the bark from contaminations by microorganisms which cause disintegration of the underlying bark and sapwood over a small area of the trunk resulting in vertical scar exposing the dead wood.

Cause

The freshly exuded sap is free from microorganisms and, therefore, the cause is not parasitic. The primary cause of water blister is high root pressure. Under conditions of high soil temperature and abundant soil moisture during the period when trees are leafless, transpiration is cut down considerably but root pressure is not reduced but builds up forcing water up the stem. In teak, internal shakes result is longitudinal split in the main bole. The sap accumulates in the split in an increasing pressure till it finds release through the rupture in the thin bark. The extractives from the wood leach into the sap which becomes light yellow in colour. The sap contains various inorganic salts. This, together with the fact that water blister is mostly restricted at lower heights of the tree where the root pressure is maximum, are good circumstantial evidences to suggest that the cause of water blister is physiological due to high root pressure (Bakshi and Boyce 1959).

Exudation of water from stems and leaves of intact plants is commonly met with under conditions that are favourable for low transpiration and high absorption of water from the soil. Bleeding is known to occur in birch (*Betula alba*) where the sap flows through a cut branch and can be collected by the litre (Ingold 1959). Such bleeding also occurs in sycamore (*Platanus* sp.) (James and Baker 1933) and in sugar maple (*Acer saccharum* Marsh.), where large quantities of sap will escape during late winter and early spring through tap holes made in the lower part of the trunk. The root pressure has been found to be as high as 9 atmospheres in *Aesculus hippocastanum*, 6 in native trees of Austria, 2-2.5 in *Betula lenta*, about 1 in sugar maple, 1.25 in grape vine and 2-4 in Monterey pine (*Pinus radiata* D. Don.) (Miller 1938).

Damage

Blister always indicates a shake or split in the main stem, causing reduction in the usable timber. Also, at the regions of water blister, secondary microorganisms including decay fungi may establish and the wood becomes decayed and sloughed off leaving large wounds.

Control

Since the cause of water blister is physiological and the phenomenon occurs on certain sites, nothing can be done with trees showing water blister or to prevent the disease. As the wood is eventually damaged in

PLATE 13

DISEASES OF TEAK

1–5. Heart rot in coppice teak: 1. Development of callus shoots. 2. Callus shoot occluding stool. 3. Decay in logs. 4. Side shoot with stool displaced. 5. Development of side shoots. 6–7. Water blister: 6. Scar due to sap flow. 7. Water spouting from stem.

PLATE 14

DISEASES OF CASUARINA

1–2. *Trichosporium vesiculosum* : 1. Plantation showing diseased trees with blistered bark and *T. vesiculosum* spore mass. 2. Infection through lopping wound. 3. Root grafts between trees. 4. A failed plantation.

diseased trees, it is desirable to remove such trees during improvement fellings and thinnings. It is advisable not to plant teak in very moist sites where the disease usually develops (Bakshi and Boyce 1959).

DYING OF TEAK

In Mandvi, Maharasthra State, extensive dying of 25 years old coppice teak occurred during 1954-57 when more than 70,000 trees were affected. The mortality later subsided. The dead and dying trees exhibited long open scars, extending from the ground level to a distance of about 2.5–6 meters up the trunk, mostly oriented on the southern face of the stem. The scars might have been caused due to excessive heat in open stands to which thin-barked teak is susceptible though wounds did not appear to be connected with widespread dying of the species. The cause of the mortality was probably a non-infectious agency (Boyce 1959).

In South Chanda Forest Division, Maharashtra State, mortality occurred in the natural teak forests over 1962 hectares when about 30,000 trees were marked for dry felling in 1968. Trees of all diameter classes were affected, mortality being higher in larger diameter classes. Trees died from top downwards and roots were the last to dry up. No pathogen was associated with dying trees. Mortality was ascribed as due to drought from October to May during 1964-67 when shallow-rooted and drought-sensitive species like teak suffered most (Bakshi, Reddy, Puri and Sujan Singh 1972).

In Bori, Hoshangabad Forest Division, Madhya Pradesh, teak mortality occurred during late sixtees and early seventees. No pathogen was present. In affected trees, growth rings during certain years were compact and consisted of early wood vessels followed immediately by late wood consisting of thick-walled fibres with small lumen. This is unlike the normal wood where development of early wood vessels is followed by a band of thin-walled fibres which gradually merge into late wood. This may suggest that trees were growing under certain abnormal conditions, possibly of drought (Bakshi, Reddy, Puri and Sujan Singh 1972).

YELLOWING IN TEAK

Abnormal yellowing in teak accompanied by pronounced curling of leaf margins and premature leaf fall was reported from Trinidad by Pawsey (1970).

17. CASUARINA EQUISETIFOLIA FORST.

GENERAL

Casuarina equisetifolia, an exotic tree on the mainland of India, has now become naturalised. The species grows naturally in the Andamans, Bangla Desh, Burma, Malaysia and Australia. *Casuarina* is extensively planted in India along the sea coast for afforesting sandy beaches, binding shifting sand dunes and adding beauty to the landscape. The species is also raised along river banks and in plantations mixed with *Eucalyptus*. *Casuarina* is fast-growing and the rotation varies from 6 to 10 years or in some cases upto about 18 years. The species is valued for fuelwood due to its high calorific value. The species thrives well in sandy loam, sand mixed with river silt or where the sand deposit covers rich alluvial arable land. Seedlings are sensitive to drought and, therefore, watering is done for the first one or two years after seedlings are transplanted in the field, usually daily for the first 6 months and less frequently thereafter. Later on, plants become drought-resistant as the species possesses xerophytic adaptations in the form of scaly leaves.

In sandy locations, *Casuarina* produces well-developed and spreading laterals (Pl. **14,** 3) which mostly remain on or near the surface and may extend upto about 3 m from the tree. Tap root is not strongly developed and may go down to about 2 m. Roots produce nodules. Lateral roots dichotomise repeatedly due to infection by bacteria (McLuckie 1923) which fix atmospheric nitrogen (Narasimhan 1918, Mowry 1933, Bond 1957). The organism in the nodules is also believed to be an *Actinomyces* (Uemura 1964). When active, the nodules are pink but become black and shrivelled when dead. In plantations raised near human habitations, the villagers indulge in illicit lopping of branches and also regularly sweep the plantation floor to remove needles for use as fuel. While this practice lessens the risk of fire to which *Casuarina* is very susceptible, it removes the only source of replenishing the soil even though needles decay slowly.

PATHOGENIC DISEASES

BACTERIAL WILT OF SEEDLINGS

Pseudomonas solanacearum (E. F. Smith) E. F. Smith causes seedling wilt in *Casuarina* in Mauritius. The pathogen occurs on a wide range of hosts. The common symptoms of the disease include sudden wilt and shrivelling of leaves. The root tissues of wilted plants show brown discolouration of vascular bundles (Browne 1968).

SEEDLING BLIGHT

Pellicularia rolfsii (Curzi) West causes seedling blight in *Casuarina* in the Philippines. Besides *Casuarina*, the pathogen attacks seedlings of many tree species in warmer climates where it is of universal occurrence (Spaulding 1961).

BLACK ROOT ROT OF SEEDLINGS

Macrophomina phaseolina (Tassi) G.Goid. causes black root rot of *Casuarina* seedlings in Karwar (Karnataka) and Thana (Maharashtra) in India (Qureshi 1956). The disease is also reported from Uganda. The pathogen attacks through fine roots spreading slowly upwards forming small black sclerotia and, rarely, fertile fruit bodies on the surface (Browne 1968).

GANODERMA ROOT ROT

Ganoderma lucidum (Leyss.) Karst. root rot in *Casuarina* is recorded from South India and West Bengal (Butler and Bisby 1960).

PHYTOPHTHORA ROOT ROT

The disease, caused by *Phytophthora cambivora* (Petri) Buisman, is recorded on *Casuarina* from Mauritius (Browne 1968). The pathogen is widely distributed both in the temperate and tropical regions. It occurs in India on ornamental plants but has not been recorded on any tree species (Butler and Bisby 1960).

ARMILLARIA ROOT ROT

Armillaria mellea (Vahl.) Quel., the root-rot pathogen, is recorded on *Casuarina* from Mauritius (Browne 1968).

STEM WILT

Occurrence

The disease, caused by *Trichosporium vesiculosum* Butler, was first recorded in India by Butler (1905) from Ganjam District, Orissa. The disease was later recorded in plantations raised in Sriharikota Island, Andhra Pradesh and along banks of river Cauvery (Bakshi 1951), from coastal plantations in Digha (West Bengal), Rameshwaram (Tamil Nadu) and Junagadh (Gujarat) (Bakshi, Reddy, Puri and Sujan Singh 1972). The disease, therefore, appears to be widely distributed, particularly in coastal *Casuarina* plantations. Outside India, the disease is recorded from Mauritius (Spaulding 1961).

Infection and damage

T. vesiculosum produces numerous spores which are dispersed by wind and may infect healthy stems usually through injury caused during pruning or illicit lopping of branches. By this means, the fungus is able to establish in plantations. After infection on the stem, the pathogen travels up the stem and also down into the roots, where it spreads mostly on the laterals upto the distal end. The secondary spread of the disease in the plantation then occurs by means of root contact and root grafting, both of which are common in view of the long spreading network of lateral root system in *Casuarina* (Pl. 14, 3). Therefore, the disease spreads rapidly in plantations. The pathogenicity of *T. vesiculosum* has been proved by Marudarajan, Ramakrishnan and Soumini (1950).

In a heavily diseased area, mortality may occur every year when dead trees are removed. In an assessmènt conducted in a 5-year old *Casuarina* plantation worked on a rotation age of 7 years, Bakshi (1951) found that 45 per cent trees had been removed earlier. In the then existing stock, about 50 per cent were healthy and the remaining dead or dying due to attack by *T. vesiculosum.*

Symptoms

Infected trees exhibit symptoms of wilt, like paling of needles which later dry up. Blisters develop on the bark due to formation of spore mass beneath the bark. The blisters eventually rupture in the form of scales or flakes and the black powdery mass of spores, appearing like charcoal dust, becomes exposed. The fungus also sporulates on the roots. Since the disease spreads by root contact, diseased trees occur in groups. The manifestation of the disease on the bark in the form of black spore masses of the pathogen is not common in plantations showing high incidence of mortality due to the disease (Butler 1905, Bakshi 1951). Dry atmosphere prevailing in plantations may possibly explain low incidence of diseased trees manifesting sporulation of the fungus.

Causal organism

T. vesiculosum produces colourless to brown hyphae in host tissues and colourless mycelial cushions in the cambium. Fertile hyphae are colourless and bear cluster of spores at or near the tip on the thickened end of the main axis or of secondary or tertiary branches. Spores are sessile, brown or black in mass, ovoid and form a dense layer in the cambium but are fewer in the tissues (Butler 1905).

Control

Illicit lopping and pruning of branches should be stopped to prevent the primary establishment of the disease in a plantation. However, if the disease has already secured a foot-hold in a plantation, its spread can be checked by trenching diseased trees which occur in groups. In diseased plantations where mortality occurs every year, the dead and diseased trees should be removed as early as possible. This will remove the source of infection and thereby minimise incidence of the disease.

PINK DISEASE

Pink disease caused by *Corticium salmonicolor* Berk. & Br. is reported from Mauritius (Browne 1968). The disease has been discussed in detail under *Eucalyptus* (Chapter 25).

HEART ROTS

Polyporus glomeratus Peck, *Fomes fastuosus* (Lev.) Cke. and *Fomes senex* (Nees & Mont.) Cke. cause decay of heartwood in standing trees (Butler and Bisby 1960, Bakshi, Reddy, Puri and Sujan Singh 1972).

LEAF SPOT

Phoma casuarina Tassi causes leaf spots and is recorded from Karwar (Karnataka) and Calcutta (West Bengal). The disease is uncommon and causes negligible damage (Browne 1968).

PHANEROGAMIC PARASITES

Three phanerogamic parasites, namely, *Cuscuta campestris* Yuncker from Fiji, *Dendrophthoe falcata* (L. f.) Ettingsh. from India and Pakistan and *D. lanosa* (Korth.) Dans. from Malaysia are recorded on *Casuarina* (Browne 1968).

WITCHES'-BROOM

In the American tropics, *Casuarina* exhibits symptoms of witches'-broom with multiple branching. The succulent green stems become irregularly mottled with yellow to yellowish green areas in the dark green twigs. The symptoms also include considerable shortening of internodes, death of leaders and branches and sometimes finally of the tree. The symptoms suggest that the disease may be of virus origin though no transmission studies were made (Wellman and Grant 1951).

PHYSIOLOGICAL DISORDER

FAILURE OF PLANTATIONS

Complete failure of *Casuarina* plantations (Pl. **14**, 4) occurs in coastal areas of different parts of the country. The plants in such locations present a bushy habit due to the development of a few weak branches from the base and from which the main leader cannot be distinguished. These plants never attain any height and keep on struggling for life.

Cause

Based on a field study, Bakshi (1951) concluded that the failure of *Casuarina* in Sriharikota Island (Andhra Pradesh) was mainly due to successive failure of monsoons combined with deficiency of soil nutrients. These places normally experience drought conditions. Even in years of normal rainfall, the precipitation is mostly restricted between July and December, while during the remaining months, the rainfall is low or absent. In *Casuarina*, most of the root system consists of superficial laterals which are well adapted to absorb any atmospheric precipitation or condensation and this is scarce in years of drought or during the months of low rainfall. In the dry season, the water-table may remain at about 4–5 m depth and may come within 1 m from the surface during the rains. Though the water-table is high, enough capillary force is not developed in sandy soils, especially when sand particles are coarse, during the dry period, to raise water from the water-table to the surface where most of the roots are located. Even though *Casuarina* is drought-resistant, in advanced stages of growth at any rate, water is essential for plants to perform the vital functions. Therefore, such prolonged periods of drought adversely affect plant growth. Due to moisture deficiency in the soil, root nodules become shrivelled and dead and, therefore, the only source of enriching the soil with nitrogen in an otherwise barren substrate is also lost. In Balukhand plantations (Orissa State) certain areas, particularly those adjoining the canal joining the Sur lake and the sea, complete failure is possibly due to the lowering of the water-table as a result of the Sur lake cut and subsequent deepening of the cut.

Control

In raising healthy stands, it is necessary to improve the vigour of plants at an early age. This is possible by manuring in pits prior to transplanting seedlings. It is a common experience that in an area where there is a general failure, plants grow healthy on sites previously occupied by nurseries where the soil is well manured. Based on their studies on nutrient requirements of *Casuarina* by foliar analysis, Qureshi, Kaul and Srivastava (1967) suggested that a balanced dose of fertilizer containing N, K and Mg may result in healthy growth of plants in areas where the growth is stunted. The seedlings are sensitive to drought and, therefore, watering should be carried out with strict vigilance during the early stages soon after tansplantation. Deep planting will encourage development of long tap root system which will be able to draw water from water-table and meet the requirement of the plant from the third year onwards when watering is discontinued. Leaf litter should not be removed from the plantation floor but allowed to decompose to form humus and enrich the soil and at the same time improve its texture (Bakshi 1951).

SALT INJURY

Coastal plantations of *Casuarina* may suffer from injury from salt-laden wind from the sea. Plants of all ages may become affected as observed in coastal plantations in Gujarat State. Symptoms include drying up of branches and sometimes the tree. Planting seedlings in pits prevents damage to plants till they reach the surface. Species resistant to salt injury may be planted to serve as a shelter-belt behind which *Casuarina* may be raised.

18. SANDAL-SANTALUM ALBUM LINN.

GENERAL

Sandal is valued for its strongly scented heartwood which yields a fragrant oil used mostly in perfumery and medicine. The timber is used for carving and other fancy work. The wood is sold by weight, on which basis, it is probably the most valuable wood in the world. During the year 1973–74, the export of sandalwood, sandalwood chips and dust and sandal oil amounted to approximately 400 tonnes, 0.4 million kg and 93,000 kg respectively earning a total foreign exchange equivalent of Rs. 43.5 millions (Anon. 1974).

All species of *Santalum* occur in the southern hemisphere where they are indigenous except *S. album* which, in addition, also occurs in South India in the northern hemisphere where sandal is believed to have been introduced from Indonesia in the remote past. By another view, sandal is believed to be indigenous to India where it occurs mainly in Karnataka and Tamil Nadu States. Sandal forests flourish best in the upland plateaux between 650 to 1150 m and rainfall between 60 to 115 cm under conditions of good drainage. Though sandal grows at its best on rich, fairly moist and fertile soil, it frequently occurs on rocky ground where trees occur mainly in open scrub forests. In such locations, trees do not attain large dimensions but possess more highly scented wood. The species does not occur in dense high forests. Being a light demander, sandal grows well under a light canopy.

Sandal is a root hemi-parasite and possesses a wide host range of nearly 150 species with the roots of which sandal forms root attachments (Rama Rao 1911). Of these, some constitute 'good' hosts on which sandal thrives, while others are 'bad' hosts on which sandal grows poorly (Venkata Rao 1938). The choice of hosts as nurse crops of sandal is, therefore, important. Barbar (1906) and later Rao (1942) studied root haustoria of sandal. Ramaiah, Parthasarathi and Rao (1962) demonstrated the presence of K, Ca, Mg, Fe, Zn and Cu in the haustoria, suggesting that these elements are taken up by the sandal from the host plant. However, sandal is also able to obtain its mineral nutrients to some extent directly from the soil as demonstrated by the high cation-exchange activity of sandal even if spiked (Parthasarathy, Gupta and Rao 1971) when the mineral content in such trees gets reduced.

PATHOGENIC DISEASES

DAMPING-OFF

Seedlings suffer from damping-off due to an excessive amount of dense, wet, herbaceous undergrowth in high rainfall locations which hinders reproduction (Troup 1921).

SPIKE DISEASE

Occurrence

The spike disease of sandal is known only from South India and has not been reported on sandal that occurs naturally in Java or on any other species of *Santalum* occurring naturally in Malaysia and Australia. The first record of occurrence of spike disease of sandal was made by McCarthy (1898) from Coorg though it is likely that the disease was in existence earlier. The disease spread rapidly with the prevailing wind from Coorg eastwards and also towards north in the sandal belt of Karnataka. The disease also spread to Tamil Nadu State. The disease is so devastating that large areas where sandal once flourished have been laid barren. Death of trees results in serious loss in the forest revenue. In view of the fact that spike disease on *Santalum* is unknown outside India and also the assumption that sandal is not indigenous to India but was introduced in the remote past, Boyce (1959) opined that spike disease which possibly occurs on indigenous hosts and on which it is harmless and unnoticed, got transferred to sandal finding the exotic highly susceptible.

Symptoms

The most striking symptom of the spike disease is the severe reduction in the size of leaves, both along length and breadth, and shortening of the internodes so that the leaves become closely crowded together. At the same time, leaves show a tendency to stand out stiffly from the branch giving a rosette appearance to the affected branches. The leaves stand out like spikes from which the disease is so named. In some cases, the infected shoots may show continuous apical growth and assume a drooping habit giving a pendulous appearance to the spike. With the progress of the disease, the leaves become yellowish and may acquire a reddish tinge later. The symptoms which initially are present only on a few branches, gradually spread to the whole tree. As the disease progresses, the new leaves become smaller and smaller. In advanced stages of the disease, the whole tree may become spiked when the size of leaves gets reduced to a mere 0.8 and 0.15 cm along length and breadth respectively as against the normal leaf size of about 7.5 cm by 2.5 cm. The ratio of length/breadth in spiked leaves is thus much higher than that in healthy leaves. The affected trees usually die within 1–2 years. In healthy sandal trees, there is a fairly well-marked periodicity in leaf development. Majority of young leaves begin to develop during spring and reach their full development during autumn. Such periodicity is, however, not noticed in spiked branches and fresh flush of new leaves are produced throughout the year (Coleman 1917). A branch or a tree showing the disease in advanced stages never bears flowers or fruit. If flowers appear before a branch becomes spiked, they may develop into fruits but then, seeds may abort. In an infected branch, flowers may exhibit phyllody. In infected trees, the haustoria and root tips die.

In some diseased trees, symptoms may be masked. However, if such trees are pollarded, the symptoms manifest in the new flush that develops. Stain tests using Mann's stain, carbol thionin and Giemsa stain distinguish spiked from healthy tissues (Parthasarathy, Gupta and Rao 1966). Such tests are, however, not positive in diseased trees showing incipient or latent infection which may be detected by the presence of a specific protein which stains blue with bromophenol (Nayar 1968).

In diseased sandal, there is a pronounced accumulation of starch in branches and leaves. Increased starch content is considered to be due to lower diastatic activity in spiked leaves as compared to that in healthy ones (Coleman 1917). The same author further showed that spiked plants contain lower amounts of calcium and total ash content and higher nitrogen content in contrast to healthy plants. The ratio of Ca/N can be used as a diagnostic feature for determining whether or not the plant is infected. Calcium deficiency in the vegetative parts is traced to non-mobility of the same from the roots where it is stored (Iyengar 1960). In spike disease where plants remain in the vegetative stage only, the nitrate reductase activity is higher than in healthy plants, thus indicating the high rate of nitrogen utilisation which is possible due to rapid synthesis of viral protein in spiked leaves (Parthasarathy, Ramaiah, Manjanatha and Rao 1962). Also, spiked plants show lower iron content resulting in chlorosis of the leaves (Parthasarathy and Rao 1962). The ascorbic acid (AA) level in the leaves of spiked sandal is higher than that of the healthy plant at all stages of development as observed by Parthasarathy, Ramaiah and Rao (1963) who further stated that since AA is known to exert an inhibitory effect on the amylolytic activity, it may possibly be a contributory factor for the abnormal accumulation of starch noticed in spiked sandal leaves. The authors analysed that the higher respiration rate in the diseased tissues cannot be ascribed to higher AA content but to other enzymatic respiratory systems. In the spiked tissues of sandal, the normal functioning of Kreb's cycle is affected with its consequential disturbance on the biosynthesis of key amino-acids (Ramaiah, Babu and Rao 1964). As compared to healthy sandal leaves, those from spiked trees show a higher level of polyphenolase activity (PPA) due to which phenolic and tannin contents rise (Parthasarathy, Babu and Rao 1970). The higher PPA, apart from iron deficiency, may be a contributory factor for the chlorotic condition in spiked leaves. These authors further showed that as compared to healthy leaves, respiration rate is higher in spiked leaves whose indoleacetic acid content is also higher in young stages and results in the generation of a large number of buds in spiked trees.

Cause

Different views were held earlier as to the cause of the disease. It was believed to be due to intensified carbon assimilation (Butler 1903), death of haustoria and root tips resulting in cutting-off of supplies of salts and water to the plant which dies of exhaustion (Barber 1903), attack by virus (Hodgson 1918) which was demonstrated to be transmissible by grafting and which, in some cases, may be spread by insects (Coleman 1917, 1923) though virus was never detected in diseased plants but for which positive serological diagnostic tests were obtained (Nayar and Padmanabha 1967), an unbalanced circulation of sap which may be caused by a number of different factors (Hole 1917), attack by a bacterial agency which is an ultra-microscopic organism (Fisher 1918), defective lime supply to affected plants barring the infective nature of the disease (Iyengar 1960).

Sandal spike belongs to the 'yellows' group of plant diseases with which there is resemblance in symptoms such as reduction of leaf size, development of witches'-brooms and phyllody and in the transmission of the disease by leaf hoppers. Recent researches (see review by Hull 1971) show that many of the 'yellows' group of diseases which were earlier considered to be due to viruses are now attributed to mycoplasma-like bodies (MLBs). Dijkstra (1968), Varma, Chenulu, Raychaudhuri, Prakash and Rao (1969) and Hull, Nayar and Horne (1969) demonstrated the presence of numerous pleomorphic bodies in infected leaves in sieve tubes only but not in phloem parenchyma or companion cells. There is no evidence of cell-wall in these bodies which are bounded by smooth unit membranes. The structures are spherical to ovoid and vary in size from 40–750 nm. In plant mycoplasmas, the ribosomes are 14–15 nm across and occur as random components or small aggregates without regular array and contain slender fibrillar-like strands corresponding to DNA component seen in animal mycoplasma. Only a few instances are known where MLB's from plants have been isolated in pure culture and Koch's postulates proved (Hull 1971). Nayar and Padmanabha (1970) claim to have isolated the mycoplasma from diseased sandal in pure culture and successfully inoculated on healthy sandal to cause little leaf in sandal and deformation, yellowing and puckering of leaves in *Stachytarpheta*. Mycoplasma was reisolated from inoculated plants and its presence in the leaf verified under electron microscope.

Disease transmission

The spike disease is likely to spread through the soil from diseased to healthy roots through haustorial connections (Coleman 1917). This may explain the spread of the disease in an infected area, though this may not explain isolated occurrence of the disease over short or long distances.

The disease can be transmitted by grafting (Coleman 1917). Besides twig grafting (Venkata Rao and Iyengar 1934), disease transmission is also possible by bud grafting (Sreenivasaya and Naidu 1928) or by means of special leaf insertion method in which, parts of fresh, diseased leaves are inserted between the wood and bark of healthy plants (Sreenivasaya 1930). The disease can be dodder-transmitted but is not transmitted through seed or soil. Transmission by mechanical sap inoculation generally yielded negative results though Nayar (1967) claimed that the disease is sap transmissible.

The disease appears to be spread by weak-flying insect vectors that may be carried to long distances by a strong wind. This is borne by the fact that sandal trees, located within a few miles from infected trees, can remain healthy for many years and scattered trees in a diseased area are not often infected (Boyce 1959). Leaf hoppers, namely, *Moonia albimaculata* Distant (Anon 1933) and **Jassus indicus* Walker, are believed to be vectors of spike disease (Rangaswamy and Griffith 1940) though insect transmission is not invariably successful under controlled conditions. In view of the fact that leaf hoppers are responsible for transmission of 'yellows' type of diseases and spike disease belongs to this group, study of insect vectors in disease transmission is important in evolving control of the disease.

Disease resistance

Two types of resistance to spike disease exist in sandal. In heavily spiked areas, it is not unusual to find healthy sandal trees which are likely to possess inherent resistance to spike. Autogenic resistance exhibited by these individuals is largely independent of the nature of the associated host plant. Acquired resistance, on the other hand, is built up by sandal if it parasitises certain species of hosts and certain environmental conditions are provided. The influence of the floristic composition of the area on the incidence and spread of the disease lends support to the existence of this type of resistance (Sreenivasaya 1933).

Spike-like symptoms on sandal hosts

Many plant species in the forests of South India exhibit spike-like symptoms (Iyengar 1960a), though the nature of the disease in these hosts is not known. MLBs were observed in *Santalum yasi* Seem. which exhibits spike-like symptoms and also in the phloem of sandal hosts, namely, *Zizyphus oenoplea* Mill. and *Dodonea viscosa* Linn. but not in *Stachytarpheta indica* Vahl. and *Lantana camara* Linn., which, however, could be infected with the organism causing spike (Nayar 1967, Nayar and Srimathi 1968, Hull, Plaskitt, Nayar and Padmanabha 1970).

Control

No effective control of spike disease has so far been evolved. However, attempts have been made on the following lines :

*Correct identity under investigation.

Sanitation. On the assumption that the incidence and spread of the disease may be lessened by eliminating the source of infection, attempts were made in the past to remove or kill the spiked sandal trees generally by arsenical poisons and also by removing diseased branches. However, this resulted in removal of large number of trees without significantly reducing the disease incidence.

Selection of sandal hosts. Different host plants of sandal impart resistance or susceptibility to spike disease in sandal. A list of such host plants was compiled by Kristensen (1960) based on field observations by earlier workers. Among the host plants through which the sandal acquires relative resistance are *Azadirachta indica* A. Juss., *Bambusa arundinacea* Retz., *Cassia siamea* Lam., *Casuarina equisetifalia* Forst., *Dalbergia sissoo* Roxb., *Dodonea viscosa* Linn., *Erythroxylum monogynum* Roxb., *Ficus benghalensis* Linn., *Murraya koenigii* Spreng., *Ruta graveolens* Linn., *Sarcostemma acidum* Voigt., *Semecarpus anacardium* Linn. f., *Strychnos nux-vomica* Linn. and *Zizyphus mauritiana* Lam. Among the host plants which render sandal susceptible to spike are species of *Acacia, Caesalpinia coriaria* Willd., *Cajanus cajan* Millsp., *Lantana camara, Ocimum sanctum* Linn. and *Pongamia pinnata* Pierre. It may, therefore, be possible to raise sandal in association with hosts known to impart disease resistance and at the same time, encourage growth and heartwood formation in sandal.

Sandal is a light-demanding species. The selection of host plants should, therefore, be confined to those which are fast-growing and possess a light crown, particularly if the sandal is to occupy the lower canopy in the forest. In nature, sandal may generally remain healthy under a top canopy of associated host species, like *Casuarina, Eucalyptus, Cassia siamea, Azadirachta indica* among others. Apart from the acquired resistance imparted to sandal by the host plants, freedom from spike disease could possibly also be due to the fact that the top canopy serves as a protective umbrella against insect vectors transmitting the disease to sandal. In diseased areas, sandal may not become spiked until it comes above the bush of its host (Boyce 1959). Based on these circumstantial evidences, Bakshi (1963) suggested a possible control of spike disease by raising sandal under a light canopy of trees.

Chemical control. Chemical control of the spike disease may be directed on diseased sandal trees or towards insect vectors transmitting the disease. Mycoplasma diseases in plants are known to be amenable to treatment with antibiotics which, however, cause temporary remission in disease symptoms as obtained for spike disease in sandal through use of tetracycline (Raychaudhuri, Chenulu, Ghosh, Varma and Rao 1972), or with benlate, a systemic fungicide (Rao, Shrimathi and Nag 1972). The recovery from the disease lasts only for some weeks after which the symptoms reappear.

While the role of insects in transmission of the disease is generally accepted, more confirmatory work is necessary in identifying the vectors, their role in disease transmission and eventually controlling insects through use of insecticides.

Genetic resistance. Spike-resistant sandal may be evolved through selection of apparently resistant sandal trees in a heavily spiked area and testing them for resistance to spike disease through grafting, dodder and insect transmission tests. Resistant individuals thus screened can be further propagated vegetatively. Sandal may be crossed with other species of *Santalum* to evolve desirable hybrids resistant to spike.

OTHER DISEASES

Leaf curl mosaic disease of sandal is noticed in some areas in Bangalore and Mysore. Leaves become wrinkled and mottled. With progress of the disease, the leaves become progressively smaller but not to the extent as occurs in spike. Finally, the leaves turn greenish yellow, become thick and brittle and fall off prematurely. The etiology of the disease is not known (Rao 1965).

A disease of sandal is recorded from Dharwar Forest Division. The disease is similar to spike disease in the reduction in leaf-size but the diminution is proportional in length and breadth, unlike that in spiked leaves. Also, Ca/N ratio is higher in diseased than in healthy leaves and in this respect differs from the spike disease, where the ratio is lower than normal. The etiology of the disease is not known (Rao 1965).

19. SEMUL-BOMBAX CEIBA LINN.

GENERAL

Semul (*Bombax ceiba*), a soft hardwood species, occurs throughout India on flat alluvial ground near river banks where it is often gregarious and attains best development. It occurs scattered in mixed deciduous forests and is also a characteristic species on grassy savannah lands. The species also occurs in Burma and Ceylon and has been introduced into Indonesia, Kenya, Nyasaland, South Africa and Uganda (Streets 1962). The species resists slight frost but is affected by severe frost and is fairly drought resistant. The wood is used for tea chests, rough planking, packing cases, match making and dug out canoes. The capsules furnish silk cotton used for the pillow. Plantations are raised by direct sowings in lines under taungya system or on prepared mounds. Stump planting of 1-2 year old nursery seedling is also practised.

PATHOGENIC DISEASES

ROOT ROTS

Ganoderma lucidum (Leyss.) Karst. causes root rot and finally death of the affected plants in Siwalik Forest Division, Uttar Pradesh. The species is fairly resistant to the disease. Vigorous trees are rarely attacked, while suppressed trees are liable to attack by the fungus. A detailed account of the disease is given in Chapter 15.

Other root rots recorded on semul are *Fomes lignosus* (Klotzsch) Bres. from Malaysia and *Fomes lamaoensis* (Murr.) Sacc. & Trott. from Ceylon (Spaulding 1961, Browne 1968.)

PHOMOPSIS DIE-BACK

Phomopsis salmalica Khan attacks young succulent shoots of semul, upto 4 years old in Pakistan. The affected bark becomes brown and pulpy and gives out an odour of fermentation. Affected plants exhibit die-back and eventually die (Browne 1968).

PINK DISEASE

Corticium salmonicolor Berk. & Br., the cause of pink disease, possesses a wide host range including semul (Browne 1968). The disease has been discussed in detail in Chapter 25.

LEAF BLIGHT

Dicellomyces bombacis Bakshi, the leaf blight fungus, causes gelatinous rusty-brown patches on lower surface of leaves. Areas of the leaf on the upper surface corresponding to infection spots are bleached and later turn yellow to dark-brown. In advanced stages of attack, the infection spots become necrotic, leaves get wrinkled, curled and are shed prematurely (Bakshi, Reddy, Puri and Sujan Singh 1972). The disease may become serious on young plants.

LEAF SPOTS

Myrothecium roridum Tode ex Fr. causes serious leaf spot on semul. The disease was recorded from Jaldapara Sanctuary, West Bengal. Severe infections may result in premature defoliation (Bakshi, Reddy, Puri and Sujan Singh 1972).

Phyllosticta bombacis Batista recorded from Bhubneshwar (Orissa State) causes necrotic spots on semul. Several spots may coalesce to cover major area of the leaf surface. The spots are brown with whitish centre and reddish-brown margin (Mahakul and Mohanty 1969).

Sphaceloma bombacis Wani & Thirum. causes circular leaf spots on upper surface of leaves of semul. The spots are slightly raised with wax-white centre. The disease is recorded from Khandala and Mahabaleshwar, Maharashtra (Wani and Thirumalachar 1969a).

Other leaf spots on semul are *Cercospora bombacina* Ramakr. T.S. & K. and *Cladotrichum follicola* (Niessl) Ferro from Kerala (Butler and Bisby 1960, Vasudeva 1963).

LEAF RUSTS

Chrysomyxa bombacis Petch attacks semul leaves in Ceylon. The rust is believed to be autoecious (Browne 1968).

Uredo bombacis Petch is recorded on semul leaves in Balehonnur, Karnataka. The rust is also reported from Ceylon (Spaulding 1961, Sarbhoy, Girdhari Lal and Varshney 1975).

Other fungi recorded on semul are *Fusarium semitectum* Berk. & Rav. causing damping-off and root rot of seedlings from Pakistan (Huque 1964), *Cochliobolus sativus* (Ito & Turib.) Drechsler & Dastur on seeds in Pakistan (Browne 1968) and a species of *Capnodium* causing sooty mould in India (Butler and Bisby 1960).

PHANEROGAMIC PARASITES

Mistletoes *Baratranthus nodiflorus* (Thw.) v. Tiegh. and *Helixanthera ensifolia* (Thw.) Dans. both from Ceylon and *Dendrophthoe falcata* (L f.) Ettingsh. from many countries are reported on semul (Browne 1968).

20. ALBIZIA

GENERAL

The genus *Albizia* is represented by 14 species in India, of which four species, namely, *A. lebbek* Benth. (kokko), *A. procera* Benth. (safed siris), *A. chinensis* (Osbeck) Merr. (=*A. stipulata* (Roxb.) Boiv.) (siris) and *A. odoratissima* (L. f.) Benth. (kala siris) are raised in plantations in India and other countries, like Fiji, Ghana, Jamaica, Mauritius, Nyasaland, Pakistan, Rhodesia, South Africa, Tanzania and Uganda. All the four species occur naturally in India (including the Andamans) and Burma, while siris and kala siris occur in Ceylon. Kokko and kala siris occur naturally in both dry and moist deciduous forests, the former even extending to moist semi-evergreen to evergreen forests. Siris and safed siris, on the other hand, grow naturally in moist to swampy locations. The wood of all the species is extensively used for buildings, furniture and agricultural implements. The species are generally frost-tender especially in the early stages. Direct sowing is generally the common method for raising plantations.

PATHOGENIC DISEASES

CLITOCYBE ROOT ROT

Clitocybe tabescens (Fr.) Bres. causes white root rot of kokko in Madagascar. The species is highly susceptible to the pathogen. The fungus attacks uninjured roots and spreads only by root contact. The pathogen is distributed in tropical and subtropical America, Europe and Asia including India (Browne 1968).

OTHER ROOT ROTS

Other root rot fungi recorded on *Albizia* are *Armillaria mellea* (Vahl.) Quel. on siris in Rhodesia, *Fomes noxius* Corner on kokko in Fiji, *Ganoderma lucidum* (Leyss.) Karst. on kokko and safed siris from India and Pakistan and *G. pseudoferreum* (Wakef.) Overeem and Stoinm. on kala siris in Malaysia (Browne 1968).

PIT CANKER

Fusarium solani (Mart.) App. et Wr. sensu Snyder et Hansen forma *albizziae* Bagchee causes grayish-black cankers on 15–20 year old trees of safed siris in India. The canker may extend 2–4 m on the stem. In severe infections, the canker may almost girdle the stem resulting in drying up of the crown. The trees may snap at the region of the canker. The pathogen initially establishes through frost canker and later develops a stratified canker due to repeated killing of the cambium. The canker has been artificially reproduced by Bagchee (1954).

NECTRIA CANKER

Nectria ditissima Tul. causes perennial canker with irregular, greatly swollen edges of callus in kokko in Madagascar (Spaulding 1961).

GIBBERELLA DIE-BACK

Gibberella pulicaris (Fr.) Sacc. causes shoot die-back in young plantations of siris in Kenya. The fungus also causes shoot die-back in *Pinus radiata* D. Don. and *Acacia melanoxylon* R. Br. in Kenya and Australia respectively (Browne 1968).

HEART ROTS

Heart rot fungi recorded on different species of *Albizia* are *Fomes fastuosus* (Lev.) Cke. on siris, *F. robiniae* (Murr.) Sacc. & D. Sacc. on kokko in Sudan, *Ganoderma applanatum* (Wallr.) Pat. on safed siris and kala siris in India and *Polyporus gilvus* Fr. on kokko and safed siris in India. All the fungi cause white rot (Browne 1968).

LEAF SPOTS

Cercospora albiziae Kar & Mandal attacks leaflets of young plants of kokko in West Bengal (India). The infection spots which occur on both sides of leaves are brownish to whitish-gray in the centre surrounded by narrow, brown to blackish ring. In some cases, the centre of the spots gives a deep grayish powdery appearance (Kar and Mandal 1969).

Endodothella albizziae (Syd.) Theiss. causes minute numerous yellow spots during the rainy season on leaflets of kala siris and other species of *Albizia* in Pakistan and Philippines. Fruit bodies of the fungus appear as minute black dots (Browne 1968).

Phyllosticta albizzinae Vasant Rao, attacks young plantations of kokko in India. In a 4 ha plot in Phillaur (Punjab), nearly all the trees were infected (Bakshi, Reddy, Puri and Sujan Singh 1972).

Other leaf spot fungi recorded on different species of *Albizia* are *Ascochyta saccardiana* Tassi on leaves and pods of kokko in India; *Camptomeris albizziae* (Petch) Mason on kokko from Ceylon and Pakistan; *Endodothella kanarensis* Ramakr. & Sund. on kala siris in Karnataka, India; *Helminthosporium albiziicolum* Thirum. & Naras. on kokko in India; *Leveillula taurica* (Lev.) Arn. on kokko in Rajasthan, India; *Microstroma albizziae* Syd. on kala siris and other species of *Albizia* in India and *Prathigada indica* Mathappa in Karnataka, India. All the fungi are uncommon and not damaging (Butler and Bisby 1960, Browne 1968, Bakshi. Reddy, Puri and Sujan Singh 1972, Prasada, Desai, Bhatnagar, Goyal and Pathak 1971, Sarbhoy, Girdhari Lal and Varshnoy 1975).

LEAF RUSTS

Ravenelia japonica Diet. & Syd. forms dark brown rust pustules occurring solitary or in groups on both surfaces of the leaflets and pods of siris and kala siris in China, Japan and India. *R. sessilis* Berk. produces yellowish-brown to black pustules on leaves and pods of kokko in Ceylon, China, India and Pakistan. *R. clemensiae* Syd. attacks leaflets of safed siris in India. Profuse development of rust pustules almost plasters the leaf surfaces and this may affect the metabolic activity of plants. The disease is severe on young plants. Seedlings when heavily attacked are deformed, crumpled and may die subsequently. *R. indica* Berk. attacks only the pods of safed siris in India but the damage is little (Bakshi and Sujan Singh 1967, Browne 1968).

A *Uredo* sp. attacks siris in the Annamalais in Tamil Nadu (India). Infected plants are rapidly defoliated, sometimes resulting in death. Infection is severe during June to October. Plants appear to develop resistance with age. $NiCl_2$ at 2500 p.p.m. kills rust pustules without any phytotoxic effect (Venkata Ram 1967).

Other records of rust fungi on species of *Albizia* include *Sphaerophragmium acaciae* (Cke.) Magn. on kokko and safed siris from Tropical Africa and Asia including India; *S. luzonicum* Yates on safed siris and other species of *Albizia* from the Philippines; *Uredo albizziae* P. Henn. on safed siris from New Guinea and *U. ngamboensis* P. Henn. on kokko from Mauritius and the Philippines (Spaulding 1961).

SOOTY MOULD

Meliola albizziae Hansford & Dieghton occurs in Assam State in India where the disease is widespread. The disease has also been reported from West Africa (Agnihothrudu 1960).

PHANEROGAMIC PARASITES

Mistletoes recorded on species of *Albizia* are *Dendrophthoe falcata* (L. f.) Ettingsh. on kala siris and a species of *Loranthus* and another of *Tapinanthus* on kokko. *Cuscuta reflexa* Roxb. attacks kokko in India and Pakistan (Browne 1968).

21. MULBERRY–MORUS

GENERAL

Six species of *Morus* are planted in various parts of India. Of these, *M. alba* Linn., a native of China, is by far the most important plantation species in north-western India and Pakistan. The species is also cultivated for rearing silk worms in Europe, U.S.A., western and central Asia, China, Japan, Nyasaland, South Africa and Tanzania. In India, the cultivation of *M. alba* extends to high elevation in the Himalayas and in parts of Tamil Nadu. The species is frost-hardy and grows well, mixed with sissoo (*Dalbergia sissoo* Roxb.) in irrigated plantations in the plains of India and Pakistan. Two other exotic species successfully cultivated in India are *M. nigra* Linn. in Kashmir and *M. australis* Poir in eastern parts of India. In addition, three indigenous species, namely, *M. serrata* Roxb., *M. indica* Linn. (close to *M. alba*) and *M. laevigata* Wall. ex Brandis are also widely planted. *M. serrata* is the common mulberry of the Himalayas from Kumaon westwards occurring at elevations between 1300 to 3000 m. *M. indica* occurs in dry forests of sub-Himalayan tracts ascending to about 1700 m. *M. laevigata* occurs in the outer Himalayas from Kumaon (Uttar Pradesh) extending eastwards to Assam and Upper Burma. Leaves of *Morus* species are important as food plants for silk worms. The timber is prized for manufacturing sports goods and is also used for furniture, turnery and boats. Fruits are edible.

PATHOGENIC DISEASES

CROWN GALL

Agrobacterium tumefaciens (E.F. Smith & Towns.) Conn., the disease-causing bacterium, is widespread on many hosts in many countries. The disease on *Morus* occurs in Russia and Transcaucasia (Spaulding 1961).

ROSELLINIA ROOT ROTS

Rosellinia necatrix (Hartig) Berl. causes white root rot of *Morus* seedlings and trees in France, Italy and Japan. The pathogen can attack intact root surface in young trees, while in older trees, infection is mainly through lenticels. The vascular tissues of the host are killed in advance before colonisation by the pathogen (Sakuri 1952). The pathogen is favoured by overcrowding of seedlings in the nursery. Infected saplings can be successfully disinfected by hot water treatment at 45°C for one hour or at 47°C for 40 minutes (Matuo and Sakuri 1954).

R. aquila (Fr.) de Not. causes root rot of young *Morus* trees in France (Spaulding 1961). The pathogen is favoured by wet soils.

HELICOBASIDIUM ROOT ROTS

Helicobasidium mompa Tanaka is one of the most serious root diseases of *Morus* reported from China, Formosa, Japan and Korea. The pathogen attacks slender roots which turn yellowish or blackish brown, become softened and get decayed. In severe attack, the cambium of the larger roots may also be killed. The pathogen produces purple rhizomorphs which on reaching the base of trees develop into mycelial mats or fruit bodies. Small purplish-brown sclerotia are also produced on affected roots. The leaves become smaller accompanied by early defoliation and subsequent death of affected trees when most of the roots are decayed. The pathogen can live as a saprophyte on plant debris in soil (Ito 1949, Spaulding 1961).

H. tanakae Miyabe forms irregular, large, thick, velvety-brown fungus felts on trunks of *Morus* in China, Formosa and Japan (Spaulding 1961).

H. purpureum (Tul.) Pat. attacks *Morus alba* seedlings in nurseries in Japan (Spaulding 1961).

All the above three species of *Helicobasidium* possess a wide host range on both conifers and broadleaved species. *H. mompa* is recorded on 104 host-species among 76 genera in 45 families (Ito 1949).

DIPLODIA ROOT DISEASE

Diplodia morina Syd. causes death of roots in *M. alba* in India. The disease was reproduced through inoculation. The pathogen is reported to kill branches also (Spaulding 1961).

Other root pathogens recorded on *Morus* are *Armillaria mellea* (Vahl.) Quel. causing root rot of debilitated trees of *M. alba* in U.S.A. and *Morus* sp. in France and Japan, *Botrytis cinerea* Pers. ex Fr. causing gray mould blight of *M. alba* seedlings in Russia, *Ganoderma lucidum* (Leyss.) Karst. causing root rot of *Morus alba* in India and Pakistan and *Macrophomina phaseolina* (Tassi) G. Goid. on *M. alba* in India (Butler and Bisby 1960, Spaulding 1961, Browne 1968, Hepting 1971). *Phomopsis moricola, Verticillium dahliae* Kleb. and *Fusarium* sp. are recorded from Russia causing infection at root collar region of *Morus* (Goncharenko 1968).

HENDERSONULA DIE-BACK

Hendersonula toruloidea Nat. causes serious die-back of *M. alba* in Pakistan. The pathogen infects branches and twigs in the crown. Infected branches are killed back and new branches develop below the dead region which are again killed back and the process continues till the pathogen travels to the base of the tree. At this stage, the affected tree gives sable hair brush-like bushy appearance. Infection may also occur directly on the main stem near the ground level. The tree is killed when girdled on the main stem. Other symptoms of the disease are appearance of water-soaked, slightly depressed brown areas on the stem below which the bark cracks and peels off. Black masses of spores appear below the affected bark. In advanced stages of the disease, the pathogen attacks the sapwood causing grayish-black discolouration. The disease spreads through air-borne spores (Khan 1955). The pathogen attacks the cambial region of a wide range of hosts often following injury by insolation.

GIBBERELLA TWIG CANKER

Twig and branch canker on *M. alba* and *M. nigra* is widespread in Australia, Formosa, Germany, Japan, Netherlands, New Zealand, U.K. and U.S.A. The causal organisms have been variously identified as *Fusarium lateritium* Nees, *F. lateritium* var. *mori* D.W. & R. and *F. mori* Desm. The perfect state of these species is also ascribed to different species of *Gibberella*, like *G. baccata* (Wallr.) Sacc., *G. baccata* var. *moricola* Boerma & Verhoeven and *G. moricola* (Ces. & de Not.) Sacc. All these species probably relate to the same fungus. The pathogen infects through wounds, causing cankers which may girdle and kill the infected branches. The conidial state appears as reddish-brown pustules followed later by the blackish perithecia of the perfect state (Spaulding 1961, Peace 1962, Hepting 1971).

Gibberella lateritia (Nees) Snyd. & Hans. causes lesions and blight of buds of *M. alba* in Japan (Spaulding 1961).

NECTRIA CANKERS

Nectria cinnabarina Fr. causes cankers in *Morus* in France, Japan and New Zealand. The pathogen is known to exist in strains which may vary from purely saprophytic to aggressively parasitic. The pathogen enters the mulberry through buds injured by frost. The hyphae grow into the dead cells and from there enter into the sound tissues (Butler 1918).

Nectria coccinea Fr. and *N. galligena* Bres. cause perennial target canker on *Morus alba* in U.S.A. Small, red perithecia of the parasite appear under bark crevices around the margins of the cankers (Hepting 1971).

SPHAEROPSIS DIE-BACK

Sphaeropsis mori Berl. causes die-back of *M. alba* in Italy and Tanzania. The fungus infects twigs causing lesions and cankers (Spaulding 1961).

THYROSTROMA TWIG BLIGHT

Thyrostroma mori (Nomura) Hoehnel (=*Coryneum mori* Nomura) causes serious cankers and die-back of young stems in *M. alba* in India and Pakistan. The fungus infects through wounds at points injured by frost and snow (Bulter 1918).

HEART ROTS

Heart rot fungi recorded on *Morus* are *Fomes fomentarius* (Fr.) Kickx on *Morus* sp. in Portugal, *F. robiniae* (Murr.) Sacc. & D. Sacc. on *M. alba* in Japan, *F. senex* (Nees & Mont.) Cke. on *Morus* sp. in India and on *M. alba* in Pakistan, *Ganoderma applanatum* (Wallr.) Pat. on *M. alba* in India and *Morus* sp. in France, *Polyporus hispidus* Fr. on *M. alba* and *Morus* sp. in France, Germany, India and Portugal and *Polyporus squamosus* Fr. on *M. alba* in Germany (Spaulding 1961, Browne 1966).

Other parasitic fungi recorded on twigs, branches and stems of *Morus* are *Claudopus nidulans* (Fr.) Pk. on *Morus* sp. in Japan causing sapwood rot, *Colletotrichum moricola* Vishwanathan on *M. alba* in India causing twig blight, *Corticium salmonicolor* B. & Br. on *M. australis* in Formosa, *M. indica* in Tanzania causing twig blight and stem cankers, *Cytodiplospora mori* I. Miyake on *M. alba* in Japan causing twig blight, *Cytospora atra* Sacc. on *M. alba* in U. K. causing die-back, *Physalospora obtusa* (Schw.) Cke. on *M. alba* in eastern United States causing cankers on stem and branches, *Sclerotinia sclerotiorum* (Lib.) de Bary which causes die-back of young shoots in Bulgaria on an epidemic scale and is possibly favoured by wet spring and summer, *Thyrococcum sirakoffii* Bub. on *Morus* sp. in Central Europe causing die-back of twigs and cankers on stem and branches and *Valsa ambiens* (Fr.) Fr. on *M. alba* in Japan causing die-back (Spaulding 1961, Peace 1962).

POPCORN DISEASE

Ciboria carunculoides (Siegler & Jenkins) Whet. causes popcorn disease of *M. alba* in U. S. A. Another species, *C. shiraiana* (P. Henn.) Whet., causes similar popcorn disease in *M. alba* and *M. australis* in China, Formosa and Japan. Both the pathogens are considered distinct from each other but produce identical symptoms. The infection results in hypertrophy of separate druplets of fruits where the fungi produce sclerotia which overwinter on the fallen fruits to become active in the next spring (Spaulding 1958, 1961).

LEAF SPOTS

Three species of *Cercospora*, namely, *C. missouriensis* Wint., *C. mori* Hara and *C. moricola* Cke. are recorded on *Morus*. *C. missouriensis* causes dull brown to almost black spots on lower surface of *M. alba* leaves in U.S.A. *C. mori* forms mould-like, olivaceous to dark fruiting patches on the lower leaf surface of *M. alba* and other species in China, Formosa, Japan, U.S.A. and Zaire. *C. moricola* causes leaf spots with gray to tan centres and dark purple to black border on both sides of leaves of *M. alba* in Brazil, India, U.K., U.S.A. and Italian Somaliland (Spaulding 1961, Vasudeva 1963, Hepting 1971).

Cylindrosporium maculans (Allesch.) Jacz. causes serious leaf spot of *M. alba* in Albania, Denmark, Poland, Russia and Yugoslavia. The disease is successfully controlled by chemicals (Spaulding 1961, Chanturiya and Gvinepadze 1972).

Mycosphaerella mori Wolf causes serious leaf and fruit spot of *M. alba*, *M. indica*, *M. nigra* and *Morus* sp. throughout the world. The fungus is known as *Septogloeum mori* Briosi & Cavara in its imperfect state. Infection manifests as small dark spots which gradually become larger, irregular and pale brown. Heavy infection may cause severe defoliation (Browne 1968). *Mycosphaerella indica* Vishwanathan causes leaf spot in *M. alba* in Poona (India) (Tandon and Chandra 1963–64).

Pseudomonas mori (Boyer & Lamb.) Stevens causes leaf spot and elongated lesions on old twigs of *M. alba*, *M. australis*, *M. indica* and *M. nigra* in Australia, Brazil, East and South Africa, Europe, Japan, Korea, North America and Turkey. Infection results in stunting and die-back of diseased plants. In wet weather, bacteria may ooze from infected twigs (Peace 1962).

Other leaf spots recorded on species of *Morus* are *Ascochyta morifolia* Saw. on *M. alba* in Formosa, *Cercosporella mori* Peck on *M. alba* in U. S. A., *Clasterosporium mori* Syd. on *M. alba* in China and Japan, *Phyllosticta kuwacola* K. Hara on *M. alba* in Japan, *P. morifolia* Pass. on *M. alba* in India and *Haplosporella cesatii* Sacc. on *M. alba* in India (Spaulding 1961, Tandon and Chandra 1963-64, Browne 1968, Sarbhoy, Girdhari Lal and Varshney 1975).

POWDERY MILDEWS

Four species of *Phyllactinia*, namely, *P. corylea* (Pers.) Karst., *P. guttata* (Fr.) Lev., *P. moricola* (P. Henn.) Homma and *P. suffulta* (Rab.) Sacc. are recorded on different *Morus* species. *P. corylea* and *P. suffulta* are reported as synonyms of *P. guttata* (Browne 1968) which forms white, powdery coatings on leaves of *Morus* sp. in Indochina, *M. alba* in China, India, Japan, Pakistan and Russia and *M. nigra* in India. *P. moricola* attacks *M. alba* and *M. australis* in China and Japan. (Spaulding 1961, Browne 1968).

Uncinula mori Miyake, the cause of powdery mildew, forms a coating on lower surface of leaves in *M. alba* in Japan and Russia and on upper surface of leaves of *Morus* sp. in Kulu (India) (Spaulding 1961, Gill, Munjal and Chona 1961).

LEAF RUSTS

Three species of rusts attacks *Morus* leaves. *Aecidium mori* Barcl. causes yellow infection spots on leaves of *M. alba* in India, China, Japan and Pakistan, *M. indica* and *M. serrata* in India and Pakistan, *M. australis* in China, Formosa and Japan and *Morus* sp. in India and Japan. In case of heavy infections, the entire leaf surface is covered but the damage is little as the defoliation due to rust occurs about the same time as the normal leaf fall (Spaulding 1961, Bakshi and Sujan Singh 1967). *Cerotelium fici* (Hoehn.) Arth. produces powdery, cinnamon-brown rust pustules scattered on lower surface of *M. alba* in Philippines, Tanzania and other species of *Morus* in Madagascar and Guatemala (Spaulding 1961). *Kuehneola moricola* P. Henn. attacks *M. alba* leaves in China (Spaulding 1961).

VIRUS DISEASES

Mulberry mosaic virus is reported on *Morus* species from China, Japan, Puerto Rico, Thailand and U.S.S.R. under various names such as mulberry dwarf, curly little leaf virus and mulberry small leaf curl. The symptoms include mottling, puckering, reduction in size of leaf blades, curling, cupping downwards of leaves, stunting of branches of the entire tree and resulting sometimes in death of affected trees. The disease has been successfully transmitted through grafting and by insect vectors (Spaulding 1961, Raychaudhuri, Chatterjee and Dhar 1961). In China, infection of 50-80 per cent of trees resulting in 70 per cent reduction in leaf yield was reported. Silk worms fed on infected leaves produce distorted cocoons and yield inferior silk (Faan, Lai, Chan and Tom 1964). A similar dwarfing disease is reported in mulberry in Japan in which case, mycoplasma-like organisms were observed in phloem tissues. Recession of disease symptoms was obtained by using antibiotics but the symptoms reappeared when the application of antibiotics was stopped (Ishiie, Doi, Yora and Asuyama 1967, Ishiie (1970).

Yellow net vein virus attacks *M. indica* in Kalimpong, India. Symptoms include chlorotic areas starting from leaf margins and yellowing of leaf veins (Raychaudhuri, Chatterjee and Dhar 1961).

Another virus disease, known as mulberry ring spot virus of mulberry leaves showing ring spot and filiform leaf symptoms is reported from Japan (Tsuchizaki, Hibino and Saito 1971).

PHANEROGAMIC PARASITES

Three mistletoes recorded on *Morus* species are *Dendrophthoe falcata* (L.f.) Ettingsh. and *Loranthus* sp. on *M. alba* and *Viscum album* L. on *M. alba* and *M. nigra* (Browne 1968).

22. RUBBER–HEVEA BRASILIENSIS MUELL.-ARG.

Hevea brasiliensis is indigenous to Brazil. It yields latex which is commercially the most important source of natural rubber. For this reason, the species was introduced in tropical Asia during the latter part of the nineteenth century. In India, rubber cultivation is mostly confined to Kerala and parts of Tamil Nadu and Karnataka States in South India. The tree is quick-growing and thrives on good, well-drained soils. Trees become leafless during December to February. The soil under rubber plantations in South India is lateritic, porus, well-drained and acidic. The area experiences heavy rainfall, between 2000-4000 mm, from southwest and northeast monsoons in the southern parts of Kerala and from southwest monsoons only in the northern parts of the state. Rubber is also raised in low undulating hills where soil is liable to erosion and leaching. Suitable measures are, therefore, adopted to prevent rapid degradation of soil by raising cover crops particularly belonging to Leguminosae and by mulching and manuring. Trees are tapped for latex by removing thin shavings of bark to open up latex vessels for trees tapped for the first time or remove the coagulum which blocks latex vessels in trees under renewed tapping. In the rainy season, polythene skirts are fixed on the stem for protection of tapping panels and collecting cups against rains in order that tapping can continue during the rains.

PATHOGENIC DISEASES

WHITE ROOT DISEASE

Fomes lignosus (Klotzsch) Bres. causes white root disease of rubber in Ceylon, India, Indonesia, Malaysia, Philippines, Ghana, Ivory Coast, Zaire, Costa Rica, Mexico and Brazil. The disease is serious on sites previously under natural forests or under tea. The fungus colonises roots and stumps of the previous crop from where it spreads to rubber through root contact or through subterranean mycelial cords which can travel through soil. The fungus forms a network of white, yellowish-white to reddish mycelial cords on surface of rubber roots and causes decay of the bark and the wood. The decayed roots become wet and sodden. The pathogen can infect undamaged host tissues. The penetration is, however, more rapid through wounds and natural openings. External symptoms of the disease include yellowing and curling of leaves and stoppage of latex flow from the panels. Infected trees die within a year. The diseased trees produce cork cambium in the cortex and phenols and resins in woody tissues as host reaction to the invading fungus (Perias and Irugalbandara 1973).

The fruit bodies of the fungus develop rarely on diseased trees but commonly on dead and wind-blown trees, logs and stumps of rubber and other species. The sporophores are reddish brown with bright yellow margin on upper surface and bright orange on hymenial surface when in a fresh condition (Bakshi, Sujan Singh and Balwant Singh 1963).

Control measures for the disease include complete removal of roots and stumps of the previous crop before planting, treatment of pits with chemicals like sulphur at the time of planting, cutting diseased roots upto healthy tissues, collar dressing with chemicals like PCNB (Pentozene) and isolation of diseased plants through trenches to check further spread of the disease.

RED ROOT ROT

Ganoderma pseudoferreum (Wakef.) v. Overeem & Steinm., the cause of red root rot of rubber, is recorded from India, Indochina, Indonesia, Malaysia and Zaire. Initial infection takes place from infected roots and stumps of the previous crop. Subsequent spread is strictly by root contact between diseased and healthy trees. The pathogen commonly occurs along with *Fomes lignosus*, both fungi sometimes attacking the same root. Progress of the disease in infected trees is usually very slow. Trees may be attacked at as early an age as 2-3 years but the symptoms develop late after 10-12 years. The affected trees produce profuse adventitious roots at the collar region as host reaction towards the invading fungus. These roots help in increased absorption of water and nutrients from the soil and thereby maintain vigour of the tree. Furthermore, the fungus attack is contained in localised areas mainly on undersurface of roots by the production of callus tissues so that the upper side of roots remains healthy and continues to function normally.

This explains the delayed manifestation of disease symptoms. In advanced stages of the disease, die-back, yellowing and withering of leaves and thinning of the crown occur. Affected trees may be wind-blown. The fungus produces thin, red, mycelial strands developing into rhizomorphs which later fuse to form a continuous blood red membrane on underside of roots. The decayed wood becomes wet and honey-combed with brown zone lines (Sharples 1936).

Fruit bodies, which develop rarely on standing trees but are common on stumps and fallen logs, are bracket or hoof-shaped or flattened. The upper surface is dull lacquered and zoned in shades of dark, tawny brown merging into black. Cracks develop on drying. Hymenial surface is white becoming yellow on bruising. Control of the disease is practised on the same lines as outlined for *Fomes lignosus*.

BROWN ROOT DISEASE

Fomes noxius Corner, the cause of brown root disease, is recorded from India, Indonesia and Malaysia. The fungus attacks roots, specially the tap root, on which it develops an ectotrophic growth when the mycelium is cemented with soil and bark to form brown to black incrustations. The wood exhibits white rot with honeycomb-like structure. Affected plants dry up in late stages. The disease spreads only by root contact.

DRY ROOT ROT

Dry root rot or stump rot, collar rot or charcoal rot, is caused by *Ustulina deusta* (Hoffm. ex Fr.) Lind., also known as *Ustulina zonata* (Lev.) Sacc. The disease is recorded on rubber from Ceylon, India, Fiji, Ghana, Malaysia, Seychelles and Sierra Leone and also occurs on tea, coffee and other perennial crops. The fungus attacks trees of all ages, particularly old ones. It is a wound parasite primarily infecting trees at the collar region under humid weather conditions. Latex is exuded from lesions. The infected bark and wood become soft and exhibit conspicuous zone lines. White, fan-shaped mycelial felt may develop underneath the bark of roots. The disease may progress up the stem upto a short distance and into the tap root and lateral roots. The progress of the disease is, however, slow and it may take some years before the affected trees are killed or wind-blown. Fructifications develop at the base of the tree as irregular to circular, grayish white to olive gray plates appressed to the bark. With age, they become hard, brittle and purple-gray to black and exhibit zonations. Sunken stromata develop with numerous dots on the surface which are the openings of the perithecia. In final stages, fructifications may crumple into black powder due to which the disease is called charcoal rot.

Injury to trees at the collar region and also accumulation of latex pads at the base of trees should be avoided to keep trees healthy. Removal of affected tissues and disinfection of exposed tissues during initial stage of attack may save the tree. Disinfection of exposed tissue with organo-mercurials, like 'ceresan wet' at 28 g in 4.5 l water and aretan at 28 g in 11.25 l water may prove effective against further progress of the disease. Vigour of trees is to be improved with adequate manuring (Ramakrishnan and Pillai).

FOMES LAMAOENSIS ROOT ROT

Fomes lamaoensis (Murr.) Sacc. & Trott. causes brown root rot of *Hevea* in Ceylon, India, Indochina, Indonesia, Malayasia, Philippines, West Equatorial Africa, Ivory coast, Ghana and Zaire. The fungus causes white pocket rot in roots. The disease spreads by root contact. The progress of the disease in the tree is very slow. Recommended control measures include removal of diseased trees and burning of stumps in situ (Young 1954). Foster (1964), however, regards *F. lamaoensis* as a harmless saprophyte.

ARMILLARIA MELLEA ROOT ROT

Armillaria mellea (Vahl.) Quel. causes root rot of rubber in Nigeria, Uganda, West Equitorial Africa and Zaire. This disease is favoured by high rainfall and poor drainage. Serious losses have been reported on stiff and clayey soils (Fox 1964).

PORIA ROOT ROTS

Poria hypobrunnea Petch causes root rot of *Hevea* in Ceylon, India, and Malaysia. The fungus establishes on stumps through wind-blown spores and then attacks rubber through stout rhizomorphs which are red when young and become black with age and are white from inside (Browne 1968). *P. hypolateritia* (Berk.) Cke. causes root rot of *Hevea* in Ceylon, India and Ghana. Fruit bodies are white above and reddish below (Browne 1968).

STINKING ROOT DISEASE

Sphaerostilbe repens Berk. & Br. causes root rot of *Hevea* in Ceylon, India, Indochina, Indonesia, Malaysia, Ghana and Zaire. The decaying roots and surrounding soil emit stinking foul smell from which the disease can be easily recognised. The disease causes extensive damage to *Hevea* in areas which remain under flood water for sometime. Clayey and stiff soils also predispose *Hevea* to the disease. The diseased trees exhibit thinning of foliage, sometimes rapid complete defoliation, gradual die-back of branches followed sometimes by death of affected trees (Sharples 1936). In the root, the fungus produces black, flattened rhizomorphic strands between the bark and the wood. The imperfect state consists of pinkish pustules representing a *Stilbum* species. The pathogen usually attacks mature trees and the disease spreads by root contact. Control is recommended through isolation of diseased patches by trenching.

Other root pathogens recorded on *Hevea* are *Fomes ligneus* (Berk.) Cke. from Brazil causing root rot, *Fusarium javanicum* Koordu from India and Ceylon, *Ganoderma lucidum* (Leyss.) Karst. from Ceylon and Indonesia causing root rot, *Helicobasidium compactum* Boed. from Indonesia, *Macrophomina phaseolina* (Tassi) G. Goid. from Ceylon, India, Indonesia, Malaysia and Uganda causing root rot, *Pellicularia rolfsii* (Curzi) West from Indonesia causing seedling mortality, *Phomopsis heaveae* from Indonesia causing seedling die-back, *Physalospora rhodina* (Berk. & Curt.) Cke. from India, Indochina, Brazil, Zaire, Indonesia, Ghana, Malaysia and Uganda causing root rot, die-back and canker, *Phytophthora cactorum* (Leb. & Cohn.) Schroet. from Ceylon, India and Indonesia causing seedling blight, *Polyporus shoreae* Wakef., from Nigeria and *P. zonalis* Berk. from Brazil causing root rot, *Thanatephorus cucumeris* (Frank) Donk (=*Pellicularia filamentosa* (Pat.) Rogers) from Bolivia, Brazil, Ceylon, Columbia, Costa Rica, Honduras and Peru causing web blight of seedlings, *Xylaria deserticola* Speg. from Malaysia causing root rot and *Xylaria thwaitesii* Cke. from Ceylon, India, Indochina, Indonesia causing root rot (Spaulding 1961, Browne 1968).

BLACK STRIPE

Phytophthora palmivora (Butl.) Butl., the cause of black stripe, black thread or bark rot disease, is widespread in rubber plantations in Burma, Ceylon, India, Java and Malayasia. It also attacks other hosts in rubber growing areas. Infection results in linear, vertical, shallow depressions above the tapping cut. Underneath the bark, dark lines extend vertically up and down from the depression. In some cases, black lines may extend into the wood which becomes discoloured. Variations in nature and intensity of attack is due to existence of different strains of the pathogen (Sharples 1936). Bark tissues are decayed and may be shed off leaving small or big open wounds in the panel. Infection is more on tapping cuts near the ground than on those away from the ground. The flow of latex is reduced or stopped in severe infection. High rainfall and high atmospheric humidity favour epidemic outbreak of the disease. If tapping is done during the rains, freshly exposed panels on which rain water accumulates form favourable places for the pathogen to sporulate and cause infection. Polythene covers maintain high humidity around the panel and favour infection.

Stoppage of tapping during heavy rainy months reduces the incidence of the disease. If, however, tapping is continued during heavy rains, a panel protectant should be used. Application of 'aretan' (28 g in 11.25 l water) and 'ceresan' (28 g in 4.5 l water) followed by a waterproof coating is effective in controlling the disease (Ramakrishnan and Pillai 1963).

MOULDY ROT

Ceratocystis fimbriata Ell. & Halst., the cause of mouldy rot, occurs in most countries where rubber is raised. In India, the disease is recorded in several plantations though the causal organism is recorded on few occasions. Infection results in the formation of depressed spots which join up to form an irregular sunken band on both sides of the tapping cut. Diseased tissues rot to form open wound exposing the discoloured wood. Bark is not renewed thus affecting further tapping of regenerated bark. The fungus develops conidia which are wind-dispersed and perithecia develop ascospores which are disseminated by insects. The disease becomes epidemic under continuous rainfall or humid weather conditions.

Control of the disease is possible by stopping tapping during rains and use of a fungicide if tapping is done during the rains. Tapping tools should be disinfected to safeguard spread of the disease (Ramakrishnan and Pillai 1963).

PATCH CANKER

This disease of the stem including the tapping panel region and rarely of roots, is caused by *Phytophthora palmivora* (Butl.) Butl.,

P. meadii McRae and *Pythium vexans* de Bary. Infection results in swellings and bark bursts from where amber-coloured liquid oozes out in some cases. Exudation of latex also occurs. The bark rots beneath which a coagulated rubber pad is present emanating a foul smell. On removing this, a wound is exposed. Removal of affected region and dressing the wound with an organo-mercurial fungicide like 'aretan' and 'ceresan' may keep the tree healthy (Anon. 1975)

MARASMIUS PANEL DISEASE

Marasmius palmivorus Sharples attacks tapping panels causing rotting of the bark and the cortical tissues of the newly tapped surface down to the wood. The disease is recorded from Malaysia. Infection prevents satisfactory healing of the panels. Symptoms of the disease include the development of a small fan of white mycelium formed just above the tapping cut. The fan enlarges until it becomes a plate of mycelium with strands spreading out of its edges (Sharples 1936).

PINK DISEASE

Corticium salmonicolor B. & Br., the cause of pink disease of rubber, also attacks many hosts including *Eucalyptus* under which the disease is fully described (Chapter 25). In rubber, the disease is more damaging on young plants between 3-12 years' age. Plants become infected with the onset of summer monsoon, infection initiating usually at the fork of the tree where the cobwebby form develops followed by pustule, necator and pink incrustation forms. Latex oozes out from the lesions and the bark cracks and dries up killing the shoot above this region. Epicormic branches develop below the attacked region.

Application of Bordeaux paste upto 10 cm above and below the affected region during early stages of infection and replicated after scraping the damaged bark in advanced stages of the disease, pruning and burning diseased and dead branches control the disease.

DIE-BACK

Die-back in young rubber plantations occurs in South India. The snag of the stock turns black and is invaded by several fungi including *Diplodia*. The disease is also reported from Brazil on grafted rubber associated with late budding and wound invasion with *Diplodia* sp. and other fungi, Cankers can reach 20-50 cm on five year old plants (Goncalves 1970). Infection occurs through wounds or dead tissue. Affected plants dry up or are broken at the region of attack. Solar insolation predisposes plants to infection. Cutting stocks at a steep angle for water to run off easily, protection of the cut surface with Bordeaux paste and protection from sun's heat, either by whitewashing of the stem or through shading in the first year keep the disease under control.

THREAD BLIGHTS

Three fungi are known to cause thread blight disease in *Hevea*. *Corticium koleroga* (Cke.) Hoehnel causes bark lesions on branches in moist locations in Argentina and Indonesia. The mycelial strands may travel from the branch to leaves to cause leaf blight. *Marasmius cyphella* Dennis & Reid causes white thread blight of rubber in Malaysia. *M. equicrinis* F. Muell. causes horse hair blight of *Hevea* in Brazil, Ceylon, Indonesia and Malaysia. Black horse hair-like round mycelial strands are attached to the stem at certain points by small brown pad of mycelium. The strands hang loosely and hop over to leaves, twigs and branches. The thread blight diseases, though conspicuous, are not damaging (Browne 1968).

Other stem diseases recorded on *Hevea* are *Calonectria rigidiuscula* (Berk. & Br.) Sacc. from west African countries, *Tryblidiella rufula* (Spreng.) Sacc. from Ghana causing twig blight, *Fusarium solani* (Mart.) Appel. & Wr. and *Botryodiplodia theobromae* Pat. from Malaysia causing bark necrosis on tapping panels (Browne 1968, Chee 1971). *Trametes corrugata* (Pers.) Bres. from Brazil and Indonesia causes heart rot (Spaulding 1961).

ABNORMAL LEAF FALL

Phytophthora palmivora (Butl.) Butl., *P. meadii* McRae, *P. nicotianae* Breda de Haan var. *parasitica* (Dastur) Waterhouse are responsible for abnormal leaf fall of rubber in South India. The disease, in one or the other type of infection, is also prevalent in many rubber growing areas in Burma, Ceylon, tropical regions of Africa and America. The disease also occurs on cocoa and causes heavy damage to fruits and on *Areca catechu* Linn. Initial symptoms appear as pod rot when pods do not yield viable seeds. Leaves fall off either green or after turning copper red with a drop of latex on lesions which develop on petiole, midrib or leaf blade. Heavy defoliation results in die-back of branches and loss in tree vigour. Different symptom expressions are governed by the intensity of rainfall, which when low, causes pod rot and when high, as during prolonged rainy days during the summer monsoon, favours defoliation.

The disease can be prevented with a prophylactic fungicidal spray on the foliage before the onset of south-west monsoon. Ground or aerial sprays with 1 per cent Bordeaux mixture using high volume sprayers or oil-based copper oxychloride prove effective (Ramakrishnan and Pillai 1961).

Besides *P. palmivora*, *P. botryosa* Chee (Chee 1969) causes severe defoliation in Malaysia under conditions of high humidity prevailing for about a week (Wastie 1973).

SECONDARY LEAF FALL

Gloeosporium alborubrum Petch causes minor leaf shedding in plants of low vigour such as those growing under water-logged conditions. The disease manifests before and after the summer monsoons. Tender leaves become infected and fall off. Selection of suitable site with good drainage, improving tree vigour with fertilizer application and spraying with copper based fungicides keep the disease under check (Anon. 1975).

Glomerella cingulata (Ston.) Spauld. & Schrenk (=*Colletotrichum gloeosporioides* Penz.) causes anthracnose of new developing leaves of *Hevea* in rubber-growing countries all over the world. Infection is more on lower than on upper surface and is favoured by wet weather. Heavy infections result in severe defoliation both in nurseries and plantations. The new flush may again be attacked after first defoliation. The disease can be effectively controlled through artificial defoliation at the outset of the normal winter season with chemical defoliants like cacodylic acid. The new flush following artificial defoliation hardens sooner and wards off the disease (Rao 1971).

BIRD'S EYE SPOT

The disease caused by *Drechslera heveae* (Petch) M.B. Ellis (=*Helminthosporium heveae* Petch) is serious and damaging on young plants in the nursery and in the field. Debilitated plants and those growing under exposed situations are more susceptible. Necrotic spots with pale centre and dark brown margins develop on leaves. Premature defoliation and die-back may occur. Spraying with Bordeaux mixture or Dithane M–4, shading of nursery and maintenance of plant vigour through nutrition reduce the disease incidence.

LEAF BLIGHT

Dothidella ulei P. Henn. occurs in Central and South America where the fungus is not of any consequence to rubber occurring wild and spora- d ally in the forest but causes heavy defoliation and seriously affects the yield of latex when rubber is raised in pure plantations. The disease was left behind when rubber was raised as exotic in the tropical region of the East.

LEAF SPOTS

The leaf spot in rubber caused by *Corynespora cassiicola* (Berk. & Curt.) Wei, is predominant in nurseries. Pale lesions with dark brown margin develop on leaves. The affected tissues fall off giving shot hole symptom. Defoliation also occurs. Spraying with Bordeaux mixture or Dithane M-4, shading of the nursery and maintaining plant vigour reduce disease incidence.

Ascochyta heveae Petch causes white to brownish leaf spots with reddish-brown margin on periphery of leaves of *Hevea* in India, Ceylon, Malaysia and West African countries. *Ascochyta heveana* Saccas forms leaf spots with concentric brown zones corresponding to the distribution of the fungus pustules. The fungus is recorded from West African countries (Spaulding 1961). *Cylindrocladium quinquiseptatum* Boed. & Reitsma attacks immature leaves, petioles and terminal growing tips of *Hevea* in Malaysia. The lesions on leaves are circular to sub-circular with a whitish papery centre and dark brown margin. The affected leaves become brittle and crack but no defoliation occurs. In bud wood nurseries, however, the disease becomes severe after pruning (Anon. 1972 a).

Other leaf spot fungi recorded on *Hevea* are *Catacauma huberi* (P. Henn.) Th. & Syd. from Brazil and Peru causing black crust on seedling leaves leading to defoliation, *Colletotrichum ficus* Koorders from Malaysia causing leaf anthracnose under water-logged conditions, *C. heveae* Petch from tropical Africa, America, Asia causing spots with stiff black hairs with white centre and purplish-brown margin, *Mycosphaerella heveae* Petch from India and Ceylon, *M. heveicola* Saccas from West African countries, *Ophiobolus heveae* P. Henn. and *Elsinoe heveae* both from Brazil, *Pestalotiopsis versicolor* from Malaysia causing a minor leaf spot disease, *Phyllosticta heveae* Zimm. from Brazil, Ceylon, India, Indonesia, Philippines, Malaysia, Uganda and Zaire causing marginal or apical brown patches on foliage of young trees and *Scolecotrichum heveae* Vincens from Brazil causing shot hole disease (Spaulding 1961, Browne 1968).

POWDERY MILDEW

Oidium heveae Steinm., the powdery mildew, attacks young leaves of *Hevea* in Ceylon, India, Indochina, Indonesia, Malaysia, Uganda, Tanzania and Zaire both in nurseries and plantations. Attacked leaves are dull in appearance, get crinckled at tips, becoming bluish to purplish-black below the crinckled region. Severe fungus attack results in defoliation. The flowers may also be attacked affecting adversely the production of seed. New flush following defoliation is repeatedly attacked several times in the season affecting plant growth. The spread of the disease is favoured by high humidity, low temperature and mist. Hot dry periods with bright sunny days are unfavourable to disease development. Disease is controlled by sulphur dust and also by artificial defoliation with chemical defoliants (Sharples 1936, Rao 1971, Satchuthananthavale 1971).

RED RUST

Cephaleuros virescens Kunze, the red rust alga, attacks *Hevea* in Ceylon, India and Malaysia causing small, circular, purple spots on leaves. Minute, erect, red hairs crowd together on spots giving a velvety appearance. In severe infections, the leaves may become curled, mottled green and yellow resulting in cracking. The damage is usually inconspicuous (Petch 1921).

PHANEROGAMIC PARASITES

Three mistletoes recorded on *Hevea* are *Dendrophthoe lanosa* (Korth.) Dans., *D. pentandra* (L.) Miq. and *Macrosolen cochinchinensis* (Laur.) v. Tiegh., all from Malaysia (Browne 1968).

NON-PATHOGENIC DISEASES

BORON TOXICITY

Hevea suffers from boron toxicity when grown on recently deposited clay soils and on thin layer of coarse sandy loams over very acid granite as such soils accumulate sufficient boron to develop toxicity symptoms. Tip and marginal leaf necrosis was successfully reproduced in sand cultures with excess of boron (Shorrocks 1964). Typical symptoms include necrotic spots in the interveinal areas and a light brown to gray scorch of the leaf tips and margins. Heavy liming of soil to increase pH is suggested to reduce uptake of boron (Anon. 1964 a). Indiscriminate use of fertiliser mixtures with boron should also be avoided (Jeevaratnam 1967).

23. CASHEW–ANACARDIUM OCCIDENTALE LINN.

GENERAL

Cashew (*Anacardium occidentale*) is a native of South America and the West Indies but has naturalised in India, Ceylon and in many countries in Africa. In India, cashew plantations are largely concentrated in states of Goa, Karnataka, Kerala and Tamil Nadu and the cultivation has been extended to other states like Andhra Pradesh, Gujarat, Maharashtra, Orissa and Rajasthan. The species is raised both under horticultural and forestry conditions. Cashew prefers sandy and lateritic soils though the growth is good on clayey soils also. The species has been raised for reclaiming sand dunes in coastal areas and to prevent soil erosion. Plantations are raised by direct sowing or through one year old nursery seedlings raised in baskets. Cashew is best known for its fruit which yields the edible kernel which is highly prized as a delicacy the world over. The yield of nuts per tree varies from about 10 to 20 kg. The stem yields an insect repelling gum, the bark an indelible ink and the shell of the nut an oil which finds many industrial uses. Wood is mainly used as fuel and to some extent for boat building.

PATHOGENIC DISEASES

GIBBERELLA SEEDLING WILT

Gibberella baccata (Wallr.) Sacc. causes severe wilt and high mortality of cashew seedlings in nurseries in Uganda. The pathogen is a wound parasite and is commonly associated with symptoms like cankers, wilt, die-back and damping-off. Characteristic symptoms on seedlings are softening of the hypocotyl which collapses under light pressure. The roots are soft to touch because of breakdown of cortical tissues (Browne 1968).

PHYTOPHTHORA DAMPING-OFF

Phytophthora palmivora (Butl.) Butl. is reported from Maharashtra as causing damping-off of nursery seedlings of cashew. The pathogen attacks the collar region resulting in water-soaked darkened areas around the stem. Affected seedlings turn pale, rot and topple over. In severe infections, leaves may also be attacked (Kumararaj and Bhide 1961).

Sclerotium rolfsii Sacc. is also reported to cause seedling blight in cashew (Browne 1968).

DECLINE OF CASHEW

Pythium spinosum Sawada causes decline of cashew grown on lateritic undulating land and hillocks in Kerala. Disease symptoms are first manifested by a few branches which become defoliated and dried during summer. Gradually the symptoms spread to other branches during subsequent summers and the affected plants may succumb to the disease in about 2-3 years. The symptoms get accentuated after monsoons and become severe during dry summer. Plants of all ages may be affected. Infection establishes through finer fibrous roots which become dark brown, rotted and dead, resulting in decrease in uptake of nutrients from the soil.

The control of the disease may be effected by drenching the soil around affected trees with cheshunt mixture comprising of two parts by weight of copper sulphate and eleven parts of ammonium carbonate. Thirty grams of this dry mixture is dissolved in 9 litres of water immediately before use. Addition of organic manure mixed with suitable quantities of ammonium sulphate and superphosphate helps in speedy recovery of diseased trees (Ramakrishnan 1955).

ANTHRACNOSE DISEASE

Cashew suffers seriously from an anthracnose disease caused by *Colletotrichum gloeosporioides* Penz. (=*Gloeosporium mangiferae* P. Henn.) in Karnataka, Kerala and Tamil Nadu, resulting in the death of tender leaves, vegetative and flowering shoots. The disease adversely affects the production of fruit. Initial symptoms include reddish-brown, shiny, water-soaked lesions followed by resin exudation on affected parts. The lesions spread radially and longitudinally killing the affected shoots. Infected leaves become crumpled while the affected nuts and apples decay and shrivel. Successive die-back of branches for a few years ultimately results in death of the tree. The pathogen perennates in dead parts and becomes active when plants put forth new flush. The fungus produces abundant conidia and ascospores of the perfect state, *Glomerella cingulata* (Ston.) Spauld. & Schrenk, through which spread of the disease is secured.

The disease may be controlled by protecting newly formed vegetative and inflorescence shoots with 3 : 3 : 50 Bordeaux mixture during the vulnerable period of attack. Ferrous sulphate may be added to Bordeaux to improve the sticking quality of the spray. Since the fungus perennates in dead twigs and inflorescence, these should be clipped along with a few cm of uninfected region during fruit collection. The clipped material should be collected and burnt in order to keep down the fungus inoculum (Sujan Singh, Sehgal, Pandey and Bakshi 1967).

PINK DISEASE

Pink disease caused by *Corticium salmonicolor* B. & Br. is a major disease of cashew in coastal areas where the incidence of the disease is high (Anon. 1960). The disease is discussed in detail in Chapter 25.

PHOMOPSIS SHOOT DRYING

Phomopsis anacardii causes shoot and leaf drying of cashew in coastal areas in South India. The pathogen is believed to infect the host through insect punctures and is commonly associated with *Colletotrichum gloeosporioides* (Anon. 1960).

LEAF SPOTS

Pseudomonas mangiferae-indicae Patel, Moniz & Kulkarni, a serious leaf spot of *Mangifera indica* L., is found pathogenic on cashew leaves in inoculation experiments (Patel, Kulkarni and Moniz 1948). The bacterium causes water-soaked, light brown to deep yellow spots with clear halos on leaves, petioles, fruits and tender stems of *M. indica.*

Pestalotia dichaeta Speg. and *P. virgatula* Kleb. cause widespread leaf spots on cashew leaves in parts of Karnataka and Kerala (Ramakrishnan 1955, Anon. 1960).

Other foliage and fruit parasites recorded on cashew are *Beltrania* sp., *Ceratophoma* sp., *Coniothyrium* sp., *Labridium* sp., *Mycosphaerella* sp., and *Sphaceloma anacardii* Wani & Thirum., all on leaves from India, *Endomyces anacardii* F. de Mello and *Parasaccharomyces giganteus* F. de Mello, on fruits from India, *Oidium* sp., causing powdery mildew on leaves from Brazil, *Cephaleuros virescens* Kunze, an algal parasite from Tanzania, *Loranthus* sp. and *Cuscuta* sp., phanerogamic parasites from India (Ramakrishnan 1955, Subramanian and Ramakrishnan 1956, Anon. 1960, Wani and Thirumalachar 1969, Golato 1970).

24. WATTLES–ACACIA SPP

GENERAL

Wattles are a source of high quality tannins which are obtained from the bark. The wood is also used as fuel or in rayon industry. *Acacia dealbata* Link, *A. mearnsii* De Wild., *A. decurrens* Wendl. known respectively as 'silver', 'black' and 'green' wattles are the three important species extensively planted outside their indigenous home—Australia. In Australia and New Zealand, the common names of wattles are used differently (Streets 1962). In India, black wattle is widely planted in the high altitude locations in the Nilgiris and Palni Hills in Tamil Nadu. The species is raised in large-scale plantations in Kenya, Rhodesia, South Africa and Tanzania and to a lesser extent in Ceylon, Cyprus, Nyasaland and Uganda. In its native habitat, the species prefers cool, moist winters, moist soils in valleys or on hill sides and grows often as an understorey of moist type of eucalypts. Planting is preferred through direct sowing in South Africa and through 6 month old nursery seedlings in India. Hail belts and frosty areas are avoided. Silver wattle was introduced in the past in the Nilgiris and Palni Hills in Tamil Nadu, where it is now naturalised. The species is also planted in the Himalayas between 2000 to 2700 m. Elsewhere, silver wattle was planted in Ceylon, Cyprus, Kenya, New Zealand, Rhodesia, South Africa, Tanzania and Uganda but now is being replaced by black wattle due to higher yield of tannin in the bark of the latter species. Silver and green wattles are, however, more frost-hardy than black wattle. Green wattle is rated second only to black wattle in quality of its bark. In India, the species was extensively planted in the early 40's but later was given up in favour of black wattle. This is also generally true in other countries. The species has been successfully raised in the Himalayas between 1700–2300 m but suffers from snow damage.

PATHOGENIC DISEASES

SEEDLING BLIGHTS

Glomerella cingulata (Ston.) Spauld. & Schrenk (Syn. *Physalospora acaciae* K. Ito and Shibukawa) causes punctate brown to dark brown lesions on above ground parts of seedlings of *A. dealbata* in Japan. Wet weather favours disease development when the lesions elongate, coalesce and often girdle the stem and petioles resulting in rapid wilt of affected seedlings. In its imperfect state, the fungus is known as *Colletotrichum acaciae* K. Ito & Shibukawa when it produces salmon pink masses of conidia on the lesions (Ito and Shibukawa 1956). The fungus overwinters on plant debris. The disease is controlled by sowing seeds treated with uspulun in soil sterilised with chloropicrin at the rate of 10 ml per m², followed by monthly sprays of Bordeaux mixture from May to August at the rate of 360 ml per m² (Terashita 1962). The fungus is seed-borne infecting seeds through hilum from the suture of the pods. Mycelium is found on seed surface, in the parenchyma and in the embryo. Hot water treatment of seeds at 20°C for 5 minutes is effective (Hashimoto 1968).

Cylindrocladium scoparium Morg. and *Fusarium oxysporum* Schlecht. cause post-emergence damping-off in *A. dealbata* (Terashita 1962). The former also attacks *A. mearnsii* in South Africa (Spaulding 1961).

Other root diseases recorded on wattles are *Macrophomina phaseolina* (Tassi) G. Goid. on *A. mearnsii* in Ceylon, *Armillaria mellea* (Vahl.) Quel. on *A. dealbata*, *A. mearnsii* and *A. decurrens* in Indonesia, Nyasaland, South Rhodesia and Tanzania, *Armillaria fuscipes* Petch on *A. decurrens* in India, *Fomes lamaoensis* (Murr.) Sacc. & Trott. on *A. decurrens* in Indonesia, *Ganoderma lucidum* (Leyss.) Karst. on *A. mearnsii* in South Africa, *Peniophora incarnata* (Fr.) Karst. on *A. dealbata* in Australia and *Poria albobrunnea* Petch on *A. decurrens* in Ceylon (Spaulding 1961, Browne 1968).

BLACK BUTT DISEASE

Phytophthora nicotianae Breda de Haan var. *parasitica* (Dastur) Waterhouse (syn. *Phytophthora parasitica* Dastur) causes a serious stem disease, commonly known as black butt of *A. mearnsii* in South Africa. The fungus attacks bark at the base of the tree resulting in cracks and development of turbid, yellow or dark-brown to black patches of gum on the inner side of the infected bark. The disease spreads both upwards and around the stem until the bark dies and falls off. Trees of any age may suffer though the disease is most common on 2-5 year old plants. Disease resistance varies between individuals. In more resistant trees, new bark is produced after the diseased bark has flaked off and the trees fully recover. In susceptible trees, partial recovery is followed by reinfection and the process may be repeated giving rise to dark and black patches of bark which are typical of the disease. Extracts from diseased bark injected into healthy trees produced far more gumming than extracts from healthy wattles (Zeijlemaker 1963). The mottled form of the disease occurs when warm days (temperature more than 30°C) are followed by rain and cold nights (temperature less than 6°C). The disease has been reproduced by inoculation (Zeijlemaker 1970). Painting lower part of the stem with Bordeaux reduces disease incidence (Zeijlemaker 1969). Biological control of the disease through *Trichoderma viride* Pers. ex. Fr. is unsuccessful (Zeijlemaker 1972). A disease with similar symptoms is also reported from Kenya though no causal organism was isolated (Nattrass 1950).

Other fungi causing stem and twig cankers in wattles are *Corticium salmonicolor* B. & Br. on *A. mearnsii* in Malaysia and Mauritius and *A. decurrens* in Mauritius, *Hypoxylon hypomiltum* Mont. and *H. rubiginosum* Fr. on *A. dealbata* in Australia and *Physalospora abdita* (Berk. & Curt.) N. E. Stevens on *A. mearnsii* and *A. decurrens* in South Africa (Spaulding 1961, Browne 1968).

HEART ROTS

Heart-rot fungi recorded on wattles are *Fomes mastoporus* (Lev.) Cke. on *A. dealbata* in New Zealand, *Ganoderma applanatum* (Wallr.) Pat. on *A. dealbata* in Australia and New Zealand, *A. mearnsii* in Australia, Ceylon and South Africa, *G. australe* (Fr.) Pat. on *A. dealbata* in New Zealand, *Stereum ostrea* (Fr.) Fr. on *A. mearnsii* in Tanzania and *Trametes tawa* G.H. Cunn. on *A. dealbata* in Australia.

BLIND POCKET DISEASE

In South Africa, *A. mearnsii* suffers from blind pocket disease characterised by development of longitudinal pits in the wood of the stem into which bark pegs proliferate (Zeijlemaker and Margot 1971). In sterile culture, bark cambium from diseased bark produces initially a certain amount of callus tissues which die quickly in contrast to the healthy bark cambium, the callus tissues from which live for a much longer period thereby suggesting that the disease may be of virus origin (Zeijlemaker 1972).

WILT DISEASE

A wilt disease characterised by reddening of bark followed by wilting of foliage is recorded on *A. mearnsii* from South Africa. The cause is yet unknown (Zeijlemaker and Margot 1971).

LEAF SPOTS

Calonectria theae Loos (imperfect state *Cercosporella theae* Petch) causes dark brown to black spots on leaves of *A. mearnsii* in Ceylon and India, *A. decurrens* in Ceylon, India and Indonesia and *A. dealbata* in Ceylon (Spaulding 1961, Browne 1968). The disease is serious in some areas. Severe infections may result in complete defoliation of affected plants. The fungus forms cankers on young plants of *A. decurrens* in Ceylon (Kerr 1965).

Camptomeris albizziae (Petch) Mason and *C. verruculosa* (Syd.) Bessey cause leaf spots on *A. decurrens* and *A. mearnsii* in South Africa (Hughes 1952, Spaulding 1961).

LEAF AND TWIG RUSTS

Uromyces phyllodiorum (Berk. & Br.) McAlpine and *Uromycladium alpinum* McAlpine attack respectively phyllodes and leaves of *A. dealbata* in Australia (Spaulding 1961). *Uromycladium acaciae* (Cke.) Syd. (syn. *U. bisporum* McAlpine) forms powdery leaf spot and dark swellings on branches of *A. mearnsii* and *A. dealbata* in Australia and New Zealand. *U. notabile* McAlpine forms large, swollen, distorted, yellow-brown to chocolate brown galls on branches, phyllodes and pods of *A. decurrens* and *A. dealbata* in Australia and New Zealand. *U. tapperianum* (Sacc.) McAlpine forms gall-like masses along the entire length of phyllodes and large rounded galls on branches covered with brown powdery spores on *A. mearnsii* in Australia and New Zealand (Spaulding 1961).

PHYSIOLOGICAL DISORDER

FROST INJURY

A 40 to 50 per cent mortality in 5-year old *A. mearnsii* plantation in Tamil Nadu, India, occurred during 1969. Main symptoms of the disease were splitting of the bark of the main stem near the ground upto approximately 1.5 m, followed generally by girdling and death of the plant above the girdled region. New coppice shoots generally developed below the dead part. No pathogen was associated with dying trees. Histo-pathological studies of the affected stem sections revealed the presence of abnormal tissues consisting of undifferentiated parenchyma in the region of the injury. The medullary rays either discontinued or the ray cells became broader and the affected rays deflected from their normal path (Pl. 6, 6). These abnormalities are suggestive of damage by frost which occurred in the affected areas and to which *A. mearnsii* is susceptible (Bakshi, Reddy, Puri and Sujan Singh 1972).

BORON DEFICIENCY

Mortality occurred in 18–22 month old *A. mearnsii* in Tanzania due to boron deficiency. Plants showed symptoms of curling over of the leader, severe cambial necrosis, distortion of root collar, leaf chlorosis and die-back. Diseased plants temporarily recovered during wet season but in the following dry period, symptoms aggravated and some plants died. Leaf analysis of the diseased and healthy plants showed only 7.2 ppm boron in the former compared to 20.4 ppm in the latter (Vail, Calton and Strong 1957). Application of borax at the rate of 75 lb. per acre spectacularly reduced disease symptoms and also increased the growth (Elmer, Gosnell and Stark 1963).

25. EUCALYPTUS

GENERAL

Eucalyptus is indigenous to Australia. Because of the wide adaptation as regards site conditions and fast growth in exotic locations, eucalypts have been introduced in many countries in South America, Africa and Asia. In India, *Eucalyptus tereticornis* Sm. (so-called Mysore hybrid) is being raised in large-scale plantations in most states since 1962 and the total area planted till about 1972 exceeded 0.3 million ha. *E. grandis* (Hill) Maiden is suited to high altitude and high rainfall locations and is the principal species raised in the Grassland Afforestation Division in the high altitudinal ranges in Kerala, where more than 10,000 ha were planted till 1971. *E. globulus* Labill, introduced in the Nilgiris in the nineteenth century, has naturalised in the area. *E. citriodora* Hook. is also planted on a large scale but is susceptible to frost. In addition, trials with a large number of other species are under progress. As fast-growing exotics, eucalypts may be worked for a short rotation for industrial wood. As a vigorous coppicer, the species of *Eucalyptus* are well suited for management on a short coppice rotation. Apart from meeting local demand for timber and fuel wood, *Eucalyptus* is planted to meet the increasing demand of industrial wood, particularly for the paper industry.

For raising plantations of *Eucalyptus*, seedling stock is raised in the nursery. Seeds are sown broadcast in boxes or beds during September/October or February/March. Germination takes place in one to two weeks. Seedlings are pricked out in polythene containers when they are 5–10 cm high. The plants are watered regularly. The polythene bags are shifted from time to time to prevent plants taking roots in the soil. Nursery seedlings are outplanted with the onset of the summer monsoon. Transplanting is completed early to enable seedlings to establish during the rains. Seedlings in polythene bags are generally used for planting though naked-root planting is practised in some states. Before planting in pits, an insecticide is generally added to soil to protect seedlings against termite damage to which *Eucalyptus* is susceptible.

The area to be planted is clear-felled a year in advance, debris burnt and stumps uprooted. The area is worked with tractor which helps in soil working and eventually results in increase of plant vigour. Mechanisation also helps in removing residual roots and stumps and thereby eliminate source of infection for root diseases. Planting may also be done in pits.

Weeding and hoeing are normally done for three years. In some states like Uttar Pradesh and Kerala, *Eucalyptus* is raised under taungya method when different agricultural crops are raised for the first 2–3 years along with *Eucalyptus*. This provides relief to the agriculturists in areas where there is an acute pressure for land. Also, the lease rent lowers the plantation cost. The dense vegetation of agricultural crops changes the microclimate of the plantation floor which becomes more humid and creates ideal glass house conditions for seedling pathogens, if present, to develop into epidemic proportions. Intercultivation, however, helps to keep down weeds and thus reduces fire hazard. However, weeds invade the plantation as soon as the cultivation is stopped. In high rainfall areas like Kerala, *Eupatorium* forms an impenetrable weed in the plantation while in certain areas in Uttar Pradesh, grasses come up. Such weed growth exposes plantations to risk of fire hazards.

PATHOGENIC DISEASES

CYLINDROCLADIUM SEEDLING BLIGHT

Seedling mortality in *Eucalyptus* is caused by different species of *Cylindrocladium*. In India, *C. scoparium* Morg. has been recorded from *Eucalyptus* nurseries in Goa and Dehra Dun (U.P.) and *C. quinqueseptatum* Boedijn & Reitsma from Dehra Dun and Kerala. Both fungi cause mortality in the nursery and in outplantings. In Kerala, high summer rainfall and intercultivation with crops, like tapioca and paddy, favour building up of high moisture around the plants thus creating a favourable microclimate for the development of the pathogen Elsewhere, damping-off and root rot due to *C. scoparium* are widely prevalent throughout Latin America and have caused total loss of seedlings in some nurseries (Reis and Hodges 1975). This pathogen has also been recorded from Costa Rica causing foliar damage and occasionally complete defoliation in several species of *Eucalyptus*, except *E. deglupta* which is highly resistant

(Bazan De Segura 1970), *C. quinqueseptatum* with its perfect state in *Calonectria quinqueseptata* Figueiredo & Namekata, obtained in culture and on inoculated leaves, causes damping-off of *Eucalyptus* seedlings in Brazil (Figueiredo and Namekata 1967). Leaf blight of *Eucalyptus* caused by a new species of *Cylindrocladium* occurs in Brazil and is most prevalent in plantations established from seed source of Australian origin than of Brazilian origin (Reis and Hodges 1975). *C. clavatum* Hodges & May, the cause of root disease in conifers in Brazil, is also recorded on seedlings of *Eucalyptus saligna* (Hodges and May 1972).

All the three fungi infect leaves, young twigs and also the main stems of one year old plants. Infection on leaves results in premature defoliation whereas infection of twigs and main stems results in wilting of plants. The initial symptoms are appearance of dot-like spots and slight discolouration which deepens to become grayish black. Necrotic tissues present a water-soaked appearance. Under warm and very humid conditions, the pathogen develops, on affected plant parts, white mycelial growth on which numerous spores are produced disseminating the disease. The incidence of the disease declines in dry weather.

The disease may be controlled through spraying with chemicals like blitox and captan.

OTHER SEEDLING DISEASES

In Brazil, *Cercospora eucalypti* Cke. & Mass. occurs commonly in *Eucalyptus* nurseries and attacks leaves but causes only slight damage. The fungus is also recorded from Argentina, Australia, India, Italy, Peru, Taiwan and Zaire. *Puccinia psidii* Wint. attacks foliage and young twigs of 3-12 months old seedlings resulting in severe losses in the crop in some provenances of *E. grandis* and *E. saligna* in Brazil (Reis and Hodges 1975).

Botrytis cinerea Fr. causes damping-off and leaf spot in *Eucalyptus* in Argentina, Australia, Brazil and Kenya. Excessive density of plants favours disease development. *Ramularia pitereka* Walker & Bertus causes leaf spot and shoot blight in Australia. *Ceuthospora innumera* Massee and *Piggotia substellata* Cke. cause necrotic spots on juvenile leaves of *Eucalyptus* in Australia. *Harknessia eucalyptina* Aniceta Santos attacks leaves and petioles of *Eucalyptus* resulting in defoliation and sometimes causes serious loss to planting stock in Portugal (Magnani 1964, Walker and Bertus 1971, Ashton and Macaulay 1972).

Oidium eucalypti Rostr. in Denmark and *Oidium* sp. in Australia, Europe and South America cause powdery mildews on *Eucalyptus* seedlings in the nursery. Affected leaves get wrinkled, blistered and drop off. Application of sulphur fungicides gives good control of the disease (Magnani 1964).

Other nursery pathogens recorded on *Eucalyptus* include *Rhizoctonia solani*, species of *Fusarium, Phytophthora* and *Pythium* from Brazil, *Alternaria tenuis* Pers. from Argentina, *A. tenuissima* (Fr.) Wilt. from Italy, *Calonectria theae* Loos from Mauritius, *Cercospora epicoccoides* Cke. & Mars. from Argentina, Australia and Taiwan, *Cytospora eucalypticola* Van Der Westhuizen from South Africa and *Macrophomina phaseolina* (Tassi) G. Goid. from Peru (Magnani 1964, Van Der Westhuizen 1965, Browne 1968, Bazan De Segura 1969, Reis and Hodges 1975).

PHYTOPHTHORA CINNAMOMI ROOT ROT

Phytophthora cinnamomi Rands, the cause of root rot of numerous broadleaved species and conifers, is widely distributed in all the continents. The fungus has caused considerable damage in dry sclerophyll forests of jarrah (*Eucalyptus marginata* Sm.) over 80,000 ha in south western Australia. The fungus can survive on small pieces of infected roots or soil particles and may be dispersed during logging, vehicle movement or shifted with gravel for road making. The pathogen is also dispersed through infected nursery stock. No economic control is available but sanitation and hygiene procedures are expected to reduce the rate of spread of the pathogen (Newhook and Podger 1972).

GANODERMA ROOT ROT

Different species of *Eucalyptus* are susceptible to attack by *Ganoderma lucidum*. Attacked seedlings are killed. In trees, the fungus causes decay in roots due to which plants loose footholds in soil and become windblown. Roots and stumps of the previous crop decayed by *G. lucidum* serve as infection centres for *Eucalyptus* planted under reforestation programmes. In mechanised plantations, roots and stumps of original forest crop are mostly removed thereby minimising the risk of *G. lucidum* root rot. If, however, such infection centres are not removed, the incidence of root rot is high which may be minimised by raising *Eucalyptus* in mixture with resistant species which will prevent rapid spread of the disease through root contact. In case mortality occurs and replacements are necessary, plants should not be put on the same place originally occupied by dead plants where disease inoculum is likely to be present. Instead, planting should be done as far away from old planting holes as possible to avoid roots coming in contact with decayed roots. The disease is discussed in detail in Chapter 15.

CROWN GALL

Agrobacterium tumefaciens (E.F.Sm. & Town.) Conn., the cause of crown gall, has been reported on eucalypts from Argentina and Columbia (Reis and Hodges 1975).

OTHER ROOT ROTS

Other root rots recorded on different species of *Eucalyptus* include *Armillaria mellea* (Vahl) Quel. from Australia, Cyprus, New Zealand, Tanzania and Tunisia; *Botryosphaeria ribis* (Fr.) Gross. & Dug. from Argentina; *B. berengeriana* de Not. from Portugal; *Clitocybe tabascens* (Fr.) Bres. from U.S.A. and Tunisia; *Helicobasidium compactum* Boedijn from Nigeria; *Ganoderma colossum* (Fr.) Bose from South Africa and *Polyporus dryadeus* Fr. from Australia (Purnell 1959, Magnani 1964, Browne 1968).

PINK DISEASE

Distribution

Corticium salmonicolor B. & Br., the cause of pink disease, is widely distributed in the tropics and subtemperate regions of Asia, Africa, West Indies, New South Wales and New Zealand and is probably indigenous in the regions where it occurs. It attacks many plants of economic importance like coffee and cinchona in Java, tea in Ceylon, cocoa in West Indies and rubber in India and Malaysia. It attacks *Eucalyptus* planted as exotic. In India, it is recorded on *E. grandis* (Hill) Maiden, *E. tereticornis* Sm., *E citriodora* Hook. and *E. globulus* Labill and probably other species. The fungus is recorded on *E. grandis* and *E. camaldulensis* Dehn. in north-east Brazil and on *E. kirtoniana* F.v.M. in Costa Rica. The fungus possesses a wide host range among woody plants numbering more than 141 species belonging to 104 genera (Sharples 1936). In India, the fungus is recorded on a wide range of forest trees including *Acacia leucophloea* Willd., *Albizia falcataria* (L.) Fosberg, *Aleurites montana* E.H. Wilson, *Anacardium occidentale* L., *Artocarpus heterophyllus* Lam., *Azadirachta indica* A. Juss., *Butea monosperma*

(Lam.) Taub., *Cassia siamea* Lam., *Casuarina equisetifolia* Forst., *C. montana* Jungh., *Eucalyptus alba* Reinw., *E. camaldulensis*, *E. globulus*, *E. saligna* Sm., *E. multiflora* Poir, *E. tereticornis*, *E. tessellaris* F. v. M., *E. trabuti* Vilmorin, *Grevillea robusta* A. Cunn., *Hevea brasiliensis* Muell.-Arg., *Leucaena glauca* (Linn.) Benth., *Mangifera indica* Linn., *Podocarpus gracilior* Pilger, *Populus* spp., *Salix daphnoides* Will., *Tamarindus indica* Linn. and *Tectona grandis* Linn. f. (Browne 1968, Bakshi, Reddy, Sujan Singh and Pandey 1970).

Damage

As exotics in India, species of *Eucalyptus* have not developed the resistance against *C. salmonicolor* which occurs naturally on a wide range of hosts in the natural forests. All the three species of *Eucalyptus*, namely, *E. tereticornis*, *E. grandis* and *E. globulus* which are widely planted are highly susceptible to the disease which becomes epidemic under conditions of high rainfall and high temperature during the summer. *Eucalyptus* is susceptible to the disease at a very early age. The disease appears sporadically during the second year but becomes epidemic in the two subsequent years when the entire plantation may become affected under ideal conditions for development of the disease. Young plants may show repeated die-back and mortality, while those which have escaped infection during the early age, may at a later period become infected resulting in development of cankers. Over extensive area in Industrial Plantation Circle, Kerala, the disease may be so devastating (Pl. **15**, 1) that *Eucalyptus* cannot be perpetuated as a commercial species.

Symptoms

In young plants, infection results in girdling of branches and the main stem towards the latter part of the rainy season. The affected plants are killed outright in *E. grandis* and *E. globulus*. In *E. tereticornis*, however, infection on the stem or branches causes girdling and death of parts above the girdled region. Simultaneously below the region of girdling, many epicormic branches develop one of which becomes the leader which again becomes infected and killed in the following year (Pl. **15**, 2) and the process continues resulting in diminution of height growth of plants. The growth increment and productivity become seriously affected in plantations where the disease is epidemic.

If plants escape infection at an early age, they put up good height and diameter growth. However, such trees may subsequently become infected on the main bole resulting in splitting and cracking of bark and exposure of wood (Pl. 15, 3, 4). Gum eventually exudes from the resultant cankers. The attack is localised and girdling is usually not complete and, therefore, the main leader continues to grow and the crown appears healthy.

Predisposing factors

In India, *C. salmonicolor* occurs naturally and is not damaging to many indigenous hosts on which the fungus is recorded. As exotics in India, the species of *Eucalyptus* have not developed resistance against attack by the fungus and are highly susceptible on the contrary. Therefore, *Eucalyptus* plantations raised in a locality where the fungus is present become diseased. High summer rainfall above 250 cm and warm tropical conditions encourage development of different growth forms of the fungus which is disseminated rapidly resulting in epidemic outbreak of the disease. The severity of the disease becomes greater with increasing rainfall above 250 cm as in the Industrial Plantation Circle, Kerala, and in Goa and western districts of Karnataka. Freedom from the disease in *E. grandis* plantations in high rainfall areas in Peermade-Pamba, Kerala, is due to the fact that the site was formerly grassland and, therefore, the natural hosts of the pathogen and the inoculum are absent.

Pathogen

C. salmonicolor occurs on the host in four growth forms as described below.

1. **Pustule.** Pustules appear on branches and stem within a few weeks after infection at the beginning of the monsoon as pink to salmon coloured sterile cellular bodies upto 1 mm in diameter (Pl. 15, 6).

2. **Cobweb.** The infected parts produce on the bark thin, white, arachnoid mycelia (Pl. 15, 5) originating from pustule or from the edges of the pink incrustation.

3. **Necator.** The fruit bodies are orange red, about 2 mm in diameter and develop on the upper side of the attacked branch exposed to sun. Conidia develop after the rains.

4. **Pink incrustation.** This comprises the perfect stage of the fungus and develops as thin, light pink incrustation on the bark (Pl. 15, 7).

It develops during autumn on stems and branches originating at the fork usually on the under surface of the stem on the shady side, but spreading around the stem later on.

The different stages of the fungus develop on the bark which usually becomes killed. The disease is possibly disseminated not only by the two sporing forms but also by the two sterile stages when the dead bark with viable portion of the fungus flakes off and is carried away by wind to infect trees.

Control

Effective control of pink disease of *Eucalyptus* has not yet been obtained though efforts are being made on the following lines.

In high rainfall areas in Kerala and Karnataka where the pink disease has limited the success of *E. grandis* and *E. tereticornis*, it may be necessary to replace them with other species of *Eucalyptus* or other fast growing species suitable for pulpwood and, at the same time, resistant to the pink disease. This may be determined from species trials when the performance of the species and their resistance to the disease may be ascertained in the field. It may be mentioned that *E. torelliana* F.v.M., growing in heavily diseased locations, exhibits certain degree of freedom from the disease. This is also borne out in artificial inoculation. The species develops a thick canopy and is, therefore, suitable in high rainfall locations to discourage weed growth. The wood is suitable for chemical pulping.

In heavily diseased plantations, some individuals remain healthy, free from infection. Such individuals may be further tested for inherent resistance by raising progenies from them and screening them against infection to provide material for developing a stock resistant to the disease.

Pink disease in rubber is effectively controlled by application of Bordeaux paste to infected trees. Such a curative treatment is justified for a greatly valued cash crop like rubber. In locations where the disease hazard due to *C. salmonicolor* is high, a prophylactic spray to *Eucalyptus* during the most vulnerable period of infection, between 2 to 4 years age, is likely to afford protection to plants against infection. Such sprays should be done at the beginning of summer monsoon and repeated 2–3 times during the rainy season. This will also reduce inoculum of the pathogen.

For this, it is necessary to work out a suitable chemical, its mode of application and economic feasibility. Beyond the fourth year of plantation age, plants resist infection and further chemical protection may not appear necessary.

DIAPORTHE CANKER

Canker in *Eucalyptus*, caused by *Diaporthe cubensis* Bruner, is recorded in Brazil, Surinam and Cuba and is likely to become important in tropical America wherever *Eucalyptus* is grown. The disease is more damaging in areas of high temperature and humidity. The fungus causes basal and trunk cankers of a large number of species which show considerable variation in the relative susceptibility to the disease both within and among the species. *E. saligna* Sm., *E. maculata* Hook., *E. grandis*, *E. alba* Reinw., *E. propinqua* Deane & Maiden and *E. tereticornis* are moderately or highly susceptible while *E. citriodora* Hook., *E. robusta* Sm., *E. microcorys* F.v.M., and *E. paniculata* Sm. moderately to highly resistant and *E. torelliana* appears to be immune. The fungus attacks plants at an early age, even less than one year, at the root collar causing discolouration, girdling and death in some cases. Young cankers develop around the base of dead branch stubs. In the initial stages, bark cracks longitudinally. Open cankers eventually develop which may reach more than a metre in length and from which exudation of gum occurs. Pycnidia and perithecia of the fungus develop profusely on dead bark. Through the open canker, decay fungi and insects establish in the wood which becomes weakened and the tree may break at the level of canker. Basal cankers cause a reduction in the number of stumps that sprout. Control of the disease is possible only by planting resistant species (Hodges and Reis 1975).

OTHER CANKER DISEASES

Hypoxylon annulatum (Schw.) Mont. and *H. howeianum* Peck cause cankers on *Eucalyptus* in Australia (Browne 1968).

JUVENILE DECAY

Butt or stem decay in *Eucalyptus* may occur when stands are young about 4 years of age or more. In young trees of *E. tereticornis* in India, heartwood does not develop and decay may occur as white spongy rot. Such decay is reported from Australia (Da Costa personal communication). In Zambia, incidence of butt decay is 15 to 40 per cent, entailing a loss in volume upto 0.2 per cent in *E. grandis*. In *E. cloeziana* F. Muell., incidence and volume of decay are low. *Poria epimiltina* Berk., *Lopharia cinerascens* (Schw.) G. H. Cunn., *Phellinus gilvus* (Schw.) Pat. and *Mycoacia uda* (Fr.) Donk are associated with decay. Incidence of decay can be reduced by modifying the pruning treatment given to trees (Ivory 1975).

OTHER DISEASES

Xanthomonas eucalypti Truman causes die-back of 3–7 m high *E. citrodora* plants in Australia. The bacterium kills terminal buds and juvenile leaves and causes complete defoliation in cases of severe infection (Truman 1974).

Fistulina hepatica Fr. from Australia and Portugal, *Fomes fomentarius* (Fr.) Kickx from Portugal, *F. robiniae* (Murr.) Sacc. & D. Sacc., *F. robustus* Karst., *Ganoderma applanatum* (Wallr.) Pat., *Pleurotus lampas* (Berk.) Sacc., *Polyporus portentosus* Berk., all from Australia and *P. sulphureus* from Argentina, Portugal, South Africa and U.S.A. cause heart rot on different species of *Eucalyptus* (Spaulding 1961, Browne 1968).

Important records of leaf spot fungi in plantations include *Alternaria alternata* (Fr.) Keissler from India, *Readeriella mirabilis* Syd. from Mauritius and Australia and *Rhytisma eucalypti* P. Henn. from Australia (Browne 1968, Bakshi, Reddy, Puri and Sujan Singh 1972).

Of the four virus diseases recorded on *Eucalyptus*, one causes chlorosis and reduced leaf development in Argentina. From India, three virus diseases, namely, mosaic causing bushy growth, little leaf causing stunting and reduction in leaf size and leaf crinckle causing upward curling of leaves are recorded (Magnani 1964, Sastri, Thakur, Gupta and Pandotra 1971).

PHYSIOLOGICAL DISORDER

YELLOWING AND DIE-BACK

Extensive yellowing occurs in plantations of *E. tereticornis* in South and North Gadgadia Blocks, Tarai and Bhabar Plantation Division (U.P.) where an area of 1400 ha was affected and also in Pipli, Bareilly Forest Division (U.P.). The symptoms exhibited by diseased plants include reduction of leaf size, yellowing of leaves and premature defoliation.

Branches and trees may eventually die. No pathogen was apparently associated with affected trees. In diseased areas, the soil is highly calcareous. Also, it has a high pH (about 8.6) which may have adversely affected the availability of plant nutrients, deficiency of one or more of which may have been responsible for large-scale yellowing in plantations.

GUMMOSIS

Exudation of kino, erroneously termed as 'gum', is a common phenomenon in *Eucalyptus*. The commonly used term 'gum tree' given to some eucalypts is a misnomer for no gum in the strict sense is obtainable from eucalypts. Gums are usually water-soluble substances of carbohydrate composition, while kino, which occurs in eucalypts, is a resinous exudation and accretion containing tannins (Penfold and Willis 1961). In some species of *Eucalyptus*, the kino zones may form in the bark but not in the wood while in others, they are formed in the wood but not in the bark. This is suggested as a systematic distinction typical of the two groups of eucalypts namely, the 'boxes' and the 'gum' (Day 1959).

Symptoms

During gummosis in eucalypts, symptoms of the disease include swelling and splitting in the bark, often with a certain amount of exudation of kino, a shining golden brown viscous liquid which hardens on exposure to air as brown or reddish lumps on bark of trees. In severe cases, the bark may die along varying lengths and widths on the main bole exposing the wood from beneath (Pl 6, 3). In the following growing season, the injury, if small, may be healed over if the cause of gummosis is no longer present. Else, the healing may be delayed or may not occur. In some cases, gummosis may occur extensively all around the stem resulting in kino oozing out from different places. If the girdling is partial, the affected trees continue to live; if complete or nearly so, the plant parts above the region of girdle usually die. Within the wood, gum veins develop in the annular zone as red brown pipes filled with kino (Pl. **6**, 3).

Pathological anatomy

Gummosis represents a wound reaction. A mild injury causes the cambium to develop undifferentiated parenchymatous tissue instead of normal cells. The medullary rays become laterally displaced and the ray cells become markedly broadened at the region of the false ring (Pl. **6**, 5). Within this tissue, kino pockets develop lysigenously or schizo-lysigenously and lining such cavities, parenchymatous tissues develop and kino is produced filling up the cavities (Pl. **6**, 4). In the next growing season, cambial activity may be renewed resulting in formation of normal wood and bark in which the parenchyma tissue gets embedded as an incomplete or a false ring (Bakshi and Sujan Singh 1964). In severe injury, the cambium is killed exposing the wood. This injury may be healed over partially or completely by callus growth from renewed cambial activity on the outer side of the undifferentiated tissue which remains embedded in the wood as in the 'gums'. In the 'boxes', such activity is renewed on the inner side of the undifferentiated tissue which is eventually flaked off with the bark.

Cause

Injury to the cambium with a consequent development of the gum veins may result from various causes, mainly physiogenic and mechanical. Gummosis may also be caused from animal damage or attack by insects and fungi. The gum zone formation in *E. gomphocephala* A. DC. and *E. camaldulensis* Dehn. in Cyprus is attributed to deficient water supply (Day 1959). Injury to the cambium resulting in development of abnormal tissue, the so called frost ring, is a common form of frost damage recorded for both conifers and hardwoods (Hemenway 1926, Harris 1934). Mortality in *E. citriodora* Hook. in plantations in Bijnor (U.P.) is ascribed to severe frost and also possibly to fire (Bakshi and Sujan Singh 1964).

Control

Plants should be protected from all avoidable injuries to minimise gummosis. Species should be planted on suitable site. Frosty locations should be avoided for planting species, like *E. citriodora*, susceptible to frost damage. Injury to plants, particularly when young, due to fire may be minimised by keeping down weed growth or through control burning.

PLATE 15

PINK DISEASE OF EUCALYPTUS

1. Mortality in a plantation. 2. Repeated die-back and epicormic branches. 3–4. Canker in early (3) and advanced (4) stages. 5–7. Growth stages of *Corticium salmonicolor* : 5. Cobwebby growth. 6. Pustules. 7. Pink incrustation (sporophore). 8. Girdling and epicormic branches after inoculation.

PLATE 16

ROOT—ROT IN CONIFERS

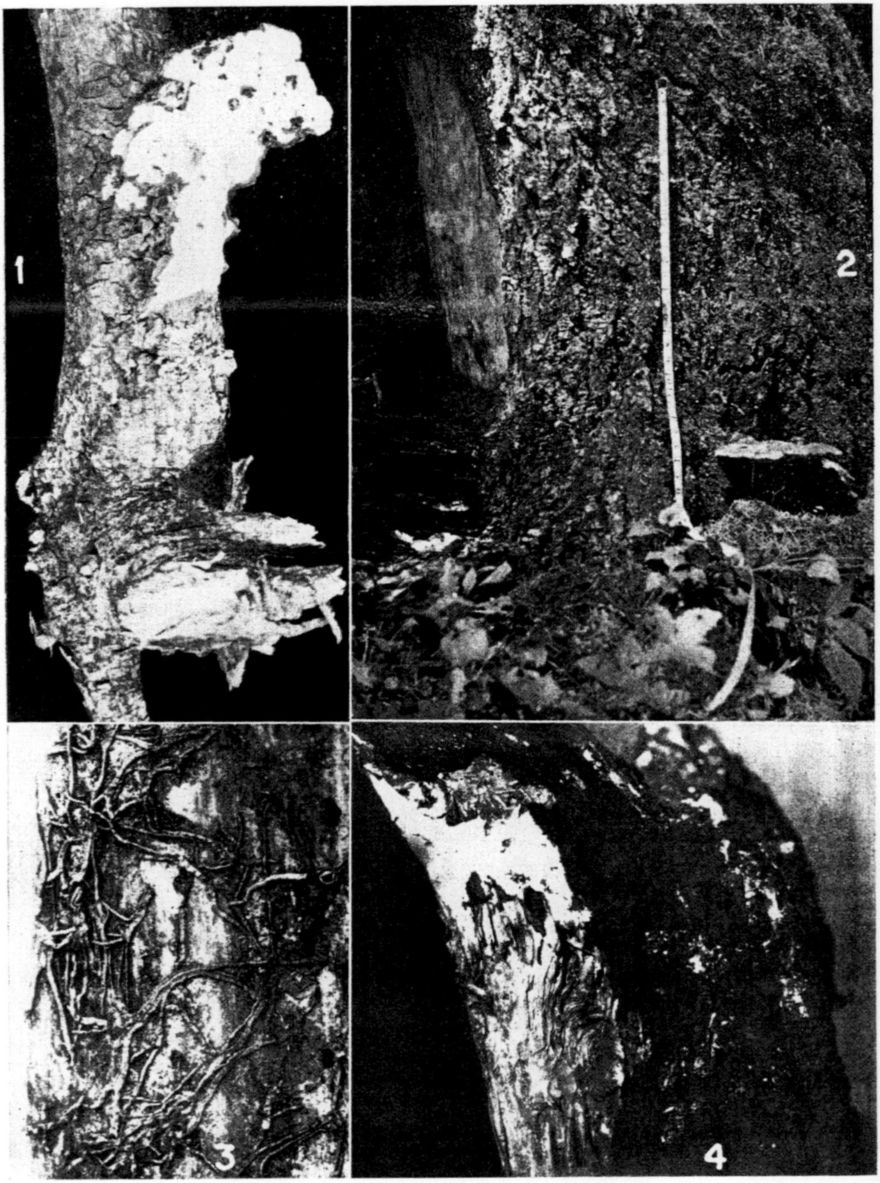

1. *Fomes annosus* sporophore on deodar. 2. *Polyporus tomentosus* var. *circinatus* sporophore on deodar. 3–4. *Armillaria mellea* on silver fir: 3. Rhizomorphs on wood. 4. Mycelial mat of the fungus below bark. (1, Photo–FRI)

26. DEODAR–CEDRUS DEODARA LOUDON

GENERAL

Deodar is typically gregarious and commonly occurs as a pure crop, though the species also occurs in mixture with both coniferous and broadleaved species. In the latter case, deodar retains its gregarious habit and often forms the bulk of the crop. Blue pine in the drier types of forests and spruce are the most important associates of deodar. Less commonly, deodar also occurs mixed with silver fir at higher elevations and with chir at the lower. Among broadleaved species, oaks besides other hardwoods are the common associates of deodar. Deodar occurs naturally throughout the western Himalayas at elevations varying from 1,000 to over 3,000 m but most commonly between 2,000 to 2,500 m. Being the most important timber species among indigenous conifers, deodar is widely planted for which site should be suitable as regards aspect, elevation and drainage. Aspect and elevation govern the distribution of deodar. At lower elevations, deodar seeks cool aspects and moist soil. At medium elevation where deodar is at its best, it flourishes on any aspect provided the soil is well-drained. Deodar is a strong light demander though it can tolerate much shade from the side. In mixture with spruce which affords dense lateral shade, deodar produces a clean bole. Deodar is fire-sensitive and is also affected by drought, chiefly in the seedling stage but is not affected by frost. Snow may cause serious injury, particularly to the pole crop.

PATHOGENIC DISEASES

FOMES ANNOSUS ROOT AND BUTT ROT

Occurrence

Fomes annosus (Fr.) Cke. is world-wide in distribution and is economically important and widespread in the temperate climatic regions of the northern hemisphere but is sporadic and of minor importance in sub-tropical climatic zones. The fungus is recorded on many conifers and also attacks some broadleaved trees. In the southern hemisphere, the fungus is recorded on dead trees and stumps of indigenous conifers, namely, *Araucaria cunninghamii* Sweet and *A. bidwillii* Hook. in Australia. The fungus has, however, not been recorded on extensive plantations of exotic conifers, except on standing dead trees of *Pinus patula* Schlecht. & Cham. (Shain and Bolland 1974).

Damage

Losses from *F. annosus* occur in three ways : through out-right killing of trees usually limited to young stands of conifers, particularly pines, douglas fir (*Pseudotsuga menziesii* (Mirb.) Franco) and deodar; through butt rot resulting in loss of most valuable portion of the tree due to decay in heartwood as in western hemlock (*Tsuga heterophylla* (Raf.) Sarg.), western red cedar (*Thuja plicata* D. Don), larch and spruce in Great Britain; and by reduction in growth due to death of roots, which may lead to greater economic loss in spruce than due to decay. Finally, the root rot renders instability to trees which become liable to wind throw. Conifers may vary in their susceptibility to root rot mortality and butt rot due to *F. annosus* (Peace 1962).

In India, the fungus occurs usually as a saprophyte and sometimes attacks living trees. Damage to deodar in plantations is reported in Kulu (Troup 1921); Seraj, Pabar and Kulu forests and Takrasi plantations (Aggarwal 1933, 1936 and 1949), and Konain, Chakrata (Bakshi 1955a). The fungus does not normally cause damage to other coniferous species in the Himalayas though group dying of blue pine due to *F. annosus* is reported in Kinnaur forests (H.P.). *F. annosus* causes serious damage to planted stands but is of minor consequence in natural forests. The disease, therefore, presents a serious problem in Europe in extensive planted coniferous stands. In North America, the fungus is becoming increasingly important with changing over from natural to artificial stands and intensification of management, involving thinnings (Boyce 1961).

Infection

The fungus becomes initially established in new areas usually through basidiospores and sometimes through conidia which infect freshly cut stumps or, in some cases, through wounds on the roots or at tree bases. The sporophores develop commonly on trees at the collar region (Pl. **16**, 1) or apparently on soil but then attached to roots or a woody food-base. They also develop on stumps, logs and litter. The basidiospores are released from sporophores in large quantity to infect freshly cut stump surfaces during thinning or harvesting of the crop (Rishbeth 1950). Conidia are rarely formed in nature (Bakshi 1952) though they develop abundantly in culture. Conidia have been used successfully as a source of inoculum in field inoculation of stumps and are reported to be the source of primary infection in plantations (Klyushnik and Morgun 1961). However, conidia may not be as effective an inoculum as basidiospores (Hodges 1969).

The freshly cut stump surface is highly selective for *F. annosus* (Meredith 1959). This selectivity decreases markedly as the interval between felling time and infection increases. Infection may not take place if more than about two weeks have elapsed after felling due to competition with other microorganisms at the stump surface (Rishbeth 1951, Meredith 1960, Yde-Anderson 1962, Cobb and Schmidt 1964, Cobb and Barber 1968, Ross 1968). After the fungus has colonised the stump surface, it spreads rapidly down the stump into roots, which serve as infection centres from where the disease may spread to adjoining trees through root contact and grafts. The amount of initial stump infection is not necessarily correlated with the number of trees infected in the residual stock. Even if *F. annosus* thoroughly colonises the stump and its roots initially, it is often replaced by other microorganisms especially *Peniophora gigantea* (Fr.) Massee and *Trichoderma* spp. (Meredith 1960, Kuhlman and Hendrix 1964). Killing due to *F. annosus* becomes serious on reforested site from the beginning if the disease was present in the earlier stand. In afforested and arable sites, the disease becomes important only after thinning and colonisation of stumps by *F. annosus*. *Fomes annosus* is unable to spread through non-sterile soil though the fungus grows readily in sterile soil or litter, indicating the antagonistic effect of microorganisms in the former (Rishbeth 1950).

Climatic conditions, like temperature and rainfall, the latter directly affecting the moisture content of the newly cut stump surface, control spore dispersal and infection. Air temperature at the time of felling is an important factor affecting conifer stump colonisation. The infection is less when the mean daily temperature is above 21°C or below 0°C. In between these two extremes, infection is favoured (Anon. 1967, Ross 1973). Attack due to *F. annosus* may occur when trees grow on suitable as well as on unsuitable sites. In Great Britain, pines may be attacked on former agricultural site or with alkaline soil reaction but not on acid soils (Low and Gladman 1960). In India, the fungus which commonly develops as a saprophyte does not cause damage to deodar in the natural forests or in plantations raised on suitable site. If the species is forced in areas unsuitable as regards aspect, elevation and drainage, roots are attacked resulting in mortality of trees.

Symptoms

F. annosus may cause mortality usually on young trees, or butt-rot in old trees. The fungus spreads from infected to the healthy plants through root contact and the diseased plants occur in groups. Decay in roots may result in wind-throw of trees. The fungus produces on the root thin, white, mycelial sheets between the bark and the wood. The mycelium may travel from the root to the butt-region where sporophores usually develop. The fungus produces white pocket rot in the wood.

Sporophore

The perennial fruit bodies develop at the base of the tree, usually hidden in the leaf litter. They are bracket shaped (Pl. **16**, 1), with hymenial surface white when fresh, turning biscuit-coloured on drying. The upper surface is light brown and lightly zoned. Basidiospores are produced in abundance. The fungus produces in culture conidia of imperfect *Oedocephalum* state.

Control

Silvicultural. The ability of the fungus to sustain itself as a saprophyte on dead roots makes practical control measures difficult in a reforested stand. In such areas, mechanised plantations may be raised where stumps and residual roots are removed, and, therefore, the disease inoculum is eliminated from the site. This operation may, however, be expensive and may be justified only where the disease is severe and where site conditions permit raising mechanised plantations. To give good control, stump removal must be fairly complete, as even small infected roots left in the site may initiate new infection centres. Severity of disease may depend on site conditions like drought or high soil moisture, species composition in the earlier and also in the present stand, soil alkalinity, spacing and thinning. In infected sites, it is possible to reduce further losses by raising resistant species. In deodar, serious attacks are reported on damp and badly drained ground and remedy lies in choice of suitable site (Aggarwal 1933). Seasonal thinning of pine plantations when temperature is lethal to *F. annosus* spores for infecting stumps (Anon. 1967) is a practical and effective means of controlling *F. annosus* (Ross 1973).

Chemical stump treatment. The risk of *F. annosus* root rot can be minimised by protecting freshly cut stump surface with chemicals which prevent stump colonisation by the fungus. Creosote and urea (20 per cent) (Sinclair and Risley 1964–65), sodium nitrite (10 per cent) and herbicides reglone (containing diquat) and gramoxone (containing paraquat) both at 5.5 per cent concentration are effective in protecting the stump surfaces from colonisation by *F. annosus* (Greig and Burdekin 1970). Complete protection of stumps by creosote, however, increases the risk of *F. annosus* becoming dominant in stumps which already have root infection (Rishbeth 1959). Stumps generally die more rapidly after treatment with urea than after creosoting (Rishbeth 1959 a). Except for creosote, a marker dye can be mixed with each of the chemicals to show up on the stump and this ensures complete coverage of exposed surfaces. Sodium nitrite is poisonous and adequate safeguards are necessary to protect lives. It is now used throughout the Forestry Commission plantations in Britain with the exception of water catchment areas and plantations from which it is impossible to exclude livestock (Greig and Burdekin 1970). Chemical treatment of stumps is effective in keeping off *F. annosus* from securing a foot-hold in the stand.

Biological. Woody tissues in the initial stages provide virgin substrates which are highly selective and, therefore, opportunities for biological control may be correspondingly greater. This principle is taken advantage of in inoculating the freshly exposed stump surface with a fungus innocuous to the standing crop thereby preventing colonisation of the stump surfaces by *F. annosus*. In Britain, inoculation of pine stumps with oidia of *Peniophora gigantea* soon after felling gives good protection against *F. annosus* (Rishbeth 1963). *P. gigantea* enters lateral roots of inoculated stumps and tends to check the advance of *F. annosus* in any tissues infected at the time of felling and also replaces the fungus to some extent. For large-scale inoculation of stump surfaces, tablets containing oidia of *P. gigantea* are dissolved in water and applied to stump surfaces (Rishbeth 1963). The marker dye, hexacol violet, is added (Anon. 1970). *P. gigantea* can colonise only pines but is incapable of colonising stumps of Douglas fir or Japanese larch (Gremmen 1963). In India, *P. gigantea* is recorded on dead stumps and logs of spruce and fir in the western Himalayas (Bagchee and Bakshi 1954). However, the role of this fungus in the biological control of *F. annosus* in India is not yet ascertained.

ARMILLARIA MELLEA ROOT ROT

Armillaria mellea (Vahl) Quel., also known as shoestring fungus and honey agaric, is world-wide in distribution and occurs both on conifers and hardwoods. In India, the fungus occurs commonly as a saprophyte on deodar, spruce, fir, blue pine and oaks, but may sometimes cause root rot and mortality in deodar planted after clear-felling original conifer forests. In the living tree, the fungus develops a white leathery mycelial felt between the bark and the wood on roots (Pl. **16**, 4) and stem bases. The mycelial felt is much thicker than the thin, paper-like mycelial sheet developed in roots on trees attacked by *Fomes annosus*. The fungus also develops dark brown to black rhizomorphs which may spread in the soil (subterranean type) or grow between the bark and the wood (subcortical type). The rhizomorphs are tough, about 2 mm broad, dark brown to black, flattened under the bark where they branch and anastomose (Pl. **16**, 3). The sporophores which develop in the rains occur singly or in clusters, possess stipe bearing annulus and a cap. The fungus attacks both heartwood and sapwood. In advance stages of decay, the wood becomes soft and spongy in which numerous black zone lines appear (Bakshi 1955).

Since *A. mellea* generally attacks trees of low vigour, maintenance of stand vigour is an effective means of controlling attack by the fungus (Boyce 1961). *A. mellea* colonises roots and stumps soon after felling due to high carbohydrate content in them. In clearing a forest land for planting, ring-barking of trees sometime prior to felling, results in lowering of carbohydrate content in roots which are invaded not by *A. mellea* but by innocuous fungi thereby reducing the inoculum potential and threat to the plantation (Leach 1939).

PENIOPHORA BUTT ROT

Peniophora luna Romell, the cause of serious butt and trunk rot, is recorded in deodar from Kulu (Himachal Pradesh) in plantations raised along river banks on a bouldery site. The trees become infected through injury occurring on the butt and roots during their passage through boulders. The sporophore and its culture are described by Bakshi, Reddy and Balwant Singh (1970). The fungus causes brown cuboidal rot in the heartwood. A correlation between diameter of decay at the stump level and the extent of decay in the bole is needed in programming cross-cutting of logs to eliminate decayed wood as worked out by Krasikova (1966).

POLYPORUS TOMENTOSUS ROOT ROT

Polyporus tomentosus Fr. var. *circinatus* (Fr.) Sartory & Maire causes root and butt rot of conifers in Canada, U.S.A. and India. In India, the fungus attacks deodar, spruce and blue pine, mainly through injury caused by fire though Gosselin (1944) suggests that the fungus forms a symbiotic mycorrhizal association with its host before becoming parasitic. The cambium is attacked and decay, in the form of elongated pockets with zone lines, develops in the heartwood. Sporophores which develop at tree bases (Pl. **16**, 2) are sessile, corky becoming hard and brittle with inrolled margin, brown with predominantly curved setae in the hymenium. Prevention of injuries, particularly those caused by fire, is important to control the disease (Bakshi 1955 a).

PERIDERMIUM WITCHES'-BROOM

Witches'-broom caused by *Peridermium cedri* (Barclay) Sacc. on deodar occurs in Kulu and Bashahr (H.P.) and elsewhere, though the incidence has declined.

Damage

On saplings and young pole crops, the attack may result in the arrest of height growth or transformation of the affected branches or the entire tree into witches'-broom and, in extreme cases, death of affected branches and the tree. A large-scale mortality in planted deodar due to *P. cedri* occurred around Manali, Kulu (Troup 1914), and in a natural pole crop of deodar in Koti Kanasar, Chakrata Div. (U.P.) where plants with otherwise promising dominant stems were badly forked and mis-shapen at the top, and had to be removed during thinnings in favour of smaller healthy trees (Troup 1912)

Symptoms

P. cedri attacks young needles of the current year's shoots. The affected needles are shorter than the normal and curve backwards. Since the affected needles are shed earlier than normal, the foliage becomes thin. Witches'-brooms are produced on the affected branches (Pl. 3, 3) and in extreme cases, the whole tree may be converted into witches'-brooms.

Causal fungus

The rust is known only in pycnial and aecial stages on deodar. Pycnia are minute, subcuticular, mostly projecting above the epidermis. Aecia are subepidermal and develop in numbers of 10-12 on each needle. The aeciospores ripen in May-June. The peridermium is delicate, consists of a single layer of cells and falls apart easily. The aeciospores are yellow, round or oval. The epispore is thick and striated. The fungus is believed to perennate as mycelium on the needles and in the shoots containing them during the winter to infect the new needles which appear in the following spring. Barclay (1886), however, states that the rust is confined entirely to the needles and does not occur on stems bearing infected needles.

Control

In badly affected areas, systematic removal of the diseased parts and trees carried out for a number of years keeps the disease under some control. Damp valleys should be avoided for deodar plantations since the disease is favoured under wet conditions.

OTHER DISEASES

Other pathogens recorded on deodar are *Phytophthora cactorum* (Leb. & Cohn) Schroet. from U.S.A. and U.S.S.R. causing seedling blight, *P. cinnamomi* Rands from U.S.A. causing root rot, *Pythium artotrogus* (Mont.) de By. from U.S.S.R. causing damping-off, *Fusarium fuliginosporum* Sibilia from Italy causing root rot, *Ganoderma lucidum* (Leyss.) Karst. from Pakistan causing root rot, *Fomes pini* (Fr.) Karst. from India, *F pinicola* (Fr.) Cke. from Pakistan, *Polyporus cuticularis* (Bull.) Fr. from India, *P. schweinitzii* Fr. from India all causing heart rot, *Panellus rupicola* (Mass.) Singer from India causing stem canker, resulting in group dying of trees, *Potebniamyces coniferarum* (Hahn) Smerlis from U.K. and Czechoslovakia causing bark canker, *Diplodia pinea* (Desm.) Kickx from U.S.A. causing canker and die-back and *Scleroderris lagerbergii* Gremmen from U.S.S.R. causing canker and die-back (Spaulding 1961, Browne 1968, Shishkina 1969, Shishkina and Tsanava 1970, Hepting 1971).

27. FIR–ABIES, SPRUCE–PICEA

GENERAL

There are two indigenous firs in India, namely, *Abies pindrow* Royle occurring at 2300 to 3600 m in western Himalayas extending eastwards upto Sikkim and *A. spectabilis* (D. Don) Spach occurring at 3600 to 4600 m in western Himalayas extending eastwards upto Sikkim and Bhutan. Firs are frost-hardy, suffer little from snow damage and are sensitive to fire. Natural regeneration takes place on newly exposed mineral soil in absence of sour humus, provided the seedlings are protected in early stages from sun, grazing and fire. Artificial regeneration is practised through nursery-raised seedlings, which take about 5–6 growing seasons to reach transplantable height of 22 cm. The period in nurseries has been successfully reduced to 4 growing seasons by application of growth promoting hormones and fortification of seedlings with mycorrhiza (Bakshi 1974). The wood is white and is used for building, planking, boxes, match-manufacture and pulp.

There are two indigenous spruces in India, namely, *Picea smithiana* (Wall.) Boiss. occurring in the western Himalayas at elevation of 2100 to 3400 m and *Picea spinulosa* (Griff.) Henry in the eastern Himalayas, extending to Sikkim and Bhutan. The western spruce occurs commonly mixed with fir, deodar, blue pine or oaks and other hardwoods. Spruce is a strong light demander. Natural regeneration establishes well in open gaps where there is full overhead light and in freshly exposed mineral soil and when the area is closed to grazing. A thick layer of dead needles is unfavourable for regeneration. Since good seed years occur at long intervals, artificial regeneration is practised.

Diseases recorded on *A. pindrow* and *P. smithiana* are described.

PATHOGENIC DISEASES

ROOT ROTS

Armillaria mellea (Vahl) Quel. causes root rot of overmature spruce and fir in India. Other root rot fungi recorded on spruce and fir in this country are *Fomes annosus* (Fr.) Cke., *Polyporus tomentosus* Fr. var. *circinatus* (Fr.) Sartory & Maire causing white pocket rot, *Polyporus schweinitzii* Fr. causing brown cuboidal rot and *Ustulina deusta* (Hoffm. ex Fr.) Lind. on spruce (Bagchee and Ujagar Singh 1960, Bakshi 1955, 1971, Spaulding 1961).

HEART ROTS

Heart rot fungi recorded on spruce and fir are *Fomes fomentarius* (Fr.) Kickx, *F. robustus* Karst., *Ganoderma applanatum* (Wallr.) Pat. and *Polyporus sulphureus* Fr. from India; *F. pini* (Thore ex Pers.) Lloyd, *F. pinicola* (Fr.) Cke., *F. igniarius* (Fr.) Kickx and *Polyporus cuticularis* Fr. from India and Pakistan (Spaulding 1961, Browne 1968).

RED SPRUCE

In spruce, the wood is normally white and heartwood is absent. Frequently the wood develops a red colour in the centre, leaving a narrow white outer zone, in roots and stems upto a height of 8 m or more. There is no decay associated with red spruce. A *Fusarium* sp., isolated from the discoloured wood is possibly the cause of the stain. Infection probably occurs through roots (Bakshi 1957). Sleepers made from red spruce are heavy and sink in water, and, therefore, cannot be transported through water courses. Such sleepers, however, can be floated down if they are air dried for some months prior to floating.

SPRUCE NEEDLE RUSTS

Peridermium thomsoni Berk. occurs in the western Himalayas (Simla, and Kulu) and is not very common. The fungus attacks current season's growth when all needles are attacked. The infected needles close together embracing the stem instead of standing out stiff as in healthy twigs. The infected shoots become longer than the healthy ones and the diseased stems and needles are appreciably thicker and assume a uniform pale yellow colour. In advanced stages, the needles curve outwards (Pl. 3, 4). The affected twigs at this stage droop down as long orange-coloured tassels. The minute pycnia are embedded and appear in large numbers at the distal end of the needle spreading on the entire surface, and becoming fewer towards its base. The necks protrude above the epidermis. Pycniospores are hyaline and globose. Aecia occur in two rows on the convex portion of the curved needles, a third row sometimes appearing near the apex on the concave side of the needle. They are flat, initially pale red, later becoming reddish orange. Aeciospores are catenulate and the successive spores are separated by an intercalary cell.

They are pale yellow, oval and beset externally with tubercles (Barclay 1886, Bakshi and Sujan Singh 1967).

Chrysomyxa deformans (Diet.) Jaczew. is a microcyclic rust, known only in its telial state. The fungus is more common than *Peridermium thomsoni* and attacks current year's shoots in which all needles get infected, turn yellow and become twisted and curved at the apices. The affected shoots have compact needles (Pl. 3, 10) and are stunted and smaller than the normal shoots. The fungus rarely attacks green cones, the scales of which become soft and swollen and covered with orange flakes (Butler 1905). The telia occupy two continuous flattened beds on the upper surface and two rows of smaller ones below. When mature, the needles are covered with a homogeneous orange-red layer of teliospores forming elevated beds. Teliospores are yellow with granular contents (Bakshi and Sujan Singh 1967).

Ceropsora piceae (Barclay) Bakshi & Singh is common on spruce. Needles of the previous year are attacked. Two to many telia develop in a single row on the upper side of each needle and the tissues around the telia become discoloured. No deformity, however, occurs. Telia are subepidermal, erumpent, waxy, deep orange-red and flat to convex. Teliospores are borne on long, serially arranged, septate, orange-red filaments forming a compact sporogenous mass. Teliospores are sessile, cylindric, one-celled, adhering laterally in young sori, separating at maturity, with orange-red contents. A few small peridial cells are present which may represent a peridium. Teliospores germinate from the tip to form a four-celled external promycelium which is stout and deep orange-red. Each cell forms a stout sterigma bearing sporidia formed basipetally. Sporidia are light orange-yellow and globose to pyriform (Bakshi and Sujan Singh 1960).

Peridermium piceae (Barclay) Sacc. is known only in the aecial stage on spruce needles from the western Himalayas (Simla, Chakrata). Only young needles of current season's growth are attacked like *P. thomsoni* but unlike the latter, all needles of an affected shoot are not attacked. Also, the affected needles do not show severe deformity as observed in other needle rust diseases but twist only slightly. The pycnia appear as minute dark spots all over the needle. Aecia are borne on the upper surface, usually 4–20 on each needle. Around each aecium, there is a yellow discolouration of the leaf tissue. Aecia are long, linear and yellow. When ripe, the peridium opens irregularly at the summit. The aeciospores are oval with orange-yellow granular contents. The epispore is warty (Barclay 1890). The infected needles are shed and when the defoliation is severe, the affected trees may die (Troup 1921).

Peridermium piceae is likely to be the imperfect state of *Chrysomyxa himalensis* Barclay on *Rhododendron arboreum* Sm. occurring in the neighbourhood of spruce infected by *P. piceae*. Spruce needles produce rust pustules when inoculated with basidiospores of *C. himalensis* (Spaulding 1961). The rust on *R. arboreum* attacks petioles and leaf blades forming rosette-like deformity of branches (Pl. 3, 8) and stunting of affected trees (Bakshi and Sujan Singh 1967).

SPRUCE CONE RUST

Pucciniastrum areolatum (Fr.) Otth is recorded on *Picea smithiana* from Pakistan. The basidiospores of the rust attack female flowers of spruce and later invade the tissues of young cones destroying the seeds and forming powdery, yellow pustules on the scales. The rust is heteroecious producing uredinia and telia on leaves of *Prunus* spp. and pycnia and aecia on spruce cones (Browne 1968).

OTHER NEEDLE DISEASES

Other needle pathogens recorded on fir and spruce are *Rehmiellopsis abietis* (E. Rostr.) O. Rostr. on *A. pindrow* from Britain causing twig die-back in severe infections, *Lophodermium piceae* (Fckl.) Hoehn. on *P. smithiana* from Denmark causing needle cast, *L. pinastri* (Fr.) Chev. on *P. smithiana* from Australia causing needle cast and *Rhytisma piceum* Berk. on *P. smithiana* from India causing tar spot (Bagchee and Ujagar Singh 1960, Spaulding 1961, Browne 1968).

28. CHIR PINE–PINUS ROXBURGHII SARGENT

GENERAL

Chir pine (*Pinus roxburghii*) occurs naturally in the outer ranges of the central and western Himalayas and the Siwaliks. Elsewhere, it is planted in Suphkar (M.P.) and Naterhat (Bihar). Being a light demander, the species regenerates in even-aged stands and, therefore, the Selection System under which it used to be worked is usually discarded in favour of system of regeneration fellings by periods with a view to produce even-aged stands. The species is frost-hardy, though it is not exposed to severe frost in its natural habitat. It is fire-resistant owing to its exceptionally thick bark. All the same, control burning is carried out early in the season in stands not under regeneration to minimise fire damage, particularly where resin tapping is in operation. Natural regeneration in chir pine is plenty. Artificial regeneration is done by raising chir pine in the nursery or preferably by direct sowing. Chir pine is characterised by twisted fibre in some localities, the twist being either left-handed or right-handed. Twisted chir, even when the degree of twist may be small, is unsuitable for sawing into planks or sleepers for which the wood is principally used. Chir pine is heavily tapped for resin.

PATHOGENIC DISEASES

ROOT ROTS

Root rot fungi recorded on chir pine are *Armillaria mellea* (Vahl) Quel. from Nyasaland and South Africa, *Fomes annosus* (Fr.) Cke. from India, *Ganoderma lucidum* (Leyss.) Karst. and *Polyporus schweinitzii* Fr. from India and Pakistan and *Helicobasidium compactum* Boed. from South Africa (Browne 1968).

PINE–SWERTIA FELT RUST

Occurrence

Cronartium himalayense Bagchee, chir pine-*Swertia* felt rust, is common in West and East Almora, Nainital and Garhwal Divisions, U.P.

(Bagchee 1929, 1933) and is also recorded from Simla Hills (Barclay 1890). The rust also attacks canary pine, *Pinus canariensis* C. Smith, the only close relative of chir pine (Critchfield and Little 1966), and an exotic raised in trial plantations in the chir pine zone in western Himalayas in India (Bakshi and Sujan Singh 1972). Outside India, the rust is reported on chir pine from Pakistan and the Philippines (Browne 1968).

Damage

Chir pine is susceptible to infection upto about 10-15 years age when infected branches or whole plants are killed. Trees develop resistance with age. Large-scale mortality in plantations in and around Almora since about 1900 was ascribed to *C. himalayense*, which was, however, of little significance in natural chir forests (Champion 1922). Later, the disease was recorded as severe in natural forests during 30's and 40's (Bagchee 1929, 1933, 1944). During 60's, however, the incidence of the rust was sporadic or absent though the states of the rust on *Swertia* were common. The decline in damage on chir pine due to the rust is possibly due to the resistance that may have developed in the pine through natural selection of resistant individuals, as those which were susceptible were killed outright. Regeneration from resistant individuals may have, therefore, led to young stands which are at present resistant to rust infection (Bakshi 1964 b). The rust attacks and kills saplings of canary pine raised as exotics in Ranikhet, U.P. (Bakshi and Sujan Singh 1972). Canary pine is widely planted in other countries (Streets 1962) where the rust, if inadvertently introduced, may become damaging on this pine provided *Swertia* is present.

Symptoms

Early symptoms of the disease are manifested by resin exudation. In late stages, bark cracks and cankers develop (Bagchee 1929). At this stage, small pycnia and later aecial cups burst through the ruptured bark. Infected saplings become pale in the needles and may be killed.

Causal fungus

Cronartium himalayense possesses 5 types of spores in 4 states namely, pycnia and aecia on chir and uredinia and telia on *Swertia*. Pycnia appear on the bark as minute blisters of the size of pin-heads. When mature, hyaline, oval pycniospores are liberated in a drop of orange-yellow fluid. Aecia appear on stems or branches as blisters on the bark

(Pl. 3, 5) about March and ripen in May-June. They are orange-yellow, cylindric, 4–12 mm long and 2–6 mm broad. They have a persistent peridium which ruptures irregularly at the sides when aecia mature liberating numerous orange-yellow aeciospores which have a thick verrucose wall. The aeciospores may be blown to long distances, 3–5 km, to infect the alternate host in three species of *Swertia*, namely, *S. alata* Royle ex C.B.Cl., *S. angustifolia* Buch.-Ham. ex D. Don and *S. cordata* Wall. ex C.B. Cl., which come up as weeds during the rains.

Uredinia which appear both on leaves and stems of *Swertia* are subepidermal, orange-yellow or rusty-yellow and about 0.2 mm broad. The peridium is delicate and degenerates. Urediniospores are orange-yellow, oval, thick-walled and spiny and infect *Swertia*. Telia develop on leaves and stems of *Swertia*. Teliospores which are light brown and cylindric are united to form filiform columns upto 3 mm long and 0.1 mm wide, which develop profusely on the host giving a felty appearance (Pl. 3, 6). Teliospores germinate *in situ* without rest producing a promycelium with four cells from each of which a sterigma appears bearing a sporidium. The sporidia are hyaline and globose. A gentle disturbance of wind dislodges sporidia, or the whole promycelium, which infects chir pine within a radius of about 200 m, thus completing the life-cycle of the rust (Pl. 2) (Bagchee 1929, 1933).

Peridermium himalayense Bagchee, the *Peridermium* stage of *C. himalayense*, is reduced by Peterson (1967) as synonym of *Peridermium pini* (Pers.) Lev. which is known to be the *Peridermium* stage of as many as four species of *Cronartium*, namely, *C. asclepiadeum* (Willd.) Fr., *C. delawayi* Pat., *C. gentianeum* Thum. and *C. ruelliae* Diet. attacking about 10 scarcely related families of Dicotyledons among which Paeoniaceae, Asclepiadaceae and Gentianaceae appear to provide the alternate hosts over the widest area. There is also a European race of *P. pini* which is autoecious on pine and never infects any telial host. *Peridermium pini* is reported to attack *Pinus kesiya*, *P. nigra*, *P. pinea*, *P. sylvestris*, *P. pinaster*, *P. densiflora*, *P. halepensis* and *P. mugo*. The first five and many other hard pines raised in a locality where chir pine was heavily infected by the stem rust were, however, found free from rust infection (Bakshi and Sujan Singh 1972) which may indicate that the rust on chir pine is distinct from *P. pini*. Morphological studies of the sporophytic phase of their life-cycles is necessary to determine the taxonomic rank of some of the *Peridermium* species (Peterson 1967).

Control

As explained earlier, damage due to the rust has declined appreciably and no control measures are called for. However, control of the rust may be possible by eradication of *Swertia* from chir pine forests upto a distance of about 300 m, to prevent infection from sporidia. Such eradication needs to be continued for about three years when infected chir pine produces crops of aecia. Being a small weed, mechanical eradication is difficult and a weedicide may be tried in heavily diseased areas.

HEART ROTS

Fomes pini (Thore) Lloyd, *F. pinicola* (Fr.) Cke. and *Ganoderma applanatum* (Wallr.) Pat. are reported to cause heart rot in chir pine in India (Browne 1968). *Poria monticola* Murr. causes brown cuboidal decay of chir pine trees in India. The decay progresses in the wood after conversion. The fungus develops thick, white mycelial mat and resupinate fruit bodies on the wood (Bagchee and Bakshi 1951). Elsewhere the fungus is recorded from Canada and U.S.A.

COLEOSPORIUM NEEDLE RUST

Coleosporium campanulae (Pers.) Lev. is common on chir pine throughout its distribution in India, in the western Himalayas and in plantations in Naterhat (Bihar). The rust occurs on trees of all age classes and stands, though specially abundant on the dense and bushy foliage of young trees. The affected needles turn pale yellow, particularly around the sori. Since needles do not become infected till late autumn and are shed along with healthy needles during April, no apparent damage is caused by the rust.

The life history of this rust has been described by Barclay (1890) and Bagchee (1941). Infection of chir pine needles takes place in autumn, by means of sporidia, which develop on *Campanula colorata* Wall. Pycnia and aecia develop on needles of chir pine. Pycnia appear within the chlorotic areas as minute, dark-orange spots, smaller than pin-heads during October-November. They are subepidermal and develop hyaline pycniospores. The aecia are usually borne on only one out of the three needles (Pl. 3, 11). They appear in November and later between February–March and are subepidermal, erumpent, with a tongue-shaped, orange-yellow peridium. Aeciospores are brownish yellow with orange tinge, oval, wall hyaline, thick and warty. Aeciospores infect *Campanula colorata*, a weed which comes up in and around chir pine forests during rains in July. Crops of uredinia develop on the leaves and stems of *C. colorata* continuously from March to September. Urediniospores develop in chains, and are orange-red, thick-walled, beset with tubercles. The urediniospores may overwinter to reinfect *C. colorata* in the next season. Telia which develop in autumn on the lower surface of the leaf are waxy, orange-red, and covered by a hyaline layer. Teliospores are orange-red, germinate by internal division into a promycelium which finally develops sporidia. It may be mentioned that a *Coleosporium* occurs on *Campanula canescens* Wall. ex A. DC., which, however, does not alternate with needle rust of chir, as shown by cross-inoculation experiments (Hafiz Khan 1928, Bagchee 1941). *Coleosporium inulae* (Kunze) E. Fisch. on *Inula cappa* DC., a common weed in the chir forests of Nainital, is also not connected with *C. campanulae* (Hafiz Khan 1928). Since no apparent damage is caused by the rust, no control measures are called for.

Other pathogens on twigs and needles of chir pine include *Dothistroma pini* Hulbary from India causing needle blight, *Lophodermium pinastri* (Schrad.) Chev. causing needle cast, *Septoria pisi* Westend, *Capnodium pini* Berk. & Curt., causing sooty mould from India, *Cercospora pini-densiflorae* Hori & Nambu from India causing needle blight, *Pestalotia funerea* Desm. from South Africa causing needle blight and *Diplodia pinea* (Desm.) Kickx from Australia causing twig blight (Anon. 1950, Bagchee and Ujagar Singh 1960, Spaulding 1961, Browne 1968, Bakshi, Reddy, Puri and Sujan Singh 1972).

Dendrophthoe falcata (L. f.) Ettingsh. the mistletoe, is recorded on chir pine (Browne 1968).

29. BLUE PINE–PINUS WALLICHIANA A.B. JACKSON

GENERAL

Blue pine (*Pinus wallichiana*) occurs naturally throughout the temperate zone of central and western Himalayas, mainly between 2000–3000 m though sometimes extending beyond these altitudinal limits. Owing to its capacity for springing up in dense even-aged crop, it may form extensive pure stands though at lower elevations, it is associated with chir pine, and at the higher, with deodar, spruce, fir and broad-leaved species. Blue pine is a strong light demander. Regeneration fails to establish in areas which are annually or periodically burnt. All the same, thorough burning of an area produces a favourable bed where regeneration is profuse (Troup 1921). Artificial regeneration through direct sowing of seeds is preferred to transplanting nursery-raised stock. Blue pine is one of the most beautiful pines in the world with its soft and tender drooping needles, five on each dwarf shoot. Apart from its forestry importance, the species adds beauty to the landscape, particularly in tourist resorts. Blue pine is not generally tapped for resin though its resin is superior to that of chir pine. The wood is normally used for buildings specially interior work, furniture and general carpentry.

PATHOGENIC DISEASES

ROOT AND BUTT ROTS

Root rot fungi recorded on blue pine are *Armillaria mellea* (Vahl) Quel. from Germany, India, Japan and Pakistan, *Fomes annosus* (Fr.) Cke. from Belgium, India, Pakistan and U. S. A. and *Polyporus tomentosus* Fr. from India (Buchanan 1964). Butt rot due to *Polyporus schweinitzii* Fr. is recorded from Pakistan (Browne 1968).

FOMES PINI HEART ROT

Occurrence

Fomes pini (Thore) Lloyd, commonly known in India as *Trametes pini* (Thore) Fr., is widely distributed throughout the world in the coniferous forests of north temperate zone. In India and Pakistan, the fungus is common on blue pine, the incidence being higher in high rainfall (100 cm and above) areas in the outer Himalayas. The incidence gradually diminishes with decrease in rainfall (below 100 cm) in the inner ranges of the Himalayas. *F. pini* rarely attacks chir pine, chilgoza pine, spruce, fir and deodar.

Infection

The fungus mainly enters through branch stubs and open knots (Boyce 1961). In India, a concession of controlled lopping of branches of blue pine is extended to villagers in some areas where the needles are used as bedding material for cattle. Branches are also lopped to allow sunlight for agriculture. However, lopping becomes uncontrolled and most branches in the tree are lopped (Pl. **17**, 3). Trunk wounds caused during lopping form infection courts for *F. pini* to establish in trees (Bakshi 1964). Infection may occur through root grafts (Hafiz Khan 1910, Sher Singh 1924, Hole 1933), which, however, may not be of much practical significance in the spread of the disease.

Damage

F. pini causes significant damage to blue pine in high rainfall areas. The fungus causes decay mainly in the heartwood. In incipient decay, the heartwood is stained light pink or brown or reddish, due to which the disease is known as 'red-rot'. In advanced stages of decay, white pocket rot develops in the heartwood (Pl. **17**, 5). If decay occurs at stump level, sporophores continue to develop on stumps. The rot is not known to develop in wood in service (Boyce 1961).

Sporophore

Fruit bodies of *F. pini* develop commonly on infected blue pine (Pl. **17**, 4) and are usually associated with punk knots. Sporophores are perennial, broadly reflexed and dull brown. The upper surface is gray, black with age, cracking and zoned. Setae are common in the hymenium (Bakshi 1971).

Control

F. pini infects blue pine mainly through trunk wounds caused during lopping which should be prohibited or controlled to minimise infection. For this, it is necessary to raise farm forests, with blue pine as a principal species, to meet the requirement of the local population so that reserved forests are not encroached.

OTHER HEART ROTS

Other heart rot fungi recorded on blue pine are *Fomes pinicola* (Fr.) Cke., *Polyporus gilvus* (Schw.) Fr., *Stereum purpureum* (Fr.).Fr., *Polyporus dryadeus* Fr. all from Pakistan, *P. cuticularis* Fr. from India and Pakistan, *Fomes conchatus* (Fr.) Gill., *F. roseus* (Fr.) Cke. and *F. fomentarius* (Fr.) Kickx all from India (Browne 1968).

STEM BLISTER RUST

Cronartium ribicola Fischer, the stem blister rust of 5–needle white pines, is Asiatic in origin and has spread gradually to Europe where it found exotic eastern white pine (*Pinus strobus* L.) as a very susceptible host. The rust was later transferred to North America during early part of this century with the seedlings of eastern white pine imported into that country from European nurseries. Since that time, the rust has become widespread on majority of 5-needle white pines on which the damage is serious (Boyce 1961). In India, the rust attacks blue pine, which, however, is resistant.

Occurrence

C. ribicola is widely distributed in Asia, Europe and North America. In India, the earliest record of the fungus is by Nisbet (1895) who collected aecia on stem of blue pine from Chakrata. The rust is recorded from Kulu, Upper and Lower Bashahr Divisions (H. P.), Hazara, Kashmir and Kagan. The *Cronartium* stage of the rust occurs on *Ribes* spp. However, the rust is not common in India.

Damage

The fungus is not as damaging on blue pine as it is on white pines in America. The fungus usually attacks young plants on branches or stems which become swollen, blistered (Pl. 3, 7) and girdled. Resin exudation may occur.

Causal organism

The fungus is a macrocyclic heteroecious rust. In India, pycnia and aecia occur on blue pine and uredinia and telia on *Ribes rubrum* Linn. and *R. orientale* Poir. Pycnia are minute and exude pycniospores in a drop of honey. Aecia emerge on infected bark (Pl. 3, 7) in spring and continue to appear till the middle of July. Aecia are tubular with persistent 3–4 celled thick peridia. Aeciospores are oval to ellipsoid, with wall partly smooth and partly marked with tubercles. The aeciospores are wind-dispersed and infect *Ribes* during May and June. Uredinia are golden-yellow and develop urediniospores which reinfect *Ribes*. Later, telial columns develop. Teliospores germinate *in situ* to produce the promycelium and sporidia which are wind-dispersed and infect blue pine. None of the spore forms appears to overwinter.

The aecial stage of *C. ribicola* on *P. wallichiana* was named *Peridermium indicum* by Colley and Taylor (1927), because it was considered distinct from *Peridermium strobi* Klebahn, the aecial stage of *C. ribicola* on *Pinus strobus*. Bagchee (1950) found two distinct types of *Peridermium* on blue pine in India. One resembled *P. indicum* of Colley and Taylor with the alternate host being *R. rubrum*, occurring in Kulu and Chakrata. The other was similar to *P. strobi* with *R. orientale* as alternate host occurring further to the west, in Kashmir. However, he considered the morphological difference inadequate for considering them as distinct. Peterson (1967) maintains separate identity of the two aecial stages occurring on *P. wallichiana* and *P. strobus* based on differences in aeciospore and peridial wall characters. In view of this, the possibility of the existence of more than one strain in *C. ribicola* may have to be examined. Until now, *C. ribicola* has been regarded as a homogenous species (Boyce 1961) and is not known to occur in more than one strain on *Ribes* (Hahn 1949). However, to confirm this, a critical morphological study of *C. ribicola* based on material collected from all over the world and cross inoculation experiments are necessary, as this may have a bearing on the relative blister rust resistance of 5-needle pines (Bakshi 1972).

Control

The rust does not cause any damage to blue pine, so that no control measures are called for, unlike in North America where the rust is destructive on white pines and control is attempted by eradication of *Ribes*, breeding resistant pines and by application of systemic fungicides (Boyce 1961). Due to its resistance to blister rust and also heterogeneity, blue pine provides a suitable material for breeding white pines resistant to the rust.

STEM CANKERS

Cenangium ferruginosum Fr. from Pakistan, *Scleroderris lagerbergii* Gremmen from Denmark, *Leucostoma kunzei* (Fr.) Munk. from U.S.A., *Trichoscyphella resinaria* Cke. & Phill. from Great Britain and *Valsa superficialis* Nits. from U.S.A. are recorded as the causes of stem and twig cankers in blue pine (Buchanan 1964, Browne 1968).

COLEOSPORIUM NEEDLE RUST

Coleosporium barclayense Bagchee is common on blue pine in its entire range of distribution in the western Himalayas. It attacks trees of all ages and only needles of the previous season. The affected needles are shed at about the same time as the healthy needles and damage is insignificant.

Causal fungus

The rust produces pycnia and aecia on blue pine and uredinia and telia on *Senecio rufinervis* DC., a weed that propagates through root suckers (Bagchee 1950 a). Pycnia are amphigenous, subepidermal, lens-shaped and dark brown. Aecia appear on the needles early in April and increase in number till June. The aecia are tongue-shaped, orange-red, upto 8–10 on each needle in 1–2 rows, 2–3 mm long, 1 mm broad, becoming wider due to adjacent ones coalescing The peridium is tough and white, 2–layered and wall finely striated along the length. The aeciospores are orange-red, catenulate, round or oval and wall coarsely verrucose. Uredinia develop on the undersurface of the leaves of *Senecio rufinervis* during June, as broad, orange-yellow patches. Urediniospores are orange, pedicillate with a hyaline, spiny wall. Telia which develop later are waxy, subepidermal and erumpent. Teliospores are 1-celled with gelatinous walls and germinate to form an internal promycelium bearing sporidia which infect blue pine.

MELAMPSORA NEEDLE RUST

Melampsora oblonga Bagchee is known only in its telial state which develops on the needles of blue pine (Bagchee 1950 a). The rust is common in the Chakrata Forest Division, U.P. Telia develop in numbers of 3-6 or more on each needle. They are pinkish red, non-erumpent and subepidermal. Teliospores are coloured, 1-celled, compressed laterally into a palisade and germinate *in situ* to produce an external 4-celled promycelium of which the upper cell produces a sterigma and a sporidium. The sporidia reinfect the needles. The rust is, therefore, microcyclic, lepto-form.

PUCCINIASTRUM NEEDLE RUSTS

Two species of *Pucciniastrum*, namely, *P. agrimoniae* (Diet.) Tranz. with alternate states on species of *Agrimonia* from India and *P. aceris* Syd. with alternate states on various species of *Acer* from India and Pakistan are recorded on blue pine (Browne 1968).

NEEDLE CASTS

Needle cast fungi recorded on blue pine are *Dothistroma pini* Hulbary from India, *Hypoderma desmazierii* Duby from France, Germany, Great Britain and U.S.A., *Leptostroma ahmadi* Petr. from Pakistan, *Lophodermium nitens* Darker from U.S.A , *L. pinastri* (Fr.) Chev. from India and *L. pini-excelsa* Ahmad from Pakistan (Anon. 1950, Buchanan 1964, Bakshi, Reddy, Puri and Sujan Singh 1972).

Other fungi recorded on blue pine are *Botryosphaeria obtusa* (Schw.) Shoemaker, *Capnodium pini* Berk. & Curt. from India causing sooty mould and *Sclerotinia fuckeliana* (de Bary) Fuckel from Japan causing seedling blight (Butler and Bisby 1960, Browne 1968).

FLOWERING PLANT PARASITE

Distribution

Arceuthobium minutissimum Hook. f., the dwarf mistletoe, is common and widespread on blue pine in low rainfall areas in the inner Himalayas but uncommon or absent in comparatively high rainfall areas. The parasite is widespread throughout Kashmir Valley in Kamraj, Pirpanjal, Sind, Kashmir and Jhelum Valley Forest Divisions. In Jammu and Chenab Circles, the parasite is common in Bhadrawah, Kishtwar and Ramban Forest Divisions. In H.P. State, the parasite is common in dry and arid regions of Kinnaur, Lahaul and Pangi Forest Divisions where the incidence of infected trees may be upto 100 per cent in certain localities.

Blue pine is managed under the Selection System on steep grounds and under the Uniform System on gentler slopes (Anon. 1966 a). The spread of dwarf mistletoe is rapid in selection felling apparently because of improved light conditions (Korstian and Long 1922) and also because it goes quickly from overstorey to understorey (Gill and Andrews 1942). Buckland and Marples (1952), however, found that incidence of dwarf mistletoe may be high in forests under all systems of management except under clear cutting.

Symptoms

The infection results in spindle-shaped or fusiform swellings on branches or on main bole causing deformity and development of witches'-brooms, top dying and death of affected branches (Pl. **17**, 1). Heavily infected stands get stunted leading eventually to death of trees. Heavy infection, besides resulting in mortality and stand openings, also reduces the aesthetic value of forests, particularly in tourist and health resorts (Bakshi and Puri 1971.)

Parasite

A. minutissimum, one of the smallest flowering plants (Pl. **17**, 2), is dioecious. Infection spreads mainly through berries which on ripening develop considerable internal pressure to eject the seeds with great force, often to a distance of about 10 metres. In course of flight, the seeds may adhere to stems of blue pine by means of the mucilage on the seeds and initiate infection. Birds and animals do not eat the seeds and, therefore, they are not significant in the spread of dwarf mistletoe (Kimmey 1957), although Gorrie (1929) suggested that *A. minutissimum* may be disseminated by birds.

Control

In heavily diseased stands where most trees are infected, it is advisable to clear-fell and regenerate the area. It may be advisable to burn the area after clear-felling as fire produces a favourable germinating bed and is often followed by profuse regeneration (Troup 1921). If, however, for some reasons natural regeneration fails to establish, as happens in the inner dry types, it may be necessary to regenerate the area artificially, despite the difficulty of the operation. Artificial regeneration by direct sowing in lines or through broadcast seeding is recommended (Troup 1921). To minimise chances of fresh infection from an adjoining infected stand, a protective zone may be delineated as discussed later. Alternatively, in severely infected forests in semi-arid locations, where blue pine forms the principal species, species substitution with *Pinus gerardiana* Wall. ex Lamb., suitable for such locations, may be considered.

In locations where infection is not widespread and is confined mainly to branches, pruning of infected branches bearing dwarf mistletoe shoots and also without them, but exhibiting symptoms of the disease, is a recognized and effective method of control. Pruning should be carried out before dispersal of seeds of the current season's mistletoe shoots, thus minimising chances of fresh infection. Stand surveillance at regular intervals is necessary to eliminate infection centres. Under Selection System, heavily broomed trees and those with infection on the main bole should be marked for sanitation cutting to remove the source of infection.

When infection is not widespread and concerns a relatively small part of the forest, it is possible to stop the spread of the parasite by establishing a protective zone of a cleared area or one containing species other than blue pine. This zone should be about 20 metres wide, which is within the effective range of seed dispersal or as wide as the height of tallest blue pine on the fringes adjoining the protective zone (Kimmey 1957).

30. EXOTIC TREES

The pattern of demand in forestry is changing rapidly in favour of industrial wood. The natural stocking may not fully meet this demand or the crop may be slow-growing. There is, therefore, a need for introducing exotics many of which have proved to be fast-growing in exotic locations and, therefore, ideally suited for industrial plantations. Most exotic conifers have their indigenous home in North America where there is a great variety of species. Success of exotics depends on choice of suitable site and of species and provenance. A knowledge of their silviculture and management (Chapter 11), introduction of mycorrhiza, particularly in afforestation projects (Chapter 4) and resistance against destructive diseases and insects (Chapter 12) is necessary. An exotic should be raised successfully in trial plantations to complete at least one rotation before a promising species may be raised in large-scale plantations. Diseases of some exotic hardwoods widely planted in India and which have since become naturalised in the country have been described in earlier chapters. The present account is confined to diseases recorded in India on exotic hard woods and conifers which are mostly in trial stages, though *Pinus patula* Schlecht & Cham., *P. caribaea* Morelet and *Cryptomeria japonica* (L.f) D. Don are fairly widely planted.

DISEASES OF PINES

DOTHISTROMA BLIGHT

Dothistroma pini Hulbary, the cause of red band disease, is widely distributed and attacks many species of *Pinus* raised as exotics in Australia, Chile, Japan, East Africa, New Zealand and elsewhere (Browne 1968, Ito, Zinno and Suto 1975). In India, the disease caused serious damage to exotic *Pinus radiata* D. Don plantations in Palni Hills, Tamil Nadu State (Bakshi and Sujan Singh 1968). The fungus has since been recorded on two indigenous pines from India, namely, chir pine from Dehra Dun Forest Division, U.P. and blue pine from J. & K. State (Bakshi, Reddy, Puri and Sujan Singh 1972), though no significant damage is reported on these species. Species of *Pinus* with their relative resistance to *Dothistroma* blight are listed by Gibson, Christensen and Munga (1964).

Symptoms

The infection appears on needles as yellow chlorotic spots which rapidly become necrotic and assume a reddish tint. Fruit bodies appear on necrotic spots. The portion of the needle above the necrotic region dies. Infection results in defoliation which, when severe, causes death of plants. *P. radiata* is susceptible when young, upto about 8-15 years, depending on infection hazard of the location. Plants develop resistance later on.

Pathogen

In the imperfect state, the fungus produces black subepidermal stromata appearing as minute dots scattered in the red band area. Conidia are straight or slightly curved (Bakshi and Sujan Singh 1968). The fungus is distinguished into 3 varieties, namely, *linearis* Thyr & Shaw, *pini* Hulbary and *keniensis* Ivory (Ivory 1967), of which damage due to var. *pini* known to occur in Chile, New Zealand and India, is most severe. The conidia are dispersed in minute drops of water by a splash mechanism. The perfect state is described as *Scirrhia pini* Funk & Parker. The spread of the pathogen is favoured in cool misty locations with high rainfall.

Control

The disease can be effectively controlled by copper-based fungicidal spray. For the control measure to be economically feasible, such sprays are recommended only when defoliation reaches 25-30 per cent level because volume increment is not significantly retarded until more than 25-30 per cent of the unsuppressed crown is destroyed (Anon. 1971). In New Zealand, one spray containing 4.48 kg of cuprous oxide in 45 l water per ha applied in November or December almost completely protects the current year's foliage for one year and results in significant gain in volume increment. Spraying only three or four times in a rotation is suggested (Anon. 1971). One pruning upto 30-40 per cent of tree height is also suggested (Anon. 1971) because such operation does not have significant effect on growth. Because of the fact that control measures form an integral part of production forestry in New Zealand, radiata pine plantations are extended in spite of the disease and more than 16,000 ha of radiata pine were planted during 1971 representing more than 90 per cent of the total area stocked during the year (Anon. 1972).

DIPLODIA TWIG BLIGHT

Twig blight caused by *Diplodia pinea* (Desm.) Kickx occurs typically in species of *Pinus* though the pathogen may attack other conifers also.

The fungus also causes sap stain in wood. The fungus is widely distributed in most countries in all the continents. In India, the disease was recorded in the 1964 exotic pine plantations of *Pinus patula* and *P. pseudostrobus* in Dehra Dun (Bakshi, Reddy, Puri and Sujan Singh 1972).

Infection

On living trees, the fungus may infect terminal buds of seedlings or young needles of current season's growth. Hail injury, insect damage or reduced host vigour by causes such as excessive competition for soil moisture facilitate infection (Gibson 1958). In the field, pines vary in their relative resistance to attack by *D. pinea* (Waterman 1943, Gibson 1958).

Symptoms

Diplodia pinea causes stunting of affected plants through killing back of the current season's growth year after year. The attack results in browning and death of needles and subsequently the entire shoot and the tree. Profuse resin exudation and formation of cankers around the nodes often accompany the death of leaders. Shoots not killed by infection bend or curl at the point of infection resulting in a permanent deformity. The fungus is also known to accelerate death of stagheaded trees growing in unsuitable environment (Birch 1936).

Pathogen

Pycnidia of the fungus, which develop on stems and bases of needles after their death, are minute, black, scattered or in linear groups, ostiolate, papillate, innate, becoming erumpent at maturity. Outer wall of the pycnidium is dark brown to black. Conidia are dark brown, obovate to elliptical with both ends rounded or sometimes with one end flattened, one to two-celled with indistinct septa.

Control

Diplodia die-back is serious in unfavourable locations and in poor soils which should be avoided for raising exotic pines. For individual trees, pruning and burning infected twigs during autumn followed by three sprays of Bordeaux mixture (4 : 4 : 50), first when new shoots just begin to emerge, second when needles begin to develop and finally when needles are partly grown control the disease (Buchanan 1967). *Diplodia pinea* may get introduced in a country by means of viable spores adhering to the surface of pine seeds or through infected timber for which adequate quarantine safeguards are necessary.

CERCOSPORA NEEDLE BLIGHT

Cercospora pini-densiflorae Hori & Nambu, the cause of needle blight of pines in nurseries, is native to Japan and Taiwan (Hodges 1964). Elsewhere, the disease is reported from Tanzania, Zambia, Rhodesia (Ito 1972) and India (Reddy and Pandey 1973).

The pathogen affects needles producing pale green lesions which later turn yellow-brown to grayish brown. The infection progresses gradually towards the tip of the needles. Blighted needles become dry and shrunken. In an epidemic outbreak of the disease, 100 per cent seedlings may be infected and 50–80 per cent killed, thus seriously affecting production of planting stock (Ito 1972). The same author also listed different species of *Pinus* according to their relative resistance to the disease, based on field observations and artificial inoculations. The fungus produces dark brown stromata which almost fill the stomatal cavities. Conidiophores are produced in dense fascicles and are olivaceous to yellowish brown in mass. Conidia are subhyaline to pale yellowish, obclavate-cylindric, straight to slightly curved, 3–7 septate, base obconically truncate and tip conic to obtuse.

Control

The disease can be successfully controlled by spraying seedlings with 0.2 per cent copper-based fungicides at fortnightly intervals during humid months (Reddy and Pandey 1973).

Intercontinental spread of the disease may occur through infected needles which may get mixed with seeds. The seeds should, therefore, be free from all the inert matter when moved from one country to another.

LOPHODERMIUM NEEDLE CAST

Needle cast of pines caused by *Lophodermium pinastri* (Schrad.) Chev., is widely distributed in North America, Europe, New Zealand and Japan. In India, the disease is recorded in exotic pine plantations of *Pinus elliottii* Engelm. and *P. radiata* D. Don (Bakshi, Reddy, Puri and Sujan Singh 1972). The disease is normally not serious as affected plants generally recover the following year though serious outbreaks of the disease have been reported from Europe and New Zealand (Peace 1962).

Symptoms

In early stages, small yellow spots appear on needles almost over the entire surface giving them a mottled appearance. The spots ultimately become brown and the affected needles dry up and fall off prematurely.

Pathogen

Pycnidia appear as minute, black spots on needles followed by development of shiny black ascocarps which occur on all surfaces of needles. Asci are cylindric-clavate and produce hyaline, filiform, unicellular ascospores surrounded by a gelatinous sheath. Ascocarps complete their development after needles are shed.

Control

The disease may be controlled in nurseries by chemical sprays e.g., with Bordeaux mixture, lime sulphur or organic zinc compounds. Three to four sprays before and during the period of infection are necessary. As infection occurs from infected fallen needles, soil mulching keeps the disease in check (Peace 1962). Quarantine safeguards are necessary against entry of the pathogen to new areas by keeping seeds to be imported free from pine needles infected by the fungus.

DISEASES OF CYPRESSES & JUNIPER

MONOCHAETIA CANKER

Monochaetia unicornis (Cke. & Ell.) Sacc., the cause of canker in *Cupressus*, *Chamaecyparis* and *Juniperus*, is recorded from the eastern United States, East Africa, New South Wales (Australia) and New Zealand (Spaulding 1961). The pathogen caused serious damage to *Cupressus macrocarpa* Hartw. in East Africa where the fungus originally occurred on the native *Juniperus procera* from which infection passed on to the cypresses (Jones 1953). In India, the fungus is recorded on *Cupressus arizonica* Greene, *C. lusitanica* Mill., *C. macrocarpa* Hartw., *C. sempervirens* L. and *Juniperus procera* Hochst., raised as exotics in trial plantations (Bakshi, Reddy, Sujan Singh and Pandey 1970). The fungus has not been recorded on the indigenous *Cupressus torulosa* D. Don, and also *C. cashmeriana* Royle, possibly indigenous to Tibet and now naturalised in the eastern Himalayas.

Symptoms

The fungus infects young plants on the stem and the rapidly growing parts of older trees through cracks and kills tissues of bark, cortex, cambium and newly formed wood resulting in formation of cankers (Jones 1953). Resin flow often precedes canker formation. The canker may girdle the stem or branch resulting in death of affected parts. The affected plants may exhibit bushy appearance and the height growth is retarded.

Pathogen

Acervuli are erumpent. Conidia are 3–5 septate, two end cells hyaline, central cells dark-coloured, with a single appendage at the apex bent at an angle at the junction with the spore. The perfect state of the fungus was identified as *Rhyncosphaeria cupressii* by Nattrase, Booth and Sutton (1963). Ascostroma are unilocular, gregarious, immersed, black, obpyriform. Asci are bitunicate, cylindrical, with light monostichus or obliquely monostichus ascospores. Ascospores are oblong to ellipsoid, brown at maturity with three transverse septa. The fungus is known to occur in three different strains which differ in their virulence on *Cupressus* and also in some morphological and physiological characters (Jones 1954, 1954 a).

Control

Systematic removal of diseased plants may lead to a significant drop in the incidence of infection (Jones 1953).

DISEASES OF CRYPTOMERIA

ARMILLARIA ROOT ROT

Armillaria mellea (Vahl) Quel. which is widely distributed around the world both on conifers and hardwoods, was recorded in the 1922 *Cryptomeria japonica* (L.) D. Don plantations at Lava, Kalimpong Forest Division, West Bengal, causing patch mortality. The infection could be traced to infected stumps of *Quercus* which formed an important component of the natural forest prior to raising *Cryptomeria*. Ring barking of hosts susceptible to *A. mellea* prior to clearing the site for raising plantations will prevent colonisation of stumps by *A. mellea*.

DISEASES OF POPLARS

GANODERMA ROOT ROT

Species of *Populus* are highly susceptible to root-rot caused by *Ganoderma lucidum* (Leyss.) Karst. In plantations of different species of *Populus* in Phillaur (Punjab), 100 per cent mortality is recorded by Bakshi, Reddy, Puri and Sujan Singh (1972). The disease is discussed in detail in Chapter 15.

STEM CANKERS

Corticium salmonicolor Berk. & Br. causes bark splitting and canker followed by twig blight and die-back of *P.* x *euramericana* cv. 'I–214' and 'I-488'. The disease is recorded from Dehra Dun (U. P.) where all the four growth forms of the pathogen, namely, pustules, cobwebby, necator and pink incrustation, develop on affected trees (Bakshi, Reddy, Puri and Sujan Singh 1972). *Botryodiplodia palmarum* (Cke.) Petrak & Syd. is recorded on *P.* x *euramericana* cv. 'I–214', 'I–215' and 'I-488' and *P. deltoides* Marsh. from Terai and Bhabar Forest Division, U.P., causing stem cankers. The fungus establishes on wounds caused by bark bursts or sun scorch and kills the bark resulting in cankers (Bakshi, Reddy, Puri and Sujan Singh 1972). A *Phomopsis* sp. and *Macrophoma bengalensis* described by Bakshi, Reddy, Puri and Sujan Singh (1972) are associated with dying and dead branches of *P.* x *euramericana* cv. 'I–214' and 'I-488' in test plantings in Dehra Dun (U.P.) and in North Bengal. Both the fungi are weak parasites and attack plants of reduced vigour.

RUST, MILDEW AND LEAF SPOTS

Foliage fungi recorded on *Populus* are *Melampsora rostrupii* G. Wagner on *P. alba* from Simla (Himachal Pradesh), *Uncinula salicis* (DC.) Wint. on *P. alba* from Manali (Himachal Pradesh) causing powdery mildew which may cover the entire foliage in cases of severe infections, *Cercospora populina* Ell. and Ev. on *P. nigra* L. cv. 'italica' from Soil Conservation Centre, Selakui (Dehra Dun) causing grayish to light brown leaf spots, *Myrothecium roridum* Tode ex Fr. on *P.* x *euramericana*, cv. 'I–488' from Dehra Dun (U.P.), *Septoria populi* Desm. on *P. nigra* L. from Kashmir causing premature defoliation, *Sphaceloma populi* (Sacc.) Jenkins on *P.* x *euramericana*, *P.* 'regenerata' and other species of *Populus* from north-western India causing premature defoliation in severe infection and

Phyllosticta adjuncta Bub. & Serebrianikow on *P.* x *euramericana* from Solan (Himachal Pradesh) causing dark brown, irregular spots with grayish white centre (Sohi and Nayar 1969 a, Bakshi, Reddy, Puri and Sujan Singh 1972).

BARK BURSTS

Bark bursts are common on *P. deltoides* and different clones of *P.* x *euramericana* in the Terai tract of Uttar Pradesh with higher water table which comes on the surface during the rains, creating water-logged condition for trees. Sap taken up by trees in large quantity under high soil temperature is released through some weak point on the stem causing bark bursts (Pl. **6,** 1). As the sap dries up, it leaves a dark brown to blackish discolouration on the stem. Later, affected areas on the bark become necrotic and flake off exposing the wood from beneath. *Botryodiplodia palmarum* establishes on the resultant wounds and kills the cambium resulting in the development of cankers (Pl. **6,** 2). The affected plants may die if the cankers girdle the stem (Bakshi, Reddy, Puri and Sujan Singh 1972).

SUN SCALD

Sun scald is recorded on *P.* x *euramericana* from Terai and Bhabar Forest Division (U.P.). The injury results in killing of the bark and exposing the wood from beneath. Sun scald injury is located mostly on southern face of trees and may extend to several feet from ground level *Botryodiplodia palmarum* may attack the affected bark and result in widening of cankers (Bakshi, Reddy, Puri and Sujan Singh 1972).

DISEASES OF ROBINIA

MILDEW

Phyllactinia guttata (Wallr. ex Fr.) Lev., the cause of powdery mildew, is the only pathogenic fungus recorded on *Robinia pseudoacacia* Linn. from Solan (Himachal Pradesh). The mycelium is evanescent but the fungus sometimes produces inconspicuous, thin, dirty white, irregular patches of powdery growth on lower surface of leaves (Sohi and Nayar 1969 a).

DISEASES OF SALIX

STEM CANKERS

Valsa salicina Fr. is recorded on *Salix alba* L. and *S. babylonica* L. from Achhabal (Kashmir) causing cankers and die-back of twigs (Butler and Bisby 1960). The fungus is widely distributed in the north temperate zone and South America and is a semi-pathogen associated with cankers and die-back of many *Salix* species (Browne 1968). *Aposphaeria salicis* Died. is recorded on *S. babylonica* from Pir Panjal Forest Division (Kashmir) causing cankers on twigs, branches and the main stem. Affected branches are killed if girdled. The fungus produces black, subcuticular, later erumpent, subglobose to lenticular pycnidia (Bakshi, Reddy, Puri and Sujan Singh 1972).

MELAMPSORA LEAF RUST

Melampsora salicis-albae Kleb. is recorded on leaves of *S. alba* from Suket (Himachal Pradesh) and *S. babylonica* from Ootacamund (Tamil Nadu) (Butler and Bisby 1960). It is a heteroecious rust with uredinia and telia on *Salix*, spermogonia and aecia on *Allium* (Browne 1968).

WITCHES'-BROOM

A witches'-broom disease is commonly recorded in avenue trees and plantations of *S. babylonica* in Jammu and Kashmir State (Bakshi, Reddy, Puri and Sujan Singh 1972). Trees of all ages are attacked and the incidence may reach upto 100 per cent in some localities. The brooms consist of clusters of short pendulous branches. The size of leaves in the brooms is reduced to about 1/6th of the normal. No pathogen is associated with the disease which does not cause any mortality.

DISEASES OF SILVER OAK

ROOT ROTS

Rosellinia bunodes (Berk. & Br.) Sacc. is recorded on living silver oak (*Grevillea robusta* A. Cunn.) from South India attacking roots and base of the stem and forming brown strands or gray mycelial sheets under the bark. The fungus is a serious pathogen of pepper (Butler 1918). Other root rot fungi recorded on *G. robusta* are *Armillaria mellea*

(Vahl) Quel. from South India (Venkata Ram 1963), Kenya, Tanzania and Union of South Africa (Spaulding 1961) and *Ganoderma australe* (Fr.) Pat. from India (Butler and Bisby 1960).

PINK DISEASE

Corticium salmonicolor Berk. and Br., the pink disease fungus, attacks and girdles stems and branches of *G. robusta* in Industrial Plantation Circle, Kerala and other parts of South India (Butler and Bisby 1960).

LEAF SPOTS

Leaf pathogens recorded on silver oak are *Botryodiplodia theobromae* Pat. from Allahabad (U. P.) causing silvery gray leaf spots separated from healthy areas by chlorotic bands and causing crumpling and detachment of affected regions (Bilgrami 1963), *Mycosphaerella grevilleae* Munjal, Chona & Kapoor from Saharanpur (U. P.) (Munjal, Chona and Kapoor 1959), *Cercospora agharkeri* from Mahabaleshwar (Maharashtra) causing irregular, brown leaf spots (Vasudeva 1963), *Cytospora grevilleae* Sahni from Jabalpur (M. P.) (Sarbhoy, Girdhari Lal and Varshney 1975), *Seimatosporium grevilleae* (Loos) Shoemaker from India, Ceylon and Australia causing leaf spots and blotches resulting sometimes in premature defoliation (Anon. 1956 a, Browne 1968), *Pestalotia banksiana* Cav. from Calicut (Kerala), *Cercospora* leaf spot from South India, *Dothiorella* leaf disease from South India, and *Phomopsis bakeri* from India (Spaulding 1961, Venkata Ram 1962, 1963, Shreemali 1970).

PHANEROGAMIC PARASITES

Three mistletoes namely *Dendrophthoe falcata* (L. f.) Ettingsh., *D. neelgherensis* (W. & A.) v. Tiegh. and a *Loranthus* sp. are recorded on silver oak from India and Ceylon (Browne 1968).

BORON DEFICIENCY

A boron deficiency disease, recorded from South India by Venkataramani (1963), exhibits symptoms like stunting and dying of the leader accompanied by abnormal development of lateral shoots giving a bushy appearance. A soil drench of 0.1 gm boric acid per litre of water applied to seedlings controls the disease.

31. OTHER INDIGENOUS TREES

Diseases of other Indian forest trees, many of which are economically important but on which diseases have not so far assumed much significance, are described below.

ACACIA CHUNDRA (ROXB.) WILLD.

A. chundra occurs in Gujarat, Rajasthan, western Karnataka and parts of Tamil Nadu. The wood is used for beams, agricultural implements and for extraction of cutch and *katha.*

Fomes badius Berk., the cause of serious heart rot in khair (*Acacia catechu* Willd.), is the only pathogen recorded on *A. chundra* (Anon. 1950).

ACACIA EBURNEA WILLD.

A. eburnea is widely distributed in the drier parts of India. In the sub-Himalayan tract, it occurs on dry river beds.

Ravenelia esculenta Narasimhan & Thirum., recorded from Poona (Maharashtra), attacks flowering twigs causing general hypertrophy, conversion of the capitate inflorescence into a spike by apostasis and pedicellation of normally sessile florets (Prain 1890). The fungus has perennial dormant mycelium which produces vigorous growth during monsoons. The axillary buds on shoots become hypertrophied, elongated and turn oranage-yellow owing to the development of numerous subcuticular pycnia. Aecia are cupulate, deep-seated and covered with a tough dome-shaped peridium. Uredinia and telia are formed on inconspicuous leaf spots (Narasimhan and Thirumalachar 1961). All the infected parts of the plant are eaten for food. They are scraped, boiled to softness and when cold rubbed into a mess with spices, warmed up and eaten as a delicacy by the local people.

ACACIA FERRUGINEA DC.

A. ferruginea occurs in Gujarat and the eastern slopes of western Ghats. The wood is used for agricultural implements. The bark and the pods possess astringent properties.

Ravenelia hansfordii Cumm. var. *ferrugineae* Ramak. & Sund. is recorded from Walayar (Kerala) producing uredinia and telia on both surfaces of leaves. Teliospores contain verruciform papillae (Ramakrishnan and Sundaram 1952).

ACACIA LEUCOPHLOEA WILLD.

A. leucophloea occurs chiefly in open scrub or thorn forests throughout drier parts of central, western and southern India. The wood is used as fuel and for agricultural implements.

Catacauma acaciae Theiss. & Syd. is recorded from Coimbatore (Tamil Nadu) producing dirty black, unilocular stroma on the upper or rarely on the lower surface of infected leaves. Asci are stipitate and contain one-celled, hyaline ascospores (Saccardo 1926).

Hapalophragmium ponderosum Syd. & Butler, the gall rust, is common on *A. leucophloea* in northern, central and southern India. Infected trees develop tumors on the stem and branch tips. Adjacent tumors coalesce to form bigger tumors which remain firmly attached to the host for sometime but eventually dry up, turn black and fall off. Tumor-like excrescences may rarely develop on peduncles and pods also. Both pycnia and telia develop on the tumors. Pycnia are hyaline to deep brown and produce oval to spherical pycniospores. Telia also develop below the epidermis but later become erumpent and produce reddish brown to yellowish brown, pedicellate, three-celled (with odd cell terminal) teliospores which germinate within 24 hours to produce oval basidiospores (Bakshi and Sujan Singh 1967). *Hapalophragmium tandonii* Mitter, recorded on leaves and petioles from Majhgawan (U.P.), produces rusty-brown telia on both surfaces of the leaves. Teliospores are deep brown, three-celled (with odd cell terminal) and densely warted (Bakshi and Sujan Singh 1967).

ACACIA MODESTA WALL.

A. modesta occurs in the sub-Himalayan tract from the Yamuna westwards ascending to 1200 m and in the northern parts of the Punjab plains. The wood is used for fuel and agricultural implements.

Ravenelia taslimii Mundkur recorded from Delhi attacks leaves, petioles and branches producing pale yellow, well-defined necrotic areas around the sori. Telia develop on both surfaces of leaves and produce chestnut-brown teliospore heads which bear brown teliospores (Bakshi and Sujan Singh 1967).

ACACIA NILOTICA (L.) DEL. SSP. INDICA (BENTH.) BRENAN

A. nilotica ssp. *indica* occurs throughout most of the drier parts of India The wood is used for buildings, agricultural implements, boat building and as fuel.

Pathogenic fungi recorded on this species are *Fomes badius* from India causing heart rot (Lloyd 1915), *Ravenelia acaciae-arabicae* Mundk. & Thirum. from Hessarghatta (Karnataka) and Coimbatore (Tamil Nadu) producing rust sori with aculeate telial heads on lower surface of leaves (Bakshi and Sujan Singh 1967) and *Cylindrosporium acaciae* Anahosur from Poona (Maharashtra) causing greenish yellow specks with tiny, black fruit bodies on leaves (Anahosur 1968).

ACACIA POLYACANTHA WILLD.

A. polyacantha occurs in Bengal, Bihar and western peninsula, preferably in moist localities. The wood is used as fuel, bark as a tan and heartwood for extraction of cutch.

Pathogenic fungi recorded on *A. polyacantha* are *Fomes badius* from India causing heart rot, *Ravenelia acaciae-sumae* Mundk. & Thirum. from Dasarhalli (Karnataka) producing rust sori with brownish yellow and paraphysate uredinia and aparaphysate telia with minutely verrucose teliospores (Bakshi and Sujan Singh 1967) and *Ravenelia hansfordii* from Walayar (Kerala), producing paraphysate uredinia with cinnamon-brown urediniospores and teliospores with a few hyaline, wart-like projections (Bakshi and Sujan Singh 1967).

ACACIA RUGATA (LAM.) MERR.

A. rugata occurs in tropical forests throughout India. Elsewhere, the species is widely distributed in Burma. The pods are extensively used as a detergent under the local name of *shikakai* as they contain upto 5 per cent saponin. Bark yields tannins.

Ravenelia acaciae-concinnae Mundk. & Thirum., is reported on *A. rugata* from Bangalore producing paraphysate uredinia with cinnamon-yellow urediniospores (Bakshi and Sujan Singh 1967). A virus causing mosaic of *A. rugata* is recorded from Mauritius (Spaulding 1961).

ACACIA SENEGAL WILLD.

A. senegal occurs in arid regions of the Punjab and Rajasthan. Elsewhere the species is abundant in Arabia and Africa. Gum arabic of commerce is obtained from wounds in the bark. Heartwood is used for manufacturing shuttles and as fuel.

Polyporus hispidus Fr., causing trunk rot, is recorded on *A. senegal* in France (Spaulding 1961). *Ravenelia acaciae-senegalae* Sanwal is recorded from Delhi producing uredinia on upper surface and telia on both surfaces of leaves. Teliospores have six, very minute warts (Bakshi and Sujan Singh 1967). *R. acacicola* Sanwal, also recorded from Delhi, is similar to *R. acaciae-senegalae* except that in the former minute warts on teliospores are absent (Bakshi and Sujan Singh 1967).

ACER CAESIUM WALL. EX BRANDIS

A. caesium occurs in the Himalayas between 1200-3000 m. The wood is used for kitchen wares and is also suitable for guns and rifle parts.

Pucciniastrum aceris Syd. is recorded on leaves from India. Alternate stage of the rust is suspected on *Pinus wallichiana* A. B. Jackson (Buchanan 1964). *Rhytisma acerinum* (Pers.) Fr., causing tar spots, is recorded from Harwan (Kashmir) and Darjeeling (West Bengal). Severe infection causes premature defoliation. The disease may be important in nurseries (Browne 1968).

ACER OBLONGUM WALL.

A. oblongum occurs in the outer Himalayas from Jhelum to Bhutan between 600-2000 m.

Foliage pathogens recorded on *A. oblongum* are *Schizothyrium annuliforme* Syd. & Butl. from Mussoorie (U. P.) causing large discoloured blotches (Spaulding 1961) and *Uncinula aceris* (DC.) Sacc. from Happy Valley, Solan (H. P.) causing powdery mildew (Sohi and Nayar 1969).

ACER PICTUM THUNB.

A. pictum occurs in the outer and middle Himalayas between 1200-2500 m. The wood is used for kitchen wares and is also suiltable for gun and rifle parts.

Foliage fungi, namely, *Pucciniastrum aceris* from Mussoorie Hills (U. P.) and *Rhytisma acerinum* are recorded on *A. pictum* (Butler and Bisby 1960).

AEGLE MARMELOS CORR.

A. marmelos occurs scattered in dry deciduous forests in the sub-Himalayan tract and central and southern India. Elsewhere, the species occurs in Burma. The wood is hard and used for agricultural implements. Fruit is edible and used as medicine in dysentery and diarrhoea.

Pathogenic fungi recorded on *A. marmelos* are *Fusarium semitectum* Berk. & Rav. from Allahabad (U. P.) attacking young shoots (Butler and Bisby 1960), *Chaetothyrium guananiticum* Speg. from Varanasi (U. P.) producing pin-head size infection spots on leaves with superficial mycelium and globose, black perithecia (Mishra 1968), *Cercospora aeglicola* Das from Cuttack (Orissa) causing leaf spots during rains (Das 1958) and *Asterina delicatula* Syd. & Bal, from Hooghly (West Bengal) causing black mildew on leaves (Butler and Bisby 1960).

AILANTHUS EXCELSA ROXB.

A. excelsa occurs in mixed deciduous forests of northern and central India. The species is extensively planted for match wood industry.

Leaf spot fungi on *A. excelsa* are *Cercospora glandulosa* Ellis & Kellerman from U. P. causing small to large, brown to pinkish buff lesions (Singh and Mathur 1963) and *Alternaria* sp., recorded from the Siwaliks (U. P.), causing defoliation in one-year old plants (Bakshi, Reddy, Puri and Sujan Singh 1972).

ANOGEISSUS LATIFOLIA (ROXB.) BEDD.

A. latifolia occurs throughout India except in some eastern parts. Silviculturally, the tree is useful in clothing dry hill sides. The wood is used for agricultural implements, furniture, poles and boats.

Pestalotiopsis versicolor (Speg.) Steyaert is recorded on leaves and petioles of *A. latifolia* from Jabalpur (M. P.) forming pinkish brown, pin-head size spots on either leaf surface. Mature lesions are dark brown with an ash-coloured centre in which acervuli of the fungus appear (Agarwal and Ganguli 1959). *Marssonina poonensis* Patwardhan

& Pande is recorded from Maharashtra producing circular, dark brown infection spots with dark black margin on upper leaf surface (Patwardhan and Pande 1969). *Monochaetia jabalpurensis* Agarwal & Hasija is recorded from Jabalpur (M. P.) causing pin-head size, circular spots with ash-coloured central region bounded by a dark brown margin on upper surface of leaves. Acervuli appear as black dots in the central region (Agarwal and Hasija 1961).

Other fungi recorded on *A. latifolia* from India are *Corticium repens* Berk. causing white thread blight of young plants, *Fomes caryophylli* (Racib.) Bres., *F. fastuosus* (Lev.) Cke., *F. robiniae* (Murr.) Sacc. & D. Sacc., *Trametes corrugata* (Pers.) Bres., *T. cubensis* (Mont.) Sacc. and *T. straminea* (Pat.) Lloyd, all causing heart rot and *Coniothyrium fuckelii* Sacc. (Sahni 1966, Browne 1968).

ARTOCARPUS CHAPLASHA ROXB.

A. chaplasha occurs scattered in moist deciduous and evergreen forests in the sub-Himalayan tract from Nepal eastwards, ascending upto 1600 m and in the Andamans. The wood is used for furniture, canoe and boat building. *Corticium salmonicolor* Berk. & Br., is the only pathogen recorded on *A. chaplasha* from India causing stem and twig cankers (Butler and Bisby 1960).

ARTOCARPUS HETEROPHYLLUS LAM.

A. heterophyllus occurs in moist evergreen forests of western Ghats ascending to 1300 m. The wood is used for furniture, cabinet work and turnery. The tree is important for its edible fruit for which it is extensively cultivated.

Root rot fungi recorded on *A. heterophyllus* are *Fomes lignosus* (Klotzsch) Bres. and *Rosellinia bunodes* (Berk. & Br.) Sacc. from Ceylon, *Ganoderma lucidum* (Leyss.) Karst., *Rosellinia arcuata* Petch and *Ustulina deusta* (Hoffm. ex Fr.) Lind. from India, *Polyporus zonalis* Berk. from Brazil and *Fomes lamaoensis* (Murr.) Sacc. & Trott. from East Indies (Spaulding 1961, Browne 1968)

Stem fungi recorded on *A. heterophyllus* are *Corticium salmonicolor* from India (Butler and Bisby 1960), *Fomes durissimus* Lloyd from India and *Ganoderma applanatum* (Pers. ex Wallr.) Pat. from Pakistan (Browne 1968), all causing heart rot.

Phyllostictina artocarpina (Syd. & Butler) Syd. is recorded from Muzaffarpur and Allahabad (U. P.) and also from Pakistan, causing circular white leaf spots with broad, dark margins (Browne 1968). Pathogenicity of the fungus was proved by Tandon and Bilgrami (1957) who observed that the disease was controlled by spraying 0.33 per cent perenox or 3 : 3 : 50 Bordeaux. Other foliage pathogens recorded on *A. heterophyllus* are *Gloeosporium artocarpi* Delacr. causing an unimportant anthracnose (Browne 1968), *Pestalotia elasticola* P. Henn. from Poona (Maharashtra) forming pustules (Butler and Bisby 1960), *Septoria artocarpi* Cke. from India associated with leaf spots (Bagchee and Ujagar Singh 1960) and *Uredo artocarpi* Berk. from India, Ceylon, Malaysia and Venezuela forming dull brown rust pustules on lower surface of leaves (Spaulding 1961, Browne 1968).

Rhizopus artocarpi Racib. is recorded on *A. heterophyllus* from Fiji, India and Pakistan. The fungus is a common saprophyte but becomes parasitic on male inflorescence and young fruits which become mummified and fall off (Browne 1968). Pathogenicity of the fungus was established by Mohanty and Patnaik (1970) who found that the disease could be controlled by sprays with captan followed by Bordeaux.

Phanerogamic parasites on the host include *Dendrophthoe falcata* (L.f.) Ettingsh., *Elytranthe capitellata* (W. & A.) Engl. from Ceylon, *Ginalloa spathulifolia* (Thw.) Oliv. from Ceylon and *Macrosolen cochinchinensis* (Laur.) v. Tiegh.

ARTOCARPUS HIRSUTUS LAM.

A. hirsutus occurs commonly in evergreen forests of western Ghats from Konkan southwards ascending to 1300 m. The wood is used for house and boat building, furniture and other purposes.

Two foliage pathogens namely *Phomatospora artocarpi* Ramakr. & Sund. from the Nilgiris (Tamil Nadu) (Subramanian and Ramakrishnan 1956) and *Phytophthora palmivora* Butler from Kerala causing severe defoliation (Ramakrishnan 1960) are recorded on *A. hirsutus*.

ARTOCARPUS LAKOOCHA ROXB.

A. lakoocha occurs in the sub-Himalayan tract from Kumaon eastwards ascending to 1300 m, southern Bihar, western Ghats and the Andamans. The wood is used for buildings, furniture and canoes.

Pestalotiopsis sp. is recorded on leaves of *A. lakoocha* from Jabalpur (M.P.) (Tandon and Chandra 1963-64).

ARUNDINARIA SUBERECTA MUNRO

A. suberecta, a gregarious bamboo, occurs along river banks in the eastern Himalayas. *Puccinia melanocephala* Syd. is the only pathogen recorded on this host from Thyria (Assam). The rust sori are produced on both surfaces of leaves. Teliospores are smooth, constricted at the septum, attenuated at the base and rounded at the apex (Bakshi and Sujan Singh 1967).

AZADIRACHTA INDICA A. JUSS.

A. indica occurs in the Siwaliks and South India. The species is widely cultivated all over India. The wood is used for house building and furniture.

Pathogens recorded on *A. indica* are *Ganoderma lucidum* causing root rot (Bakshi 1957 a), *Corticium salmonicolor* from South India causing stem and twig blight (Subba Rao 1936), *Oidium* sp. from Poona (Maharashtra) causing powdery mildew (Butler and Bisby 1960), *Cercospora subsessilis* H. & P. Sydow from Karnataka causing white to pale yellow leaf spots on both leaf surfaces (Vasudeva 1963) and *Xanthomonas azadirachtii* Moniz & Raj, a host specific bacterium from Poona (Maharashtra) causing leaf spots (Moniz and Raj 1967). *Cercospora meliae* El. & Ev. (= *C. leucosticta* Ell. & Ev.) recorded from Poona (Maharashtra) belongs to *C. subsessilis* (Vasudeva 1963). *Pseudomonas azadirachtae* Desai, Gandhi, Patel & Kotasthanae, recorded from Gujarat and Rajasthan, causes water-soaked, irregular leaf spots around veins. The areas surrounding the spots become yellow, necrotic and fall off leaving a shot hole symptom (Desai, Gandhi, Patel and Kotasthanae 1966). Disease development is maximum between 22–27°C air temperature, while the spread of the disease is favoured during rains. The leaf becomes susceptible to infection when it is half the normal size and remains so till senescence. The pathogen is host specific (Srivastava and Patel 1969).

Dendrophthoe falcata and *Tapinanthus* sp. are the two mistletoes recorded on *A. indica* (Browne 1968).

BAMBUSA

Two species of *Bambusa*, *B. tulda* Roxb. and *B. arundinacea* Willd., are economically important. *B. tulda* occurs in Bengal, Assam, Bangla Desh and Burma. *B. arundinacea* occurs throughout greater part of India except in the driest regions and is by far the commonest bamboo used for paper pulp. Most of the disease records are on *Bambusa* sp. and are included here.

Decay fungi recorded on living bamboo are *Polyporus durus* Jungh., *P. friabilis* Bose, *Ganoderma lucidum*, *Amauroderma rugosus* Nees, *Trametes corrugata*, *Merulius similis* B. & Br. and *Stereum percome* B. & Br. all from Calcutta and its suburbs in West Bengal (Banerjee and Ghosh 1942).

Foliage rusts recorded on *Bambusa* spp. are *Puccinia gracilenta* Syd. & Butler on *Bambusa* sp. from Darjeeling (West Bengal) producing dirty brown telia and teliospores not constricted in the middle (Bakshi and Sujan Singh 1967), *P. xanthosperma* Syd. on *Bambusa* sp. from Mussoorie (U.P.) producing yellowish brown to dark brown telia on lower leaf surface (Bakshi and Sujan Singh 1967), *P. longicornis* Pat. & Har. on *B. arundinacea* from China forming yellowish brown to dark brown or black pustules (Spaulding 1961), *P. melanocephala* on *Bambusa* sp. from Poona (Maharashtra) (Butler and Bisby 1960), *Uredo ignava* Arth. on *B. arundinacea* from China forming brown, powdery pustules (Spaulding 1961) and *Dasturella bambusina* Mundkur & Kheswala on *Bambusa* sp. from Bombay (Maharashtra) (Mundkur and Kheswala 1943). Bakshi and Sujan Singh (1967) later observed that telia of *D. divina* (Syd). Mundkur & Kheswala and those of *D. bambusina* are indistinguishable.

Ustilago shiraiana P. Henn. attacks branches of *Bambusa* spp. and many other bamboos in India, Japan, China and Formosa. The rust got introduced into U.S.A. but was eradicated. The rust causes witches'-broom effect and swelling of shoots which stop growth. Dark brown, powdery spore masses burst through leaf sheaths (Spaulding 1958).

Diplozythiella bambusina Died. is recorded on fading leaves of *Bambusa* sp. from Dehra Dun, U.P. (Butler and Bisby 1960). Three species of *Phyllachora*, namely *P. bambusae* Syd., *P. malabarensis* Syd. & Butl. and *P. shiraiana* Syd., all causing leaf spots and *Meliola bambusicola* Hansf. causing sooty mould are recorded in India on *Bambusa* sp. (Browne 1968). *Balladyna butleri* Syd. causes black leaf spots on culms of *Bambusa* sp. in Khasi Hills (Assam) (Butler and Bisby 1960).

BAUHINIA PURPUREA LINN.

B. purpurea occurs scattered in mixed deciduous and sal forests. It is frequently cultivated for handsome flowers. The species is frost-hardy and is useful for stocking frosty blanks. The wood is used for agricultural implements and the bark for tanning.

Two root and butt rot fungi namely *Polyporus zonalis* and *Hypoxylon ustulatum* Bulliard frequently occur together on *B. purpurea* in New Forest, Dehra Dun (U. P.). Both the pathogens cause decay of roots resulting in windthrow of affected trees. *P. zonalis* causes white rot with pockets distinct when fresh, becoming indistinct on drying. *H. ustulatum* causes white rot with black zone lines (Bakshi, Puri, Sharma, Sujan Singh and Balwant Singh 1956, Bakshi, Sujan Singh and Balwant Singh 1963).

Foliage pathogens recorded on *B. purpurea* are *Cercospora bauhinae* H. & P. Sydow from Bangalore (Karnataka) causing round to irregular, grayish to pale brown leaf spots with dark brown margins and pale to dark brown stroma with dense fascicles (Lucy Channama and Rangaswami 1969), *Phyllosticta bauhinae* Cke. from Pusa (Bihar) and Bandra (Maharashtra) causing irregular brown spots (Butler and Bisby 1960) and *Exosporium bauhiniae* Garud from Dapoli (Maharashtra) causing irregular, dark brown infection spots (Garud 1970).

BOSWELLIA SERRATA ROXB.

B. serrata occurs commonly on dry hills throughout greater part of India. The gum, oozing at wounds, is commonly used as incense and in medicine. The wood is used for paper pulp. Silviculturally the species is most useful for clothing dry, hot and barren hills.

Fomes robiniae, *F. fastuosus*, *F. senex* (Nees & Mont.) Cke., *Hexagonia sulcata* Berk. and *Trametes incerta* (Currey) Cke. are reported to cause heart rot in *B. serrata* in India. The fungi enter the trees through fire scars. The external indicators include bulbous and swollen bases and bottle-shaped stems (Anon. 1956).

Dasturella boswelliae Patel, Payak & Kulkarni causes severe rust spots on leaves of *B. serrata* in Poona (Maharashtra). Infection spots are brownish black, angular, minute and gregarious. Urediniosori are mostly produced on lower surface of leaves and are minute, reddish-brown and paraphysate (Patel, Payak and Kulkarni 1951).

BUTEA MONOSPERMA (LAM.) TAUB.

B. monosperma occurs commonly throughout greater part of India ascending upto 1000 m in the outer Himalayas. The species is mainly used for propagation of lac. Red juice exuding from wounds hardens into Bengal kino used in medicine. Flowers are used for dye and tan and leaves are used as fodder for cattle. The wood is used in well curbs and water scoops.

Xanthomonas buteae Bhatt, Patel & Thirum. produces a few to numerous minute, water-soaked areas with brown centre and pale yellow halo on leaves, injured stems and rachii of *B. monosperma* in Thana (Maharashtra). The spots become round to angular and get black with the progress of the disease. In severe infections, the entire leaf surface may be covered by the bacterium resulting in premature defoliation (Bhatt, Patel and Thirumalachar 1955).

Other fungi recorded on *B. monosperma* are *Corticium salmonicolor* on stem and branches from East Indies, *Phomopsis buteae* Sahni and *Pseudodiplodia buteae* Sahni both on leaves from Jabalpur (India) (Spaulding 1961, Sarbhoy, Girdhari Lal and Varshney 1975).

CASSIA FISTULA LINN.

C. fistula occurs commonly in deciduous forests throughout greater part of India and ascending upto 1300 m in the Himalayas. The wood is used for house posts, carts and agricultural implements. The pulp of the pods is a strong purgative. Bark is used for tanning.

Phleospora cassiae Thirum. & Narasimhan causes circular to irregular, brown to tawny-olive spots on both surfaces of leaves of *C. fistula* in Bangalore (Karnataka). The pycnidia are subepidermal and are arranged in concentric rings (Thirumalachar 1950, Browne 1968). An *Oidium* sp. produces white coating on upper surface of *C. fistula* leaves in Narsapur forests (Andhra Pradesh). The fungus produces superficial, creeping and hyaline mycelium on leaf surface (Salam and Rao 1958). *Cylindrosporium cassiae* Chiplonkar is recorded on leaves of *C. fistula* from Maharashtra. The fungus produces pale brown and necrotic infection spots on both surfaces of the leaves. Dark brown to black acervuli develop on the infection spots (Chiplonkar 1969).

Other pathogens and diseases recorded on *C. fistula* are *Cercospora angustata* Chupp & Solheim from Malaysia (Heath 1956), *Colletotrichum lindemuthianum* (Sacc. & Magn.) Broisi from Khandagir (Orissa) causing serious leaf spot (Mohanty and Beraha 1962), spike disease, showing symptoms similar to those manifested by sandal, from South India (Iyengar 1960 a), *Polyporus bicolor* Jungh. (Bakshi 1971) and *Trametes incerta* from India, both causing heart rot (Browne 1968) and three phanerogamic parasites namely *Cuscuta reflexa* Roxb. and *Loranthus* sp. from India and *Viscum articulatum* Burm. f. from Ceylon. (Browne 1968).

CASSIA SIAMEA LAM.

C. siamea occurs in southern parts of Tamil Nadu. The wood is used as fuel. The leaves are used as green manure for wet cultivation.

Cercospora cassiae-siameae Chiddarwar is recorded on both leaf surfaces of *C. siamea* from Poona (Maharashtra) causing pale brown to brown leaf spots surrounded by indistinct brown margin (Chiddarwar 1959). *Cochliobolus nodulosus* Lutrell produces brown spots at the tip or margin of leaves in Jodhpur (Rajasthan). Severe infections result in defoliation. Imperfect state of the fungus is known as *Helminthosporium nodulosum* Berk. & Curt. (Reddy 1969). An *Oidium* sp. and *Cercospora cassiae* P. Henn. were associated with defoliation of *C. siamea* in Tanzania (Etheridge 1965).

CASTANOPSIS TRIBULOIDES A. DC.

C. tribuloides occurs in the outer Himalayas from the Ganges eastwards, ascending to 2000 m. The nut is edible.

Erysiphe sikkimensis Chona, Kapoor & Gill attacks leaves of *C. tribuloides* in eastern Sikkim forming white irregular patches often covering the entire upper surface of leaves (Chona, Kapoor and Gill 1960). *Microstroma castanopsis* Muller & Sanwal is recorded on leaves of *C. tribuloides* in Bindsar, Almora (U. P.). The fungus causes witches'-broom and leaf spots on which fructifications develop (Muller and Sanwal 1959). *Palawaniella castanopsidis* Kapoor attacks leaves forming black shiny stromata on the upper surface (Kapoor 1968).

CELTIS AUSTRALIS LINN.

C. australis occurs in the western Himalayas and is often cultivated in the plains. The wood is used for oars, tool handles and ploughs. The tree is commonly lopped for fodder.

Pleochaeta shiraiana (Henn.) Kimbr. & Korf attacks leaves of *C. australis* in Harwan (Kashmir) (Pandotra, Kachroo and Sastry 1968). This fungus is previously recorded as *Uncinula polychaeta* (Berk. & Curt.) Wei (Butler and Bisby 1960). Other fungi recorded are *Thyrostroma mori* (Nomura) Hohnel from India and Pakistan, *Helicoceras celtidis* (Biv.) Linder from Algeria, Israel and Italy attacking leaves, *Phyllosticta destruens* Desm. from Denmark attacking leaves, *Phytophthora citrophthora* (R.E. & E.H. Sm.) Leonian from Argentina attacking green fruits, *Taphrina celtis* Sadeb. from France, Algeria and U.S.S.R, causing leaf blisters, *Fomes fomentarius* (Fr.) Kickx from India causing heart rot, and *Polyporus adustus* Fr. from India and Pakistan (Spaulding 1961, Browne 1968).

CUPRESSUS TORULOSA D. DON

C. torulosa occurs scattered or isolated in the western and central Himalayas between 2000–3000 m. The wood is used for furniture and construction work.

Gymnosporangium cunninghamianum Barclay, a heteroecious rust, attacks *C. torulosa* in Simla (H. P.), Chakrata, Kumaon, Mussoorie and Dehra Dun (U. P.). The rust produces pycnia and roestelioid aecia (Pl. 3, 2) on leaves of *Pyrus pashia* Ham. The infected leaves become pale and may be shed off prematurely. Telia on *C. torulosa* (Pl. 3, 1) are dark brown, round or elongate, developing on ultimate branches or green twigs. No significant damage due to the rust, is, however, reported on *C. torulosa* (Bakshi and Sujan Singh 1967). *Gymnosporangium distortum* Arth & Cumm., another heteroecious rust, is recorded on *C. torulosa* from Chakrata (U. P.) (Bagchee and Ujagar Singh 1960). Aecia and pycnia are reported on leaves and twigs of *Cotoneaster bacillaris* Wall. from Tragbal (Kashmir) (Arthur and Cummins 1933). Infection on *Cotoneaster* results in distortion of infected leaves. The pathogen is not important on *Cupressus* (Browne 1968). *Rhynchosphaeria cupressi* Nattrass, Booth & Sutton, with conidial state in *Monochaetia unicornis* (Cke. & Ellis) Sacc., is recorded on *C. torulosa* from Rhodesia causing branch cankers (Browne 1968).

DENDROCALAMUS STRICTUS (ROXB.) NEES

D. strictus is a densely tufted deciduous bamboo occurring in deciduous forests throughout India excepting in the eastern parts of the country. The species is extensively used for furniture, basket making and pulp.

Dasturella divina, a heteroecious rust, attacks leaves of *D. strictus* throughout India where bamboo occurs. Aecia and pycnia are produced on leaves and fruits of *Xeromphis spinosa* (Thunb.) Keay and *Randia candolleana* W. & A. during May–June when nearly every leaf on the tree is attacked and the infected trees stand out sharply in the forest. Uredinia and telia are produced on *D. strictus*. Telia appear on leaves as black spots arranged in linear rows. The infected leaves fall off along with other leaves during normal leaf fall and, therefore, the damage to bamboo is neglegible.

Tunicopsora bagchii Sujan Singh & Pandey attacks leaf sheaths of *D. strictus* in New Forest, Dehra Dun (U.P.). The leaves and the twigs above the infected region are killed and the dead twigs stand out distinctly in the clumps. Uredinia appear during September. Telia also develop along with uredinia and continue their growth upto July and August next year when they form continuous and concentric rings all along the leaf sheaths (Sujan Singh and Pandey 1971). A witches'-broom disease is reported from Ranchi (Bihar) (Bakshi, Reddy, Puri and Sujan Singh 1972).

DIOSPYROS MELANOXYLON ROXB.

D. melanoxylon occurs in central and southern India. The wood is hard, reddish brown with an irregular black heartwood and used for buildings, shafts, shoulder-poles and also carved into many fancy articles. The leaves are used in the manufacture of *bidis*.

Pathogenic fungi recorded on *D. melanoxylon* are *Cercospora kaki* Ellis & Everhart from Bangalore and Nandi Hills (Karnataka) producing leaf spots with reddish brown centre and gray-black margin (Vasudeva 1963), *Aecidium rhytismoideum* Berk. & Br. from Sirsi (Karnataka) and *A. miliare* Berk. & Br. from Chanda (Maharashtra) (Butler and Bisby 1960, Sarbhoy, Girdhari Lal and Varshney 1975).

DIOSPYROS TOMENTOSA ROXB.

D. tomentosa occurs in the sub-Himalayan tract upto Nepal in the east, Bihar, Orissa and Madhya Pradesh. Uses of wood are the same as for that of *D. melanoxylon*.

Pathogenic fungi recorded on *D. tomentosa* are *Cercospora kaki* from Saharanpur (U. P.) causing infection spots with grayish white centre on upper surface of leaves (Munjal, Lall and Chona 1961) and *Aecidium rhytismoideum* from Yalwal (Karnataka) and Tinnevelli Hills (Tamil Nadu) (Butler and Bisby 1960)

EMBLICA OFFICINALIS GAERTN.

E. officinalis occurs throughout greater part of India in mixed deciduous forests ascending the Himalayas upto 1500 m. The wood is used for inferior furniture and agricultural implements. Bark, leaves and fruits are used for tannins. Fruits are also used in medicine.

Pathogenic fungi recorded on *E. officinalis* are *Ravenelia emblicae* Syd. from Dehra Dun (U. P.), Nagpur (Maharashtra), Tanjore (Tamil Nadu) and Bihar causing reddish infection spots on leaves (Bakshi and Sujan Singh 1967). *Phakopsora phyllanthi* Diet. from Poona (Maharashtra) attacking leaves (Butler and Bisby 1960) and *Pestalotia cruenta* Syd. causing irregular brown spots on fruits (Tandon and Srivastava 1964).

GMELINA ARBOREA ROXB.

G. arborea occurs scattered throughout greater part of India in deciduous and moist deciduous forests. The wood is used for construction, furniture and ship building.

Poria rhizomorpha Bagchee, the cause of root rot of *G. arborea* in North Bengal and Assam, is normally a saprophyte but becomes pathogenic when the species is raised on clayey soils which become periodically water-logged. The pathogen causes brown cuboidal rot in the roots resulting in die-back and death of affected plants (Bagchee 1953 a). *Phyllactinia suffulta* (Rabenh.) Sacc. var. *gmelinae* Patil, recorded on leaves of *G. arborea* from Chikhalda (Maharashtra), produces infection spots on lower surface of the leaves with corresponding pale yellow colour on the upper surface. Cleistothecia are golden-yellow to dark brown and bear a whorl of 5-10 appendages (Patil 1961). *Cercospora ranjita* Chowdhury is recorded on leaves of *G. arborea* from Lahing (Assam) causing leaf spots (Tandon and Chandra 1963–64).

Other pathogens recorded on *G. arborea* are *Armillaria mellea* (Vahl) Quel. from Nigeria causing root rot, *Sclerotinia rolfsii* Sacc. from Nigeria causing seedling wilt, *Fomes roseus* (Fr.) Cke. from Pakistan, *Polyporus baudoni* Pat. from parts of Africa, *Trametes straminea* from India, all causing heart rot and *Tapinanthus* sp., a mistletoe, from Nigeria (Browne 1968).

GREWIA TILIIFOLIA VAHL

G. tiliifolia occurs commonly in central and southern India and in the sub-Himalayan tract. The wood is tough, elastic and is used for furniture, construction and for agricultural implements.

Pericladium tiliacearum Mundk. & Thirum. attacks petioles and stem of *G. tiliifolia* in Santaver (Karnataka) and Vendalur (Tamil Nadu) (Butler and Bisby 1960). The fungus produces crowded, cinnamon-brown pustules, appearing like mustard seeds, enclosed by a hard woody covering and forming a long crack with 2–3 locules (Spaulding 1961). *Puccinia tiliaefolia* Ramakr. & Sund., a leaf rust, causes brown infection spots on leaves of *G. tiliifolia* in Ambalávayal (Kerala) (Bakshi and Sujan Singh 1967). Other fungi recorded on *G. tiliifolia* are *Ganoderma applanatum* from India causing heart rot, *Myriangium parasiticum* Mont. & Berk. from Maharashtra attacking living stem and *Phyllosticta sedgwickii* da Costa & Mund. from Dharwar and Karwar (Karnataka) causing leaf spots (Butler and Bisby 1960, Sarbhoy, Girdhari Lal and Varshney 1975).

HOLOPTELEA INTEGRIFOLIA PLANCH

H. integrifolia occurs in the sub-Himalayan tract ascending upto 600 m, central India, Chhota Nagpur and the Indian Peninsula. The wood is used for agricultural implements and brush backs.

Cercospora holopteleae Chiddarwar attacks leaves of *H. integrifolia* in Khandla (Maharashtra). The fungus produces round to ovoid, pale brown to dull brown leaf spots on both surfaces of the infected leaves (Chiddarwar 1959 a). Other foliage fungi recorded on *H. integrifolia* are *Colletotrichum holopteleae* Hasija from Jabalpur (M.P.) and *Mycosphaerella holopteleae* Naphade from Poona(Maharashtra) producing on leaves uniloculate pseudothecia containing bitunicate asci (Tandon and Chandra 1963–64, Naphade 1970).

HOPEA PARVIFLORA BEDD.

H. parviflora occurs in the moist tropical evergreen forests of Karnataka and Kerala States. The wood is hard, heavy and durable and is used for building and boats. Only two decay fungi are recorded on this species. *Fomes lamaoensis* causes root rot and *Trametes spongipellis* (L.) Lloyd causes heart rot (Butler and Bisby 1960).

HOPEA WIGHTIANA WALL. EX W. & A.

H. wightiana occurs in the evergreen forests of western Peninsula from Konkan southwards to Tinnevelli. The wood is used in buildings. Pathogenic fungi recorded on *H. wightiana* are *Fomes badius* causing heart rot (Butler and Bisby 1960) and *Phyllachora kanarensis* Ramakr. & Sund. from Mangalore (Karnataka) causing leaf spots on both leaf surfaces (Ramakrishnan and Sundaram 1953).

JUGLANS REGIA LINN.

J. regia occurs in the Himalayas at 1500 to 4000 m from Kashmir in the west extending upto Bhutan in the east. Elsewhere, the species occurs in Afghanistan, the Balkans, Greece, the Caucasus, and Iran. The wood is used for furniture, cabinet making and gunstocks. Fruit is edible and provides the well-known dry nut.

Among important root diseases recorded on *J. regia* are *Agrobacterium tumefaciens* (E.F. Smith & Towns.) Conn. from Italy causing crown gall, *Phytophthora cambivora* (Petri) Buism. from Italy causing destructive root rot resulting in black exudation from cankered roots and bases of diseased trees, *P. cinnamomi* Rands from Portugal and Spain causing root rot, *Chalaropsis thielavioides* Peyron. from Britain causing graft disease, *Armillaria mellea* from France and Italy causing root rot, *Rosellinia necatrix* Prill. from Italy causing white root rot and *Poria ambigua* Bres. from U.S.A. causing root rot (Spaulding 1961, Browne 1968, De Vay, Sinden, Lukezic, Werenfels and Backman 1968).

Erwinia nigrifluens Wilson, Starr & Berger causes bark canker of *J.regia* in California (U.S.A.). The symptoms include irregular dark brown necrotic areas stained with dark coloured watery material, in the bark of the trunk and scaffold branches. The cankers expand in summer but become inactive in winter (Wilson, Starr and Berger 1957). Another bacterial disease of this host, caused by *Erwinia rubrifaciens* Wilson, Zeitoun & Fredrickson is recorded from California (U.S.A.) resulting in phloem cankers of trunk and branches. The symptoms include dark streaks and necrotic bands along the inner bark and the cambium accompanied by exudation of gray slime. The bacterium produces numerous black pits in the outer layers in the sapwood. Black streaks may also be produced in xylem tissues immediately below the cambium. The bacteria move faster in the xylem elements. Slow lateral movement also occurs between xylem and phloem. The pathogen produces water-soluble red pigment in which character it differs from *E. nigrifluens* which does not produce any water-soluble red pigment (Wilson, Zeitoun and Fredrickson 1967, Schaad and Wilson 1968). According to Holdeman (1970), both the diseases are problems on specific varieties but bark canker is superficial and insignificant, while phloem canker impairs vigour and in many cases causes decline.

Hendersonula toruloidea Nattrass causes branch wilt of *J. regia* in California (U. S. A.). The fungus is a weak parasite and is able to infect sunburnt branches only. Control of the disease may be secured by white washing the stems (Sommer 1955), annual removal of diseased branches, by the use of nitrogenous fertilizers and by irrigation during growing season (Foott, Hendrickson and Wilson 1955). *Melanconis carthusiana* Tul., a weak parasite, causes die-back of young grafts in many areas in the north temperate zone (Browne 1968). *M. juglandis* (Ellis & Everh.) Graves, also a weak parasite, causes die-back of twigs resulting in stagheadedness of the affected trees in North America and Europe (Browne 1968).

Other stem fungi recorded on *J. regia* are *Cytospora juglandina* Sacc. from Europe causing die-back, *Nectria cinnabarina* Fr. from Britain and New Zealand causing cankers, *Phytophthora parasitica* Dastur from Australia, and heart rot fungi, namely, *Fomes fomentarius* from Russia, *F. igniarius* (Fr.) Kickx from Siberia, *F. robustus* Karst. from India, *Polyporus hispidus* from Italy and Russia, *P. squamosus* Fr. from Denmark and Italy and *P. sulphureus* Fr. from Germany and Italy (Spaulding 1961, Browne 1968).

Xanthomonas juglandis (Pierce) Dows. causes spots on leaves, green shoots and nuts of *J. regia* in England, Iran and Italy resulting in significant damage to the nut crop (Spaulding 1961). *Gnomonia leptostyla* (Fr.) Ces. & de Not. causes a serious leaf spot in Denmark, France, Germany, Great Britain, India, Italy and Russia. The conidial state is *Marssonina juglandis* (Lib.) P. Magn. The spots are white to grayish yellow with brown edges, irregular and are restricted to the intervenal areas of the lower surface of leaves. The ascospores and conidia are disseminated during the rains. The fungus overwinters on fallen leaves (Browne 1968). Three powdery mildews, namely, *Microsphaera alni* DC. ex Wint. from Italy and India, *Phyllactinia guttata* (Fr.) Lév. from India and *Erysiphe polygoni* DC. are recorded on *J. regia*. Other leaf spot fungi recorded on this host are *Ascochyta juglandis* Boltsh. from Bulgaria, Great Britain, Italy and the U. S. A., *Cercospora forsteriana* Chupp & Viegas from Brazil causing angular, vein-limited spots, *Microstroma juglandis* (Bereng.) Sacc. from Bulgaria, Canada, Denmark, Germany, India, Italy, Pakistan and Russia forming powdery white conidial fungus pustules; *Mycosphaerella woronowii* Jacz. from the Caucasus occurring on ochraceous spots with dark edges and *Septoria nigro-maculans* Thüm. from Europe (Spaulding 1961, Browne 1968). *Viscum album* Linn., the mistletoe, is recorded on *J. regia* in the north temperate zone (Browne 1968).

LAGERSTROEMIA LANCEOLATA WALL. EX W. & A.

L. lanceolata occurs in the mixed deciduous forests of the western coast ascending upto 1300 m altitude. The wood is used in ships and buildings.

Pathogenic fungi recorded on *L. lanceolata* are *Fomes durissimus* causing heart rot, *Cercospora lythracearum* Heald & Wolf var. *macrophora* Chiddarwar from Maharashtra causing irregular pale brown leaf spots surrounded by dark brown border on both leaf surfaces and *Rhytisma lagerstroemiae* Rabenh. from Belgaum (Karnataka) and Coimbatore (Tamil Nadu) causing tar spots on upper surface of the leaves (Chiddarwar 1959 a, Butler and Bisby 1960).

LAGERSTROEMIA PARVIFLORA ROXB.

L. parviflora occurs throughout India in the plains ascending upto 1000 m altitude. Silviculturally, the species is important as a common companion of sal, teak and other valuable trees. The leaves of *L. parviflora* are used for rearing *tassar* silk worms. The wood is used for construction work.

Pathogenic fungi recorded on *L. parviflora* are *Fomes durissimus*, *F. fastuosus*, *F. robiniae*, *Trametes incerta*, all from India causing heart-rot, *Cercospora lythracearum* from Khandla (Maharashtra) causing circular, pale yellow to dull brown leaf spots with a green or yellowish border, *Cercospora paramignyae* Thirum. & Chupp from Jabalpur (M.P.) causing circular to angular, brown leaf spots with a bright yellow halo, *Sphaceloma lagerstroemiae* Wani & Thirum. from Poona (Maharashtra) attacking leaves, petioles and young stem, *Pestalotiopsis guepini* (Desm.) Steyaert from Dharwar (Karnataka) attacking leaves, *Rhytisma lagerstroemiae* from India attacking leaves and *Gloeosporium* sp. from Coorg (Karnataka) (Butler and Bisby 1960, Agarwal and Hasija 1961 a, Garud 1970, Wani and Thirumalachar 1970, Sarbhoy; Girdhari Lal and Varshney 1975). *Helixanthera ligustrina* (Wall.) Dans., the mistletoe, is recorded from India killing branches and sometimes the tree (Browne 1968).

LAGERSTROEMIA SPECIOSA (L.) PERS.

L. speciosa occurs along river banks in moist flats in eastern and southern India. The wood is used in ship building and construction.

Xeromphalina campanella (Fr.) Maire is recorded on *L. speciosa* from India. The fungus is usually saprophytic on decaying stumps and logs but also occurs as a wound parasite causing root and butt rot in many tree species, principally of conifers, in North and Central America, Asia and Europe (Browne 1968). Stem fungi recorded on *L. speciosa* are *Corticium koleroga* (Cke.) Höhnel, an important thread blight fungus from Calcutta (West Bengal) causing leaf blight and bark lesions, *Marasmius campanella* Holterm. from Calcutta (West Bengal) causing bark cracking and die-back of branches and *Hypoxylon bovei* Speg. var. *microsporum* Miller from Darjeeling (West Bengal) attacking stem (Banerjee and Mukherjee 1954, Butler and Bisby 1960, Browne 1968, Kar and Maity 1971).

Leaf spot and leaf blight fungi recorded on *L. speciosa* are *Cercospora lythracearum* from the Philippines, *Haplosporella lagerstromiae* Mhaiskar, *Pestalotia lagerstromiae* Mhaiskar, *Phyllosticta lagerstromiae* var. *microspora* Mhaiskar, *Glomerella cingulata* (Ston.) Spauld. & Schrenk, all from India and *Mollisia ravida* Syd. from the Philippines (Spaulding 1961, Mhaiskar 1969).

Three phanerogamic parasites, namely, *Cuscuta reflexa*, *Macrosolen cochinchinensis* and *M. psilanthus* (Hook. f.) Dans. are recorded on *L. speciosa* (Browne 1968).

LANNEA COROMANDELICA (HOUTT.) MERR.

L. coromandelica occurs commonly in mixed deciduous forests throughout greater part of India and ascending upto 1300 m in the Himalayas. The wood is used for planking and agricultural and domestic implements. The bark yields a useful gum.

Two foliage fungi, namely, *Meliola geniculata* Syd. from Pulliyanur (Kerala) and *Cercospora lanneae* Kar & Mandal from Midnapur (West Bengal) causing leaf spots on both leaf surfaces are recorded on *L. coromandelica* (Butler and Bisby 1960, Kar and Mandal 1969).

LOPHOPETALUM WIGHTIANUM ARN.

L. wightianum occurs on the west coast ascending upto 1000 m altitude in evergreen forests. The timber is used for construction and plywood.

Aecidium lophopetali Wakefield is the only pathogen recorded on this host from Dehra Dun (U.P.). The rust produces globose to irregular tumours on which aecia are produced (Bakshi and Sujan Singh 1967).

MADHUCA LATIFOLIA ROXB.

M. latifolia occurs in northern and central India and northern parts of the Indian Peninsula. This is one of the most important forest trees of India. The flowers are eaten or distilled into spirit. An oil, extracted from seeds, is used for manufacture of margarine, soap and glycerine. The wood is used in bridges.

Pathogenic fungi recorded on *M. latifolia* are *Fomes caryophylli* from India causing heart rot, *Scopella echinulata* (Niessl) Mains, a foliage rust, from Calcutta, Palamau, Pusa, Tholkobad, Varanasi and Bangalore in India and also from Ceylon and Pakistan producing chocolate-brown pustules on leaves and *Pestalotiopsis* sp. from Allahabad (U. P.) attacking leaves (Tandon and Chandra 1963–64, Bakshi and Sujan Singh 1967, Browne 1968). The mistletoe, *Dendrophthoe falcata*, is recorded on twigs and branches (Browne 1968).

MALLOTUS PHILIPPENSIS MUELL.-ARG.

M. philippensis is common in the sub-Himalayan tract ascending sometimes to 1500 m altitude. Silviculturally, the species is a useful nurse crop of sal and improves the soil. The wood is used for bobbins. A dye, *kamala*, used for dyeing silk, is obtained from red glands on the surface of the capules.

Phyllachora malloticola Seshadri causes tar spots on leaves of *M. philippensis* in India. Infection spots aggregate to form necrotic spots on both leaf surfaces (Seshadri 1968). Other pathogenic fungi recorded on

M. philippensis are *Fomes conchatus* (Fr.) Gill., *F. connatus* (Weinm.) Gill., *F. robiniae*, *Ganoderma applanatum*, all from India causing heart rot and *Phyllosticta marmorata* Cke. from Dehra Dun (U.P.) attacking leaves (Butler and Bisby 1960, Balwant Singh and Bakshi 1961, Spaulding 1961).

MELIA AZEDARACH LINN.

M. azedarach is indigenous to Baluchistan and Kashmir but is cultivated all over India. The wood is used for furniture and agricultural implements.

Rosellinia bunodes causes root rot of *M. azedarach* in Argentina resulting in death of trees within 5-6 years of infection. The other root rot fungi recorded on *M. azedarach* are *Ganoderma lucidum* from Bihar (India), *G. pseudoferreum* (Wakef.) v. Overeem & Steinm. from East Indies, *Fomes lamaoensis* from Formosa causing brown root rot, *Helicobasidium purpureum* (Tul.) Pat. from U.S.A. and *Phymatotrichum ominivorum* (Shear) Dug. from U.S.A. (Spaulding 1958, 1961, Bagchee and Ujagar Singh 1960).

Corticium salmonicolor causes stem and twig cankers and girdling of trees. The species is highly susceptible to the disease in high rainfall locations. Other canker and twig blight fungi recorded are *Phytophthora parasitica* from Argentina, *Hendersonula toruloidea* from Pakistan, *Nectria cinnabarina*, *N. coccinea* Pers. ex. Fr., *Physalospora abdita* (Berk. & Curt.) N.E. Stevens, *P. obtusa* (Schw.) Cke., *P. rhodina* (Berk. & Curt.) Cke., *Botryosphaeria obtusa* (Schw.) Shoemaker, *B. ribis* (Tode ex. Fr.) Gross & Dug., *Fusarium lateritium* Nees causing leaf and twig blight, *Glomerella cingulata* causing anthracnose and *Corticium koleroga* causing thread blight, all from U.S.A. (Spaulding 1958, 1961, Browne 1968). *Fomes conchatus* from Union of South Africa and *F. senex* from Formosa and India are reported to cause heart rot (Spaulding 1961).

Three species of *Cercospora*, namely, *C. leucosticta* from India, *C. meliae* from China, India and Formosa and *C. subsessilis* from Sudan, India, Philippines and Dominican Republic are recorded on *M. azedarach* (Spaulding 1961, Browne 1968). *Phyllactinia guttata*, the powdery mildew, is recorded from China (Spaulding 1961). *Pseudoperonospora portoricensis* (Lamkey) Seaver & Chardon, the downy mildew, forms irregular, downy, diffuse and yellow areas on lower leaf surface causing premature defoliation in Puerto Rico and Dominican Republic (Spaulding 1961). Other foliage fungi recorded on *M. azedarach* are *Gloeosporium meliicola* Speg. from Argentina causing anthracnose, *Phyllosticta azedarachis* Thüm. and *P. meliae* Ell. & Everh. both from U.S.A. causing leaf spots, *Septoria meliae* Syd. from India and Tanzania causing circular white spots and *Sphaceloma meliae* from Argentina causing leaf spots and branch lesions (Fresa 1958, Spaulding 1961, Sohi and Nayar 1971).

Six species of phanerogamic parasites, namely, *Dendrophthoe falcata*, *Helixanthera ligustrina* from India killing branches and sometimes the tree, *Loranthus corynitis* Spreng., *L. umbellifer* Schult. f. from India, *Phragmanthera incana* (Schum. & Thonn.) Balle from Nigeria and *Tapinanthus pentagonia* (DC.) v. Tiegh. from Nigeria are recorded on *M. azedarach* (Browne 1968).

MESUA FERREA LINN.

M. ferrea occurs in Assam, Khasi Hills and the western Ghats. The wood is very durable and is mainly used for railway sleepers, bridge construction and many other purposes.

Pathogenic fungi recorded on *M. ferrea* are *Ganoderma lucidum* from Wynaad (Kerala) causing root rot, *Diaporthe curvatispora* Wakef. from Sibsagar (Assam) and *Rosenscheldiella indica* Anahosur from Coorg (Karnataka) forming black, superstomatal ascostroma on lower surface of leaves (Bagchee and Ujagar Singh 1960, Butler and Bisby 1960, Anahsour 1970 a).

PINUS KESIYA ROYLE EX GORD.

Pinus kesiya occurs in Khasi and Naga Hills and Manipur in eastern India at 800 to 2100 m. Outside India, the species occurs in Burma, Thailand and Indochina. The tree is a light demander, wind-firm and fire-tender in early age. Larger trees resist fire due to thick bark. Timber is hard and used for building and fuel. The species yields high quality resin.

Damping-off fungi recorded on *P. kesiya* are species of *Fusarium* from India and Malaysia, *Pythium debaryanum* Hesse from the Philippines and *Rhizoctonia solani* Kühn from India (Spaulding 1961, Browne 1968, Bakshi, Reddy, Puri and Sujan Singh 1972).

Cronartium quercuum (Berk.) Miyabe, the gall rust, is recorded on *P. kesiya* from Assam (India) causing galls of various sizes on branches and the main stem. The rust produces pycnia and aecia on *P. kesiya* and uredinia and telia on *Quercus griffithii* Hook. f. Elsewhere, the rust is recorded on *Pinus contorta* Dougl. ex. Loud., *P. banksiana* Lamb. and *P. ponderosa* Dougl. ex Laws. from Canada, *P. densiflora* Sieb. & Zucc. from Japan, *P. massoniana* Lamb. from China, *P. nigra* Arn. and *P. nigra* Arn. var. *maritima* (Ait.) Melville from U.S.A., *P. thunbergii* Parl. from Japan and U.S.A., *P. sylvestris* Linn. from Canada, Japan, Russia and U.S.A. and *P. kesiya* from the Philippines. Recorded uredinial hosts are *Quercus coccinea* Muenchh. from Germany, *Q. ilex* L. *Q. palustris* Muenchh. and *Q. petraea* (Mattuschka) L. ex Liebl. from Japan, *Q. rubra* L. emend. Du Roi from Canada and various species of *Castanea*, *Castanopsis*, *Fagus* and *Pasania* (Bagchee 1950 b, Browne 1968).

Other pathogens recorded on *P. kesiya* are *Pellicularia filamentosa* (Pat.) Rogers from the Philippines causing web blight, *Armillaria mellea* from Nyasaland causing root rot, *Diplodia pinea* (Desm.) Kickx from Australia causing canker and die-back, *Fomes pinicola* (Fr.) Cke., *Polyporus sulphureus* both from India causing heart rot, *Dothistroma pini* Hulbary from East Africa causing needle cast and *Coleosporium tussilaginis* (Pers.) Lev. from India (Spaulding 1961, Browne 1968).

PONGAMIA PINNATA (L.) PIERRE

P. pinnata occurs throughout India, chiefly along streams, rivers, sea coasts and in tidal forests. The species is planted in avenues and gardens. The wood is used as fuel. The seeds of *P. pinnata* yield a thick oil used for burning and for skin diseases.

Pathogenic fungi recorded on *P. pinnata* are *Ganoderma lucidum* from Bangalore (Karnataka) causing root-rot, *Fomes merillii* (Murrill) Sacc. & Trott. from Calcutta (West Bengal) attacking base of living trees, *Ravenelia hobsoni* Cke. from Calcutta (West Bengal), Madras (Tamil Nadu), Nagina (U.P.) and Bandra (Maharashtra) and from Ceylon attacking lower leaf surface and producing chestnut brown teliospore heads with yellowish brown, simple appendages, *R. stictica* Berk. & Br. from Ceylon and India, *Microstroma pongamiae* Munjal, Chona & Kapoor from Delhi causing white to cream coloured spots giving a yellowish appearance to the infected leaves, *Cercospora pongamiae* Kar & Mandal from Bauria, Howrah (West Bengal), *Sphaceloma pongamiae* Wani & Thirum. from Pimpri, Poona

(Maharashtra) causing anthracnose spots on leaves, tender shoots and pods resulting in severe damage and early defoliation in young seedlings and trees, *Oidium* sp. from Poona (Maharashtra), *Fusicladium pongamiae* Syd. from Nasik (Maharashtra), Coimbatore and Chalisseri (Tamil Nadu) and from Ceylon associated with blight and *Phyllachora pongamiae* (Berk. & Br.) P. Henn. from Islampur (Maharashtra), Madras (Tamil Nadu) and from Ceylon causing black spots (Munjal, Chona and Kapoor 1959, Butler and Bisby 1960, Bakshi and Sujan Singh 1967, Browne 1968, Kar and Mandal 1969, Wani and Thirumalachar 1970).

Cuscuta reflexa and a species of *Loranthus* are the phanerogamic parasites recorded on *P. pinnata* from India (Browne 1968).

POPULUS CILIATA WALL. EX ROYLE

P. ciliata occurs in the Himalayas from Kashmir to Bhutan at 1300 to 3300 m. The wood is excellent for match manufacture. The species is useful for afforestation of unstable hill sides.

Valsa sordida Nits. is recorded on *P. ciliata* from Kashmir (India) causing death of bark in patches and development of cankers. The imperfect state of the fungus, *Cytospora chrysosperma* Fr., is characterised by pin-head size fruit bodies from which orange-pink spores are liberated in a tendril. The trees weakened by drought are more susceptible to attack by *V. sordida* (Browne 1968).

Two leaf rusts, namely, *Melampsora ciliata* Barclay and *M. populnea* (Pers.) Karst. are recorded on *P. ciliata* from India and Pakistan causing premature defoliation during heavy infection. *M. ciliata* is an autoecious rust, while *M. populnea* is heteroecious producing spermogonia and aecia on species of *Larix*, *Mercurialis* and *Pinus* and uredinial and telial states on *P. ciliata* and other species of *Populus* (Bakshi and Sujan Singh 1967, Browne 1968, Quraishi and Jamal 1969). *Uncinula salicis* (DC.) Wint. is recorded from Simla (H. P.), Chakrata and Mussoorie (U. P.) and also from Pakistan causing white powdery growth on leaves (Bakshi and Sujan Singh 1961, Browne 1968). *Taphrina aurea* (Pers.) Fr. causes blistering, wrinkling and curling of leaves of *P. ciliata* in India and Pakistan. When the asci are fully developed, blisters are brilliant yellow or orange on the lower concave side but later turn brown (Browne 1968, Quraishi and Jamal 1969).

Other pathogens recorded on *P. ciliata* are *Xylaria polymorpha* (Fr.) Grev. from Pakistan causing root rot, *Fomes fomentarius* from India and Pakistan, *Pleurotus ostreatus* (Fr.) Kummer and *Fistulina* sp. from Pakistan, all causing heart rot, *Nectria cinnabarina* from India causing stem cankers, *Linospora populina* (Pers.) Schroet. from India and Pakistan causing leaf spots and *Titaeosporina tremulae* (Lib.) van Luijk. from India and Pakistan causing anthracnose (Butler and Bisby 1960, Browne 1968, Quraishi and Jamal 1969).

POPULUS EUPHRATICA OLIVER

P. euphratica occurs in southern Punjab and Ladakh. Elsewhere, the species occurs in North Africa and extends through Middle East to Central Asia. The wood is used for building and turnery.

The pathogenic fungi recorded on *P. euphratica* are *Melampsora populnea* producing dark coloured telial sori and yellow, powdery uredinial sori on leaves and *Cladosporium herbarum* Fr. causing velvety, blackish olive effused patches on bark and leaves, both from Pakistan (Quraishi 1967).

PROSOPIS CINERARIA DRUCE

P. cineraria occurs scattered in the Punjab, Rajasthan and South India.

Ravenelia spicigerae Patil & Thirum. is the only pathogen recorded on the species from Nasik (Maharashtra) (Sarbhoy, Girdhari Lal and Varshney 1975).

PTEROCARPUS DALBERGIOIDES ROXB.

P. dalbergioides occurs scattered in the mixed deciduous or semi-evergreen forests in the Andaman Islands. The red heartwood is highly prized for furniture, railway carriages, panelling and interior decorative work.

Fomes fastuosus is the only decay fungus recorded on the species from South Andamans (Sujan Singh, Balwant Singh and Bakshi 1961).

PTEROCARPUS MARSUPIUM ROXB.

P. marsupium occurs scattered in mixed deciduous forests of central and southern India. The wood is used in buildings and for agricultural implements. The red gum exuding from wounds is used in medicine.

Maravalia pterocarpi (Thirum.) Thirum., recorded on leaves from Karnataka and Kerala, may cause severe defoliation in heavy infections (Bakshi and Sujan Singh 1967). Other fungi recorded on the host from India are *Ganoderma lucidum* causing root-rot, *Polyporus gilvus* Fr. and *Fomes fastuosus* both causing heart rot and *Cercospora canescens* Ell. & Mart. on leaves from Balaghat (M. P.) causing leaf spot (Browne 1968, Bakshi, Reddy, Puri and Sujan Singh 1972, Sarbhoy, Girdhari Lal and Varshney 1975). *Loranthus scurrula* Linn. is the mistletoe recorded on the host from India (Browne 1968).

QUERCUS GRIFFITHII HOOK. F.

Q. griffithii occurs commonly in eastern Himalayas at 1300 to 2600 m. The wood is very hard and is used in buildings and as fuel.

Cronartium quercuum, the gall rust, is the only pathogen recorded on the host from Shillong Range (Assam). The rust is heteroecious with pycnia and aecia on *Pinus kesiya* and telia on *Q. griffithii* (Bakshi and Sujan Singh 1967).

QUERCUS HIMALAYANA BAHADUR

Q. himalayana occurs in the western Himalayas at 2300 to 3000 m, occasionally descending to 1800 m in cool and moist situations. The wood is mainly used as fuel.

Sphaerotheca lanestris Harkn., the powdery mildew, causes inconspicuous powdery patches on lower surface of leaves in Batote (Kashmir) Globose, reddish brown cleistothecia develop embedded in the mycelium and contain only one ascus with eight ascospores (Pandotra, Kachroo and Sastry 1968). Other pathogenic fungi recorded on *Q. himalayana* are *Fomes fomentarius*, *Polyporus obtusus* Berk. both from India and *Trametes dickinsii* Berk. from India and Pakistan, all causing heart rot. *Dendrophthoe falcata*, the mistletoe, is also recorded on *Q. himalayana* (Browne 1968).

QUERCUS ILEX LINN.

Q. ilex occurs in the inner arid tract of the Himalayas from Sutlej Valley westwards. Elsewhere, the species occurs in the Mediterranean region. The wood is very hard and is used for agricultural implements and tool handles. Branches are lopped for cattle fodder.

Epidochium ilicinum Goid. & Camici causes small cankers on branches of *Q. ilex* in Italy resulting in necrosis of phloem and cambium (Spaulding 1961). Other stem fungi recorded on *Q. ilex* are *Endothia parasitica* (Murr.) P.J. & H.W. And. from Italy and Switzerland causing stem cankers, *Hypoxylon mediterraneum* (de Not.) J.H. Miller from France and Morocco causing cankers, *Polyporus cuticularis* Fr. from Portugal, *P. dryadeus* Fr. from Britain and *Stereum gausapatum* (Fr.) Fr. from Italy, all causing heart rot (Spaulding 1961, Browne 1968).

Two leaf rusts, *Uredo quercus* Brond. from Great Britain and *Cronartium quercuum*, the pine-oak gall rust, from Japan are recorded on *Q. ilex* (Spaulding 1961, Browne 1968). Other foliage pathogens recorded on the host are *Microsphaera extensa* Cke. & Peck from Austria and Italy causing a white powdery coating on leaves and shoots, *Sphaerotheca lanestris* from Pakistan causing powdery mildew, *Ascochyta quercus* Sacc. & Speg. and *Gnomonia quercus-ilicis* Berl. from Italy causing leaf spots (Spaulding 1961, Browne 1968.) *Taphrina kruchii* (Vuill.) Sacc. causes witches'-broom of *Q. ilex* in Algeria, Corsica, Italy, Sardinia and Sicily (Spaulding 1961). *Viscum album* is the mistletoe recorded on *Q. ilex* (Browne 1968).

QUERCUS LEUCOTRICHOPHORA A. CAMUS EX BAHADUR

Q. leucotrichophora occurs in the western Himalayas extending in the east upto Nepal at 1000 to 2000 m. Wood is used for agricultural implements, fuel and charcoal.

Two root rot fungi, namely, *Armillaria mellea* from India, and *Fomes annosus* (Fr.) Cke. from India and Pakistan are recorded on *Q. leucotrichophora* (Browne 1968). The heart rot fungi recorded on the species are *F. conchatus*, *F. fastuosus* and *F. senex* all from India and Pakistan, *Fomes caryophylli*, *Polyporus cuticularis*, *P. delectans* Peck and *P. gilvus* all from India (Butler and Bisby 1960, Browne 1968, Thind and Rattan 1971). *Bulgaria inquinans* Fr. and *Diplodiella crustacea* Karst. are reported from India causing die-back in the species (Butler and Bisby 1960, Bagchee and Ujagar Singh 1960).

Sphaerotheca lanestris is recorded on *Q. leucotrichophora* from India causing powdery mildew. Infection of young trees may result in stunting and sometimes in the development of witches'-brooms (Browne 1968). The leaf spot fungi recorded on the host are *Aulographum maculare* Berk. & Br. from Darjeeling (India), *Monochaetia monochaeta* (Desm.) Allesch. from

India, *Pestalotiopsis clavispora* (Atk.) Steyaert from India and Alabama (U.S.A.) and *P. darjeelingensis* Dube, Bilgrami & Srivastava from India (Spaulding 1961, Browne 1968, Pavgi and Singh 1970).

QUERCUS SEMECARPIFOLIA SMITH

Q. semecarpifolia occurs throughout the Himalayas from Bhutan westwards at 2600–4000 m. The wood is used for building, agricultural implements, fuel and charcoal.

Pathogenic fungi recorded on *Q. semecarpifolia* are *Ganoderma lucidum* from India causing root rot, *Fomes conchatus*, *F. robiniae*, *F. senex*, *Hymenochaete rubiginosa* (Fr.) Lev., *Polyporus cuticularis*, *P. gilvus*, *P. obtusus* and *P. sulphureus*, all from India causing heart rot (Butler and Bisby 1960, Spaulding 1961).

SALIX DAPHNOIDES VILL.

S. daphnoides occurs in the western Himalayas at 800 to 5000 m. The wood is used in buildings and for utensils. Twigs are used for baskets and branches are lopped for fodder.

Cryptomyces maximus (Fr.) Rehm is recorded from Denmark causing bark cankers on stems and shoots of *S. daphnoides*. The fungus forms thick hyphal cushion below the bark which may be shed in large patches by continued growth pressure of the fungus (Spaulding 1961). Other stem canker pathogens recorded are *Corticium salmonicolor*, the pink disease fungus, from India (Bakshi, Reddy, Puri and Sujan Singh 1972) and *Scleroderris fuliginosa* (Fr.) Karst. from Denmark and Germany (Spaulding 1961).

Melampsora capraearum (DC.) Thüm. attacks *S. daphnoides* in India. It is a heteroecious rust with pycnial and aecial states on *Larix decidua* Mill. and *L. leptolepis* (Sieb. & Zucc.) Rehd. and uredinial and telial states on *S. daphnoides* and other species of the genus. *M. larici-capraearum* Kleb. also recorded on *S. daphnoides* (Spaulding 1961) is reduced as synonym of *M. capraearum* (Wilson and Henderson 1966). *M. epitea* Thüm., another foliage rust, is recorded on *S. daphnoides* from Austria and India. The rust produces pycnial and aecial states on species of *Abies* (Spaulding 1961). *M. larici-epitea* Kleb. recorded from Germany and Switzerland and alternating on species of *Larix* is regarded as a variety of *M. epitea* (Browne 1968). *M. salicina* Lev. forms black rust fungous masses on leaves of

S. daphnoides in Germany (Spaulding 1961). *Uncinula salicis* causing powdery mildew, is recorded on *S. daphnoides* from Japan (Spaulding 1961).

SALIX TETRASPERMA ROXB.

S. tetrasperma occurs throughout India ascending to 2000 m in the Himalayas and 2300 m in the Nilgiris. The wood is used for planking, well construction and match manufacture.

Two species of *Melampsora*, namely, *M. capraearum* from Ranikhet, Mussoorie, Bhowali, Bageshwar and Kumaon (U.P.) and *M. salicis-albae* Kleb. from Coimbatore (Tamil Nadu) are recorded on *S. tetrasperma* (Butler and Bisby 1960). Other pathogens recorded are *Phyllactinia guttata* from Pakistan causing powdery mildew, *Cercospora salicina* Ell. & Ev. from Maharashtra, India, *Drepanopeziza triandrae* Rimpau from Egypt and *Rhytisma salicinum* Fr. from Simla, India, all causing leaf spots (Butler and Bisby 1960, Tandon and Chandra 1963–64, Browne 1968).

SAPINDUS EMARGINATUS VAHL

S. emarginatus occurs in southern India. The pulp of the fruit is largely used as soap.

Cercospora sapindi-emarginati Ramakrishnan, T.S. & K. forms yellowish leaf spots studded with black dots in Anakapalle (A.P.) (Butler and Bisby 1960, Vasudeva 1963). *Microascus cinereus* (Emile-Weil et Gaudin) Curzi is recorded on leaves of *S. emarginatus* from Secundrabad (A.P.) causing cream coloured leaf spots. The imperfect state of the fungus is described as *Scopulariopsis cinerea* Emile-Weil et Gaudin (Reddy 1969). Other fungi recorded on leaves are *Alternaria sapindi* Rajderkar from India causing ashy black to dark brown leaf spots, *Macrophoma sapindi* Ramakr. & Sund. from Kallar (Tamil Nadu) causing grayish brown circular spots surrounded by a deeper coloured ring, *Mycosphaerella sapindicola* Pai from Poona (Maharashtra) causing circular spots with ashy white to brown centre surrounded by a dark brown margin and often followed by shot holes (Ramakrishnan and Sundaram 1955, Rao 1962, Rajderkar 1966, Pai 1968).

SYZYGIUM CUMINI (LINN.) SKEELS

S. cumini occurs throughout India (except in the most arid regions) ascending upto 1300 m in the Himalayas and 2000 m in the Nilgiris. The wood is used in buildings, for agricultural implements and well curbs. Fruit is edible.

Elsinoe kamatii Tewari is recorded on leaves of *S. cumini* from Mahabaleshwar (Maharashtra). Severe infections sometimes cause drying and withering of the leaves (Tewari 1968). *Kamatella indica* Anahosur causes well-marked erumpent leafspots scattered or aggregated on upper surface of leaves of *S. cumini* in Coorg (Karnataka) (Anahosur 1969). *Maublancia indica* Pande & Patwardhan is recorded on leaves from Maharashtra causing circular infection spots on lower surface (Pande and Patwardhan 1967). Other leaf-spot fungi recorded on *S. cumini* are *Corynespora cassiicola* (Berk. & Curt.) Wei from India, *Didymosporella* sp. from Nandi Hills (Karnataka), *Neobarclaya congesta* (Berk. & Br.) Petch from Coimbatore (Tamil Nadu), *Pestalotiopsis versicolor* from Allahabad (U.P.), *Phyllachora eugeniae* Anahosur from Coorg (Karnataka) causing tar spots with black, shining carbonaceus stromata, *P. ambigua* Syd. from Koppa (Karnataka), *Polystigma eugeniae* Anahosur from Coorg (Karnataka) and *Selenophoma kamati* Kalani from Poona (Maharashtra) causing brick red infection spots on both leaf surfaces resulting in excessive shedding of leaves in young trees (Thirumalachar 1951, Subramanian and Ramakrishnan 1956, Butler and Bisby 1960, Tandon and Bilgrami 1961, Anahosur 1968, 1969, Kalani 1969, Sarma and Nayudu 1970).

Other diseases recorded on *S. cumini* are *Fomes caryophylli* from India, causing heart rot, *Meliolina cladotricha* (Lev.) Syd. from India, *M. mollis* (Berk. & Br.) Hohnel from Ceylon, *Asterina jambolanae* Kar & Maity from India, all causing sooty moulds and the mistletoe, *Dendrophthoe falcata* from India (Butler and Bisby 1960, Browne 1968, Kar and Maity 1970).

TERMINALIA ARJUNA W. & A.

T. arjuna occurs commonly throughout greater part of the Indian Peninsula along rivers, streams, ravines and dry water courses. The wood is used for building, agricultural implements, carts and boats. The bark is used for tanning.

The pathogens recorded on *T. arjuna* are *Armillaria mellea* from Mauritius causing root rot, *Colletotrichum arjunae* Sahni from Jabalpur (M.P.) and *Sphaceloma terminaliae* Bitanc. from Khandala (Maharashtra) causing small spots on leaves and elongated, closely grouped, crustose spots on petioles and stems (Spaulding 1961, Wani and Thirumalachar 1969 a, Sarbhoy, Girdhari Lal and Varshney 1975).

TERMINALIA BELLIRICA ROXB.

T. bellirica occurs throughout India scattered in mixed deciduous forests. The wood is used for planking, packing cases and boats. The fruit is used for tanning and in medicines.

Puccinia terminaliae (Ramakr., T.S. & K.) Ramakr. & Sund. is recorded from Valparai (Tamil Nadu) causing spherical, woody galls on branches and both surfaces of leaves. The rust fructifications develop sunken in woody galls. The rust is autoecious with all the spore forms produced on *T. bellirica* (Ramakrishnan and Sundaram 1954). *Phyllactinia terminaliae* T.S. Ramakrishnan causes powdery mildew on lower surface of older leaves of *T. bellirica* in Khandala (Maharashtra). The tender leaves remain free from attack (Ramakrishnan 1957). *Phyllachora terminaliae* Seshadri & Patwardhan causes tar spots surrounded by necrotic areas on upper surface of the leaves in Mahabaleshwar (Maharashtra). Stroma are black and shiny and produce dumb-bellshaped ascospores with constriction in the centre (Seshadri 1968). *Pestalotiopsis terminaliae* Agarwal & Hasija is recorded from Jabalpur (M.P.) causing circular to irregular infection spots with light brown central regions and dark brown margin (Agarwal and Hasija 1961).

Other pathogens recorded on *T. bellirica* are *Fomes fastuosus*, *F. robineae*, *F. senex*, *Ganoderma applanatum*, all from India causing heart rot, *Colletotrichum terminaliae* Sahni from Jabalpur (M.P.), *Corticium repens* from India causing thread blight and *Sphaceloma terminaliae* from Karnataka. *Dendrophthoe falcata*, the mistletoe, is recorded from India (Browne 1968, Sarbhoy, Girdhari Lal and Varshney 1975).

TERMINALIA CATAPPA LINN.

T. catappa occurs in the Andamans and adjacent islands. The wood is used for furniture, bark yields tannins and fruits are edible.

Pathogenic fungi recorded on *T. catappa* are *Sclerotium rolfsii* Sacc. from the Philippines causing seedling blight, *Diplodia catappae* Cke. from India causing die-back, *Cercospora catappae* P. Henn. from Burma and East Africa causing pale brown to dingy gray spots with wine-coloured to black margins, *C. geraisensis* Chupp from Brazil causing pale brown to gray spots with purple to dark brown border, *Myxormia terminaliae* (Syd. & Butl.) Arx., *Phyllosticta catappae* Syd., *Polyrhizon terminaliae* Syd., all from

India causing leaf spots, *Ramularia catappae* Rac. from Java causing round, ashen spots with dark coloured margins, *Sphaceloma terminaliae* from Brazil causing scabby lesions on leaves and fruits and *Meliola amphitricha* Fr. from Wynaad, Kerala (India) causing sooty mould. *Dendrophthoe falcata*, the mistletoe, is also recorded on *T. catappa* (Butler and Bisby 1960, Spaulding 1961, Vasudeva 1963, Browne 1968).

TERMINALIA CHEBULA RETZ.

T. chebula occurs throughout India scattered in mixed deciduous forests. The fruits are the best of commercial myrabolans used in medicine and for tannins.

Pathogenic fungi recorded on *T. chebula* are *Uredo terminaliae* P. Henn. from Mahabaleshwar (Maharashtra) attacking leaves, *Phyllactinia terminaliae* from Ranchi (Bihar) and Mahabaleshwar causing powdery mildew and *Cercospora catappae* from Mahabaleshwar causing leaf spots (Ramakrishnan 1957, Chiddarwar 1959, Butler and Bisby 1960).

TERMINALIA CRENULATA HEYNE EX ROTH

T. crenulata occurs in the western and south-western India in mixed deciduous forests. The species is suitable for afforesting clayey grounds. The wood is used for buildings, carts, railway wagons, mine props and various other purposes.

Pathogenic fungi recorded on *T. crenulata* are *Ganoderma lucidum* causing root rot, *Corticium repens* causing thread blight, *Phyllactinia terminaliae* from Sinhgad (Maharashtra) causing powdery mildew and *Cercospora catappae* from Bannerghatta, Bangalore (Karnataka) and Mahabaleshwar (Maharashtra) causing leaf spots (Ramakrishnan 1957, Chiddarwar 1959, Butler and Bisby 1960, Spaulding 1961, Vasudeva 1963).

TERMINALIA PANICULATA ROTH

T. paniculata occurs along west coast from Bombay southwards. The wood is used for planking, agricultural implements and canoes.

Cerotelium terminaliae-paniculatae Nagraj, Govindu & Thirum. from Walayar (Kerala), Balehonnur (Karnataka) and *Pestalotia* sp. from Sirsi, North Kanara (Karnataka) are the only two pathogenic fungi recorded on leaves of *T. paniculata* (Butler and Bisby 1960, Sarbhoy, Girdhari Lal and Varshney 1975).

TOONA CILIATA ROEM.

T. ciliata occurs in the sub-Himalayan tract and valleys of the outer Himalayas ascending upto 1300 m, in Assam, Bihar and the western Ghats. The species is widely planted in avenues and is gaining importance as a plantation species. Outside India, *T. ciliata* occurs in Pakistan, Bangla Desh, Burma and Thailand and is planted in avenues and plantations in North and South Rhodesia, Uganda, Nyasaland, Tanzania and South Africa. The wood is used for furniture, tea and cigar boxes. Branches are lopped for fodder.

Armillaria mellea attacks *T. ciliata* in Nyasaland, Tanzania, and South Africa. The infection is serious in South Africa and Nyasaland where it probably originates from infected stumps of a local species, *Parinarium mobola* Oliver (Butler 1928, Pole Evans 1932). Other root rot fungi recorded on *T. ciliata* are *Ganoderma lucidum* from India and Pakistan and *Polyporus zonalis* from India (Browne 1968).

Toona ciliata suffers from serious heart rot in avenues in India and Pakistan. The incidence of the disease may sometimes rise upto 100 per cent as in some avenues in the Doon Valley in Uttar Pradesh. The species is subjected to various abuses like lopping of branches for fuel, fodder and for making way for communication and transmission wires and blazing of bark for putting numbers on avenue trees (Pl. 5, 1). The resultant injuries serve as infection courts for heart rot fungi. *Fomes senex* is by far the commonest and the most important fungus causing heart rot in avenue trees. The pathogen causes white stringy rot and develops abundant perennial fruit bodies on wounds and branch stubs. Other heart rot fungi occurring less commonly are *Polyporus sulphureus* causing brown cuboidal rot, *Fomes fastuosus* causing white pocket rot and *Ganoderma applanatum* and *Trametes straminea* both causing white fibrous rot. Control of heart rot in *T. ciliata* in avenues may be effectively achieved by avoiding all possible injuries. Tree number should be written on tin plates and nailed on the stem. Illicit lopping should be checked. Where pruning may be necessary, the pruning wounds should be given a chemical dressing immediately after cutting.

Discella cedrelae Ramakr. T.S. & K. causes circular to irregular spots on both surfaces of leaves of *T. ciliata* in Tamil Nadu (India). The pathogen produces shiny black flattened or discoid subepidermal pycnidia which are arranged around a central straw-coloured zone (Ramakrishnan and Ramakrishnan 1950). *Marssonina* sp. is recorded on *T. ciliata* in avenues from Dehra Dun (India) causing heavy infection (Bakshi, Reddy, Puri and Sujan Singh 1972). *Phyllactinia suffulta* causes inconspicuous patches of powdery mildew on under surface of *T. ciliata* leaves in Himachal Pradesh (India). Conidia produced by the fungus are of *Ovulariopsis* type. The cleistothecia bear upto 16 appendages (Lall and Gupta 1965). Saharan (1971) observed hyper-plastic symptoms in the form of warts associated with and originating from infection by *P. suffulta* on *T. ciliata*. The surface of such excrescences is smooth during earlier stages but becomes rough later.

Two phanerogamic parasites, *Cuscuta reflexa* and a species of *Loranthus* are recorded on *T. ciliata* from India (Browne 1968).

TOONA SERRATA (ROYLE) ROEM.

T. serrata occurs in the western Himalayas (upto 2500 m), Manipur and Burma. The species is planted as a shade tree in tea and coffee plantations in Ceylon and Indonesia and also successfully introduced in Kenya, Nyasaland, Uganda and South Africa. The timber is used in buildings and for planking and general carpentry.

The pathogenic fungi recorded on *T. serrata* are *Fomes conchatus* from Pakistan causing heart rot, *Nyssopsora cedrelae* (Hori) Arth. from Himachal Pradesh producing rust pustules surrounded by reddish brown areas on leaves and *Uromyces cedrelae* (P. Henn.) P. Henn. from Indonesia producing reddish brown rust pustules on lower surface of leaves (Spaulding 1961, Bakshi and Sujan Singh 1967).

DISEASES OF BLUE PINE

1–2. Dwarf mistletoe : 1. Witches'-broom and mortality in blue pine. 2. Dwarf mistletoe shoots. 3. Heavy lopping in blue pine. 4. *Fomes pini* sporophores on tree. 5. Decay in wood due to *F. pini*.

PLATE 18

MICROBIAL DEGRADATION IN WOOD

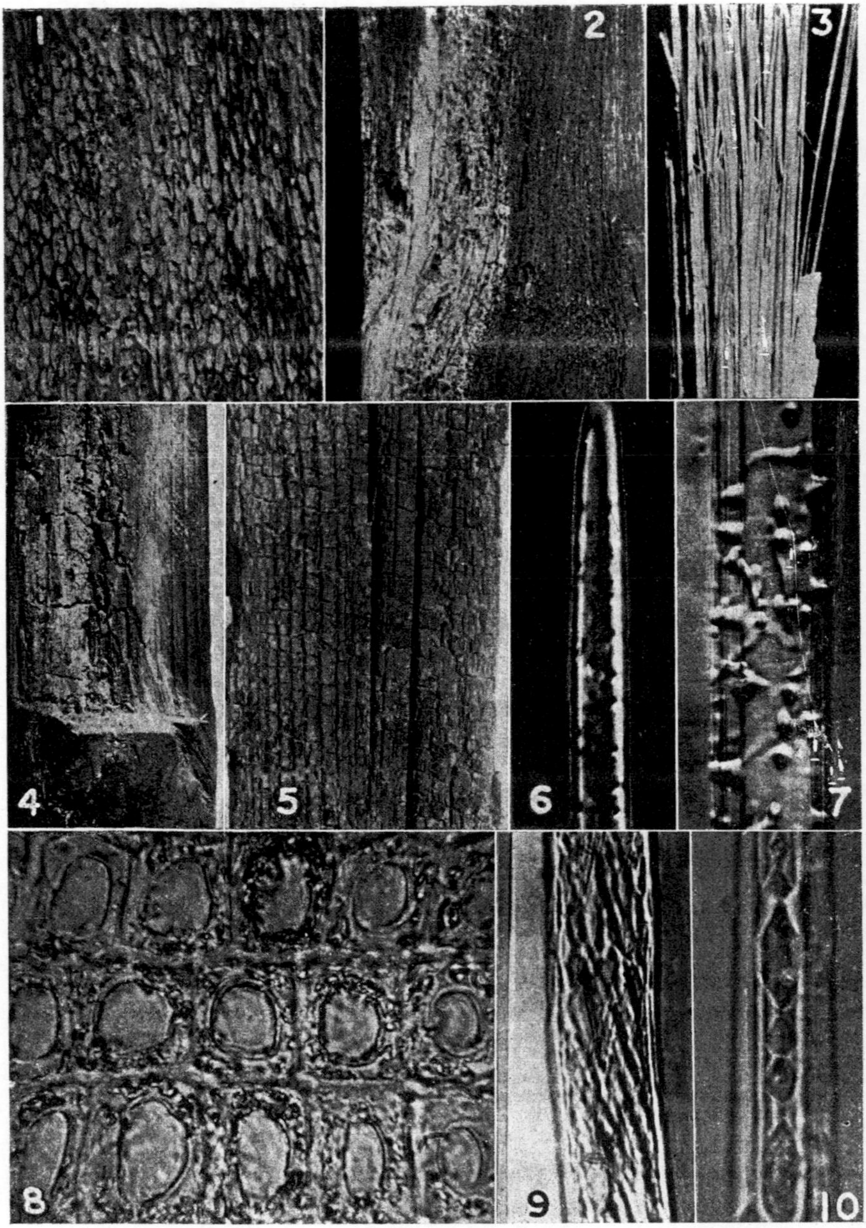

1–3. White rot : 1. Pocket rot. 2. Fibrous (spongy) rot. 3. Fibrous rot in bamboo. 4. Brown rot. 5. Soft rot. 6–7. Bore holes in fibre caused by decay fungus. 8. Cavities in secondary walls caused by soft rot fungus. 9. Soft rot hyphae in secondary wall of tracheid. 10. Diamond-shaped cavities in secondary wall of fibre caused by soft rot hyphae.

PART III

DECAYS IN TIMBER

INTRODUCTION

Wood has many exceptional properties like shock absorption, low thermal and electrical conductivity, plasticity which renders it relatively easy to work and bend to different shapes, relatively high strength with respect to weight, resistance to chemicals among others. The fibres in the wood make it unique for use as raw material for pulp and paper industry. Timber is a renewable resource available at a low cost, unlike its substitutes, like steel, the supply of which is exhaustible and which also has to be mined and smelted thus making the production cost very high. Due to all its qualities, wood has successfully withstood competition with its substitutes to fill much of the construction needs in buildings, railways, telephone and telegraphic communication, ship-building and agricultural industries. Due to a great variety of colour shading and grain pattern, wood is ideally suited for furniture making. Wood has, therefore, played a significant role in the industrial development of countries throughout the world.

As an organic substrate, wood is, however, liable to degradation due to microbial agencies causing significant losses. In the U. S. A., an annual overcut of about 10 per cent was necessary to replace wood decayed in service (Hartley and Boyce 1933). The repair of damage caused by dry rot in building timber in U. K. was estimated at £ 10 million annually (Savory 1954). Estimation of losses in wood after conversion and in use through microbial degradation has not been assessed in India though conditions obtainable in the tropics are generally more congenial for rapid biological deterioration of timber as compared to countries in the temperate regions. With modern industrial development, the requirement of wood is increasing rapidly. Since the supply of wood is limited, it is necessary to protect the wood in service from biological deterioration and thus effect economy in utilization.

DAMAGE

Dead wood does not make any pathological response to an invading organism. Many microorganisms are involved in the colonisation of wood and these include bacteria, moulds and fungi causing stain, soft rot, brown rot and white rot. One or more of these organisms follow in succession. The colonisation by the microorganisms depends on the time, place, temperature, moisture within and on wood surface and the wood species involved among other factors. Three principal causes that determine succession are competition among microorganisms for colonising the substrate, toxic metabolites which may exert either antagonistic or synergistic effects and microbial parasitism. Fungi attacking wood are classified into two broad groups—those causing decay in wood and others causing stain in the sapwood, based on whether or not they possess an enzyme system for attacking the cell wall substance.

STAINS

Sap stain in wood caused by moulds which develop as dark-coloured growth is generally superficial and may be easily brushed or planed off. Stain penetrating deep into the wood is caused by fungi which belong to Ascomycetes or Fungi Imperfecti. The wood may take different shades of blue, black and brown. Blue stain is the most common form of discolouration and occurs commonly in soft woods but less so in hardwoods. Stain in wood may also be caused by decay fungi during incipient stages of attack, though with progress of decay, typical rots develop, unlike sap staining fungi which do not cause any wood decay.

DECAYS

Fungi causing decay can be diagnosed from fruiting bodies and cultures as discussed in Chapter 3. Different types of decays are described below.

Brown rots

In this type of decay, cellulose and its associated pentosans are attacked and lignin remains unaffected. The lignin-rich residue left after decay imparts brown colour to decayed wood. Brown rotted wood shrinks abnormally when dried giving rise to characteristic cubical cracking. Brown rot occurring in building timber is referred to as dry rot. The frequency of brown rots is considerably higher in soft woods than in hardwoods.

White rots

In white rots, both lignin and cellulose are attacked. White rot is characterised by bleached appearance of the rotted wood due to utilisation or modification of chromogenic materials in the wood. White rotted wood shows normal shrinkage. A distinction is made as to whether the decomposition is successive where lignin is attacked initially and cellulose at a later stage or where lignin and cellulose are attacked simultaneously. White rots are more common in hardwoods than in soft woods. Fungi causing white and brown rots mostly belong to the Hymenomycetes in which members of the Polyporaceae are important.

Soft rots

In soft rot, cellulose is removed like brown rots but the mechanism of action on cell wall is different. The fungi causing soft rot, like those causing sap stain, belong to the Ascomycetes and Fungi Imperfecti, though the species involved and mechanism of action are different in the two. Soft rot is usually superficial and generally occurs in thin wooden pieces exposed to situations of high moisture as in cooling towers. Soft rotted wood shows typical cross checking at the surface. Soft rot occurs predominantly on hardwoods.

Bacterial rot

Bacterial degradation of wood is a slow process unlike that caused by fungi. Many bacteria may attack cellulose, while some may attack lignified cell wall.

FACTORS AFFECTING DECAY

Factors governing initiation and progress of microbial degradation of wood may be external or infernal, the latter inherent in the wood.

EXTERNAL FACTORS

Oxygen in the wood, as a constituent of air, is necessary for micro-organisms to perform the function of respiration. Insufficient air in the wood may limit the progress of degradation and if air is removed by saturation of wood with water, the organism may die.

Moisture is another controlling factor. Not only there should be adequate moisture in the wood but the relative humidity of the air should be sufficiently high, nearly 100 per cent, for the infection to establish.

Inside the wood, the spread of microorganisms depends on a suitable air-moisture balance. Fungi cannot grow if the moisture in the wood is below about 20 per cent based on oven-dry weight and this is the guiding principle in controlling degradation in wood during seasoning. If the infected wood is air-dried, the organism may not die but its active growth ceases. If such wood is able to imbibe moisture later, the organism may revive and continue its active growth in the wood. The progress of decay becomes severe when the wood reaches the fibre saturation point which varies from 25 to 32 per cent based on its oven-dry weight. At higher moisture levels in the wood, there is a progressive reduction in deterioration which ceases when the wood is fully saturated. Soft rot fungi, however, can grow in the wood with high moisture content and, therefore, possess a higher moisture tolerance than white and brown rot fungi.

With increasing specific gravity of wood, the upper limit of moisture content in the wood at which degradation is inhibited decreases. This is explained by the fact that with increasing specific gravity, there is more wood substance in a given volume of wood so that it will require less moisture to limit the air supply below which the organism will not grow.

In service, wood will not decay if kept air-dry as normally happens in wood kept under indoor conditions, like furniture or building timber not in contact with the ground or not exposed to rain. In outdoors, wood is exposed to rain and thus gets wet when decay may set in and progress. Wood on the ground imbibes moisture at the region of contact with soil and decays most at the ground level as in poles. At lower soil depths, oxygen becomes a limiting factor. Submerged piling supporting the masonry foundation of bridges or large buildings may, therefore, remain free from decay for an indefinite period.

Decay fungi differ in their temperature requirement for growth, as determined from pure culture tests on agar (Sehgal, Sen and Bakshi 1966), though the range of temperatures at which fungi may grow (minima and maxima) may be fairly wide. The temperature requirements of fungi on wood are usually fairly close to those of pure culture. Fungi can grow in wood at a fairly wide range of temperature, about 15° to 30°C. The activity decreases at temperatures above and below this range and effective growth ceases at about 5° to 10°C at the lower limit and 35° to 40°C at the higher. Fungi in wood may be killed at very low or high temperature which is obtainable in seasoning kilns and also during preservative treatment of wood under pressure when fungi in the wood may be killed.

As regards growth of fungi in relation to pH, all decay fungi causing white rot, brown rot and soft rot produce optimum growth at about pH 6, though soft rot fungi are capable of growing at pH 8 or 9 (Duncan 1960). In the wood, decay fungi during metabolism produce a certain amount of acid which renders the substrate distinctly acidic.

INTERNAL FACTORS

Natural resistance of wood to decay is defined as the inherent quality in wood to resist attack by fungi. This and natural resistance against termites constitute the major factors imparting natural durability to timber. Natural decay resistance of wood is a purely relative term since under ideal conditions of degradation, like wood in contact with the ground or under prolonged exposure to damp conditions, all species will eventually perish. Toxic natural extractives in the cell walls are responsible for imparting natural decay resistance and the subject is discussed in Chapter 7.

EVALUATION OF DECAY RESISTANCE

A knowledge of the natural durability and natural decay resistance of timber is necessary for its proper, rational and judicious utilization in sites of varying decay hazards. It is inadvisable and expensive to use a highly decay resistant wood on a dry site having less decay hazard as it is to use a less resistant wood on a damp site where the risk of decay is high. Also it gives information : whether it is safe to use wood in a location in its natural state or only after treatment with a wood preservative. Dwindling supply of conventional timbers in India has led to use of secondary timbers whose field performance is mostly not known. Laboratory tests reliably indicate the 'relative' decay resistance of timbers : whether the timber of one species is likely to last longer than another in the same situation. While laboratory tests are not intended to predict service life of timbers, a close approximation of service life may be obtained by a comparison with a timber of identical performance in laboratory tests and whose average life in service is known.

Natural durability of timber can be determined from the field tests. The data thus available are supplemented with records as to the reputation of timber from service tests to arrive at the expected service life.

Field trials, though realistic, are necessarily elaborate involving field tests and where test yards may have to be selected at many locations representing different biological hazards. The assessment of damage is qualitative and collective for all kinds of deterioration due to biological, chemical and weathering agencies so that any single cause cannot be isolated and assessed. As a supplement to field trials, therefore, accelerated laboratory tests of timbers are conducted to ascertain natural decay resistance of timbers. Details of sampling timbers and test procedure are outlined by Bakshi, Puri and Sujan Singh (1967). The tests are conducted under conditions ideal for decay to develop and are, therefore, rigorous. Also, the tests are standardised and, therefore, reproducible. Furthermore, the results are assessed quantitatively. However, laboratory tests introduce a factor of artificiality that precludes the application of results to quantitative prediction of service life though these results may supplement data from field trials in obtaining a closer approximation to service life of timbers.

Based on accelerated laboratory tests conducted at the Forest Research Institute, Dehra Dun (Bakshi 1967 a), some commercial timbers of India have been categorised into 4 decay resistant classes, namely, I (very resistant), II (resistant), III (moderately resistant) and IV (non-resistant) based on weight loss of wood in laboratory tests. For each species, both outer and inner heartwood have been tested and results presented in the Table 1. The field service class for each timber species is also indicated wherein the average life of timbers for class I is 120 months and over, for Class II, 60–119 months and for class III, 59 months and under (Anon. 1962).

TABLE 1

Decay and Field Resistance of some Indian Timbers

Timber species	Trade Name	Decay resistance class of heartwood		Field service class
		Outer	Inner	
1	2	3	4	5
Albizia odoratissima Benth.	Kala Siris	I	I–II	I
Artocarpus lakoocha Roxb.	Lakooch	I	—	I
Cassia fistula Linn.	Amaltas	I	I	II
Dalbergia latifolia Roxb.	Rosewood	I	I	I

1	2	3	4	5
Dalbergia sissoo Roxb.	Sissoo	I	I–II	I–II
Hardwickia binata Roxb.	Anjan	I	I	I
Madhuca longifolia (Koenig) MacBride var. *latifolia* (Roxb.) Chev.	Mahua	I	I–II	I
Mesua ferrea Linn.	Mesua	I	I	I
Pterocarpus marsupium Roxb.	Bijasal	I	I–II	I
Shorea robusta Gaertn. f.	Sal	I	I–II	I
Terminalia paniculata Roth	Kindal	I	I	III
Xylia xylocarpa Taub.	Irul	I	I	I
Albizia procera Benth.	Safed Siris	I–II	II–III	II
Bridelia retusa Spreng.	Kasi	I–II	II	II
Grewia tiliifolia Vahl	Dhaman	I–II	II–III	II
Lagerstroemia parviflora Roxb.	Lendi	I–II	II–III	III
Ougeinia oojeinensis (Roxb.) Hochr.	Sandan	I–II	I–IV	I
Toona ciliata Roem.	Toon	I–III	II–IV	III
Lagerstroemia speciosa (Linn.) Pers.	Jarul	I–III	II–III	II
Tectona grandis Linn. f.	Teak	I–III	II–III	I
Pinus wallichiana A.B. Jacks.	Kail	I–IV	III–IV	III
Aphanamixis polystachya (Wall.) Parker	Pitraj	II	—	I
Dipterocarpus turbinatus Gaertn. f.	Gurjan	II	—	III
Lagerstroemia lanceolata Wall. ex W. & A.	Benteak	II	II	I
Acacia nilotica (Linn.) Willd. ex Del. ssp. *indica* (Benth.) Brenan	Babul	II–III	III	III
Amoora wallichii King	Amari	II–III	—	II
Palaquium ellipiticum (Dalz.) Baillon	Pali	II–III	III	II
Terminalia alata Heyne ex Roth	Laurel	II–III	III–IV	II

1	2	3	4	5
Adina cordifolia (Roxb.) Hook. f. ex Brand.	Haldu	III	III	III
Anogeissus latifolia (Roxb.) Bedd.	Axlewood	III	II–III	III
Albizia chinensis (Osbeck) Merr.	Siris	III	III–IV	III
Chukrasia velutina W. & A.	Chickrassy	III	III–IV	III
Maniltoa polyandra (Roxb.) Harms	Ping	III	—	III
**Dillenia indica* Linn.	Dillenia	III	III	III
**Gmelina arborea* Roxb.	Gamari	III	III	I
Melia composita Willd.	Malabar Neem	III	—	—
Michelia champaca Linn.	Champ	III	II–III	III
Pinus roxburghii Sargent	Chir	III	—	III
Schima wallichii (DC.) Choisy	Chilauni	III	—	III
Acrocarpus fraxinifolius W. & A.	Mundani	III–IV	III–IV	III
**Terminalia bellirica* Roxb.	Bahera	III–IV	III–IV	III
Acer spp.	Maple	IV	IV	III
†*Boswellia serrata* Roxb.	Salai	IV	IV	III
**Casuarina junghuhniana* Miq.	Casuarina	IV	IV	—
†*Diospyros melanoxylon* Roxb.	Ebony	IV	IV	III
Juglans regia Linn.	Walnut	IV	IV	III
**Mangifera indica* Linn.	Mango	IV	IV	III
Bombax ceiba Linn.	Semul	IV	IV	III

*Species in which sapwood and heartwood are not distinguishable on the basis of colour.

†Species in which heartwood is rarely formed and that too in very large trees. Sapwood of such species are tested.

EFFECTS OF DECAY AND DISCOLOURATION

GROSS CHARACTERS

When timber is attacked by a fungus, it passes through various stages of degradation. In the early stage, the wood may appear normal except in a few cases where a change in colour may occur. This stage which can be detected by microscopic or cultural diagnosis, may be easily overlooked. In later stages, however, physical characters of the wood are changed and then the decay in wood may be identified with the unaided eye.

Colour

Change in colour of the wood from normal is usually associated with fungus attack. In sap stain, change of colour in wood usually occurs within a few days after attack. The usual shade of discolouration is blue though the affected wood may take shades of brown and black. In the incipient stage of decay, wood may sometimes change in colour to a darker shade depending on the species of the fungus and the species of wood attacked. In advanced decay, wood is bleached in white rot and darkened in brown rot.

Luminescence

In some cases, decayed wood may become strongly luminescent. The luminescence persists as long as the wood is moist and the fungus is active e.g., wood decayed by *Armillaria mellea*.

Odour

Advanced decay, particularly from freshly opened surface, may possess an odour which is sometimes characteristic of pure cultures of decay fungi.

Zone lines

White rots in wood may be associated with narrow, black to dark brown zone lines where the host cells are filled with dark brown swollen hyphae. Zone lines characterise decays due to many fungi like *Amillaria mellea*, many species of *Polyporus* and *Fomes* like, *F. fomentarius*, *F. igniarius*, *F. pini* and *Xylaria* among others (Bakshi 1971).

Features of decay

In white rots, the decay may be 'stringy' (also termed 'spongy' or 'fibrous') (Pl. **18**, 2) and the wood comes off in splinters. White rots may also occur in pockets within which tissues are usually bleached white and become completely decayed. The pockets are lined by more firm and less decayed wood (Pl. **18**, 1). The pattern of pockets, their shape and size are characteristic of the species of fungi which may, therefore, be identified from the rot characters (Bakshi 1971). In brown rots, the decay is characterised by cubical cracking in the wood, in transverse and longitudinal directions both in the tangential and radial faces, resulting in a tendency of the wood to separate into brown cubical portions (Pl. **18**, 4). The wood shrinks and becomes crumbly and powdery under pressure of the thumb. In soft rots, the wood becomes darkened and softened only at the surface which may be scraped off easily to open up relatively sound wood from beneath. Such abrupt change from heavily decayed surface wood to sound wood beneath is characteristic of soft rot. Soft-rotted wood shows bluish gray to brown discolouration, failure with a brittle fracture and softening at the surface showing characteristic cross checking (Pl. **18**, 5). Soft rotted wood usually retains its shape unlike in brown rot.

Strength

In sap-stained wood, the strength properties are not significantly reduced, except in toughness where the reduction may amount to 25 per cent or more (Chapman and Scheffer 1940, Pettifor and Findlay 1946). Sap-stained wood could, therefore, be used safely for all purposes except where toughness is an important consideration, as in timber used for aircraft.

In incipient decays, strength properties of white-rotted wood are not greatly affected, while those of brown rotted wood are seriously weakened. In advanced decays, all strength properties are affected, the reduction being more in a brown than in white rotted wood at comparable weight loss. In both decays, tensile strength is rapidly affected and the wood breaks with a brash fracture abruptly across the grain. Other strength properties that are less affected include bending strength, compressive strength, hardness and elasticity arranged in decreasing order of susceptibility (Cartwright and Findlay 1958). Unless, therefore, it is definitely known that the wood is not weakened in incipient decay, such wood should not be used for purposes where high strength is required. In advanced decays, all mechanical properties of wood are seriously affected to render such wood worthless for any purpose.

Specific gravity

Decay results in decrease in specific gravity of wood when the amount of wood substance per unit volume gets less. Decayed wood based on volume has, therefore, less calorific value and yields low grade fuel. Decayed wood absorbs water and becomes water-logged and also loses water much more rapidly than the sound wood.

MICROSCOPIC CHARACTERS

Decay fungi attacking wood utilise the cell wall substance for food. Also, fungi pass from cell to cell through the cell wall. A knowledge of the structure of cell wall is, therefore, important to understand the microscopic features of fungus attack in wood.

In the wood, each cell is separated from the adjacent one by a thin layer of inter-cellular substance called the middle lamella, composed almost entirely of lignin. This is flanked on either side by the extremely thin primary wall. Between 60 per cent (in hardwoods) and 90 per cent (in conifers) of all the lignin in wood is concentrated in the compound middle lamella comprising middle lamella and primary wall (Meier 1964). The secondary wall is laid down on the primary wall as thickening. Based on the differences in the helical angles of the microfibrils constituting them, the secondary wall is distinguished into three layers, namely, S_1 (outer layer) and S_3 (inner layer towards lumen) which are usually thin with microfibrils oriented more or less perpendicular to those of S_2 (middle layer) with fibrillar orientation nearly parallel to the long axis of the cell. The middle layer contributes most to the bulk of cell wall material as well as to its physical properties. It consists largely of cellulose although some lignin is present. In the angiospermous wood, the secondary wall is believed to contain very little lignin and consists almost entirely of carbohydrates. The cell wall, however, is not uniformly thickened leaving small openings, the pits.

Sap-stained wood

Staining fungi attack mainly the sapwood in which the hyphae initially concentrate within ray cells. The fungi derive nutrition from cell contents. Sap stain fungi do not possess an enzyme system for degrading cell wall components though certain ectoenzymes have been detected in liquid culture of a few blue staining fungi by Rosch, Liese and Berndt (1969). Therefore, when hyphae grow through the lumen, the inner cell wall does not show any deterioration though a limited attack on cell wall may occur (Scheffer and Lindgren 1940). The passage of staining hyphae from cell to cell is through pits, the pit membrane offering little resistance. The penetration occurs while the closing membranes of the pits are still soft and pliable. After the sapwood is transformed into heartwood or after the sapwood is partially dried, the membranes of the pits offer the same resistance to penetration as do the cell walls. If hyphae penetrate through bordered pits, they pass the torus in full width without constriction, thus breaking through the membrane apparently by aid of mechanical pressure. Hyphae may penetrate the cell wall directly by means of bore holes and are considerably constricted at the point of passage which does not widen subsequently. Under electron microscope, the apex of hyphae shows a special appendix, the transpressorium, consisting of a stalk and head part with a small pointed tip which serves to penetrate the cell wall. Enzymes may be secreted here locally softening the cell wall layers for the subsequent penetration of the transpressorium (Liese 1970 b). The growth of transpressorium is passive and the tip is pushed into the cell wall through intercalary growth at the joint of the mother cell and by its hydrostatic pressure. The cell wall around the penetration hole does not show any enzymatic changes. The openings produced facilitate the penetration of preservatives if the wood is dry enough (Liese 1970 a).

Decayed wood

Fungi grow rapidly through the largest available voids like the pits in the wood and through ray tissue. Decay fungi produce many ectoenzymes which act upon the cell wall substance. The mechanism in different groups of fungi is dealt with separately.

Decay due to Soft Rot

Hardwoods are usually more susceptible to soft rot attack than soft woods. Also, the resistance to soft rot does not always parallel resistance to decay. Soft rot fungi possess preferential growth within the cell wall (Pl. **18**, 9). Two forms of cell degradation occur depending whether the fungi attack hardwoods or soft woods (Corbett 1965). In both groups, the ray cells are initially colonised from where the hyphae pass on to the lumen of the fibre or tracheid via pits. Penetration between tracheids occurs through bordered pits and between fibres through simple pits or through the wall itself. There is no apparent degradation of wood up to this state. From the lumen, the hyphae give rise to short side branches at right angles to the long axis of the tracheid. The branches penetrate into the wall where the hypha becomes considerably slender than lumen hyphae and remains so. The initiation of the penetration hypha may result from a response to a biochemical stimulus from the cell wall (Levy 1965). The penetration hole widens in course of attack. Also, a lytic change of the cell wall around the hole suggests penetration by enzymatic action (Liese 1970 b). In soft woods, the penetration hypha traverses only upto the S_2 layer of the secondary wall where the hypha branches at right angles in T-shape (Levy and Stevens 1966). The paths of the branches in the S_2 layer correspond closely to the direction of orientation of the cellulose mycellae (Bailey and Vestal 1937) though Corbett (1965) held that the arms were placed at a slight but definite angle to the long axis of the tracheids. The action of ectoenzymes leads to the formation of chains of rhomboidal cavities with pointed ends, usually aligned one after another like a string of pearls (Pl. **18**, 10). Generally, a fine hypha termed proboscis hypha (Corbett 1963) grows out from the apex of cavity giving rise to chains of cavities. In transverse section, the cavities appear as circular areas of erosion within the S_2 layer of the secondary wall (Pl. **18**, 8). The size and shape of the cavity are determined by duration of attack, species of wood and fungus, cell type and moisture content (Liese 1970 b). In advanced stages of decay, the adjacent cavities tend to coalesce resulting finally in the general dissolution of the S_2 layer. However, the more resistant S_3 layer remains unattacked.

In hardwoods, the hyphae growing inside the lumen erode the S_3 layer, resulting in production of V-shaped notches in the wall (Levy and Stevens 1966). Erosion of the wall also occurs along the length of the hypha lying in the lumen and in contact with the wall leading to a sculpturing of the wall (Liese 1970 b). Both notches and erosion extend into secondary wall as far as the S_1 and primary wall—middle lamella complex. With progress of decay, cavities appear in S_2 layer though such cavities are less pronounced than in soft woods. Dissolution of the S_2 layer occurs earlier than in soft woods. Soft rot fungi attack fibres in preference to vessels and early wood rather than late wood (Greaves and Levy 1965).

Decay due to Brown Rot

Inside brown-rotted wood, only a few hyphae are visible. The hyphae pass from cell to cell usually through pits, though direct penetration of the cell wall by means of bore holes becomes common with progress of decay. In penetrating the wall, the hypha becomes considerably slender and the bore hole results from the enzymatic dissolution of cell wall substances. Local dissolution of the cell wall in advance of the actual passage also supports this view (Proctor 1941). Mechanical force of the advancing hyphal tip does not come into play. The bore holes eventually enlarge due to lateral secretion and diffusion of extracellular enzymes and so too the penetrating hypha.

In brown rot, the width of the cell wall is not reduced though under electron microscope, Meier (1955) showed that the apparently intact secondary walls of spruce wood decayed by brown rot fungi actually contain only a delicate but coherent lattice, presumably consisting of lignin that remains after selective removal of wood carbohydrates. The degradation of cell walls occurs first in S_2 layer causing the S_3 layer to become separated so that it lies free in the lumen. Cavities may appear in the S_2 layer. The S_1 layer may be attacked next and in advanced stages of attack, S_3 layer may be destroyed. The primary wall and middle lamella remain unaffected (Meier 1955, Wilcox 1968). Brown rotted wood shrinks abnormally on drying due to extensive cell wall shrinkage and wall collapse (Wilcox 1968) and exhibits deep cubical cracking (Pl. **18**, 4).

Decay due to White Rot

In white rots, hyphae are widespread and also more numerous than in brown rots, initially colonising vessels or rays and later the fibres. Passage from cell to cell occurs through pits and also by direct penetration by means of fine perforation hypha from the apex of which enzymes diffuse out resulting in a cone-shaped area of degradation. In course of attack, the canals become progressively larger resulting in large openings.

In white rot, the thickness of the cell wall at first remains stable, while in simultaneous rot, there is a gradual loss of cell wall thickness as the decay proceeds (Liese 1970 b). The degradation of cell layers occurs progressively from the lumen (S_3 layer) outwards (Wilcox 1968). Generally, no shrinkage and cell collapse occur and the original shape and the outward appearance of the wood are maintained.

Bacterial Decay

Degradation of cell wall by bacteria is a slow and prolonged process unlike that due to decay fungi. Bacteria initially colonise the ray parenchyma cells of the sapwood whose contents serve as source of nutrition. When this is depleted, the walls of ray cells are attacked when the pectic substances of the pit membranes are decomposed by means of pectinase. The ray parenchyma pits in pines are decomposed leading to enlargement of openings. Also, the pit membrane of the bordered pits between tracheids is decomposed and as a result, the whole sapwood becomes easily permeable to preservatives (Liese 1970 b) though Greaves and Levy (1965) suggest that permeability is increased not by breakdown of pits but by a change in the structural components of the ray cell walls and adjoining tracheids as evidenced by a loss of birefringence indicating lack of crystallinity in the rays. Permeability is not increased in the heartwood where pit membrane is lignified. Movement of bacteria into adjoining cells possibly occurs through pits, though evidence of this is lacking (Greaves and Levy 1955).

Many wood-inhabiting bacteria are able to attack cellulose, while some attack lignified cell wall, particularly when wood is exposed for long to wet condition as in piles and cooling towers in which situations, the surface of the wall may show numerous pits, some of which may perforate the walls (Savory 1954). Cell wall may be degraded by depolymerisation or by direct corrosion from the cell lumen (Liese 1970 b).

CHEMISTRY OF DECAY

Microorganisms causing wood decay produce different enzymes, each of which possesses specific effect on the cell to cause different types of decay. Enzyme production is greatest in brown rots and least in soft rots in which cellulase production is very low (Lyr 1962). Enzymes are water soluble protein molecules of high molecular weight. They are active in low concentrations and catalyze a particular type of chemical reaction without being permanently altered in the process. The moisture in the cell lumen swells the cellulose thus rendering the internal surface more accessible for the extracellular enzymes to diffuse into the cell wall. These enzymes catalyze the depolymerization of the different cell wall constituents into water soluble smaller molecular fragments which diffuse back to the fungal hyphae where they are assimilated by the fungus (Cowling 1961).

In studies on Indian Polyporaceae, Bose and Sarkar (1937) found the amount of extracellular enzymes to be much larger than the intracellular ones. Also, the activity of all the enzymes is greater in the mycelium in the vegetative stage than in the mycelium which is in advanced growth and in the fruiting stage. In *Polystictus sanguineus* and *Fomes lamaoensis*, dialysis enhances the enzymatic activity, optimum pH for which with both fungi is about 5. The optimum temperature for enzymatic activity with *P. sanguineus* is about 35°C in malt cultures and with *F. lamaoensis* 50°C in peptone and 45°–50°C in malt solution. Furthermore, the extracellular enzymes are more active than intracellular. Also, amylase of *P. sanguineus* remains stable for a week in aqueous solution at room temperature and after that its activity declines; that of *F. lamaoensis* is more stable (Nayaranamurti, George and Verma 1952 a). On wood blocks of *Shorea robusta* and *Pinus roxburghii*, the optimum pH and temperature for invertase activity of *P. sanguineus* are 4.7 and 36°C respectively for reaction time of 24 hours and for cellulase activity for the same fungus 5.2-5.4 and 33°C respectively for 14-day reaction (Nayaranamurti and Verma 1953). In *P. sanguineus*, *F. lamaoensis* and *Lenzites sepiaria* cultivated on different media, the pH values after an initial fall rise though not to their original level. *P. sanguineus* produces formic and butyric acids, while acetic acid and ethyl alcohol are also detected (Narayanamurti, George and Verma 1952).

Ascu, copper sulphate, coal tar oil and arsenic pentoxide depress enzymatic activity except in minimal concentrations when they are stimulatory. Also, water extracts from sound wood reduces enzymatic activity more than the water extract from decayed wood but in case of acetic acid extracts reverse trend is noticed (Narayanamurti, George and Verma 1952, Narayanamurti and Verma 1953).

Though fungi causing soft rot and brown rot are both brown-rotters, the difference between them in their action on the cell wall is attributed to a weaker production of enzymes by the soft rot fungi whose hyphae have to penetrate the cell wall and the lysis appears localised around the hyphae. In brown rot fungi, the hydrolytic enzymes decompose the polysaccharides of the secondary wall by diffusion into the wall. Among the ectoenzymes, cellulase is most important and which acts on the polysaccharide. The end product of degradation is glucose (Gadd 1957, Levi 1964) which is ultimately utilised by the fungus and at the same time oxidised into acids, possibly acetic acid and oxalic acid (Creamer 1950), the presence of which makes the decayed wood more acidic. Mannanase, xylanase and glycosidase are other enzymes known to degrade polysccharides (Keilich, Bailey and Liese 1970).

In white rot fungi, oxidating enzymes are the phenolases, the most important of which are thyrosinase, an intracellular and laccase, an extracellular enzyme. Thyrosinase oxidises both mono and polyphenols, while laccase only the polyphenols. The presence of these two enzymes, the polyphenoloxidases, are responsible for the production of a brown dark coloured zone around the mycelial mat of a white rot fungus when cultivated on agar containing phenolic compounds. Brown rot fungi do not form such coloured zones. Some soft rot fungi produce dark zones while others do not. This is one of the most useful methods developed by Bavendamm (1928) in distinguishing white rotters from brown rotters, when the former group causes oxidative browning of tannic or gallic acid agar, while the latter produces no such reaction. Preston and McLennan (1948) distinguished white rotters from brown rotters by growing fungi on media containing gentian violet or neutral red as dyes when white rot fungi discolour the dyes while the brown rot fungi do not. The discolouration of the dyes is due to extracellular enzymes in fungi causing white rots of the nature of phenol oxidase similar to laccase. Nobles (1958) developed a rapid test for extracellular oxidase when alcoholic gum guaiac solution added to actively growing cultures on malt agar develops a blue colour in 2–3 minutes with white rot fungi and not with brown rot fungi.

Brown-rotted wood shows a high degree of solubility in 1 per cent aqueous sodium hydroxide attributed largely to accumulation of carbohydrate degradation products as compared to white-rotted and soft-rotted wood where the alkali solubility is only slightly greater than that of sound wood (Campbell 1952, Savory and Pinion 1958).

DECAY IN STORAGE AND SERVICE AND ITS CONTROL

After conversion, decay, if present in the living tree, may continue to progress in the wood or may become inactive depending on the species of the fungus. When a healthy tree is converted, the natural protection to the tree afforded by the bark is lost and fungi may establish *de novo* in the wood. Fungi are most active under hot and humid conditions while during the winter, the activity is minimum. Therefore, winter felling is preferred and is actually practised in the plains in India though in the hills, where most places remain snow bound, felling is done during spring and early summer.

DECAY AFTER FELLING

The heartwood in different species may vary in its natural decay resistance. However, the sapwood of all species, timber of fir and spruce among conifers and soft hardwoods like semul, salai, white chuglam, papita and dhup which develop little or no heartwood and species with less durable heartwood may become decayed within a few months, unless adequate precautions are taken. There should be minimum delay in removing logs from the forest and this is achieved by improved and mechanised logging practices, opening up of forest roads and facilities for quick transport to the saw-mill. In case delay is anticipated, exposed ends should be treated with a preservative, followed by an end-coating to control drying checks. Fungi establish more readily at the ends and also progress more rapidly into the wood along the grain than from the sides and across the grain. In spruce and fir logs, decay may extend upto 1.3 m from the ends and only upto 2–3 cm from the sides. In these timbers, decay established through cut ends after conversion, since the causal fungi observed in the wood do not attack trees (Puri and Taneja 1967).

During transport through water courses, logs may be transported singly, usually in up-streams or built up into rafts in the plains. While the submerged portion of logs remains saturated with water and, therefore, remains free from decay, the exposed portion above the water line loses moisture and becomes vulnerable to decay. During drive through water where logs may be held up for long period, non-durable timbers like semul and fir may become severely decayed (Bakshi 1953, Puri and Taneja 1967). In such cases, debarking and spraying with a 'fixed' type of preservative within 48 hours after felling may be practised for perishable timbers.

DECAY IN STORAGE

In the storage yard, timber has to be stored for varying periods if conversion is delayed or the timber is not utilised early. Logs with intact bark do not readily lose moisture. The bark in some species is naturally resistant to fungus and insect attack and, therefore, the ends may be treated with a preservative. If the bark is not resistant to fungus and beetle attack, any chemical protection including coverage of the bark is ineffective (Scheffer 1969). In such cases, logs may be debarked facilitating rapid loss of moisture from the wood, particularly from the surface, thus minimising risk of fungus infection. Thus peeling of jack pine reduced the volume of visibly decayed wood to 2 per cent during 26 months of storage as compared with 72 per cent in rough bolts (Pascoe and Scheffer 1950).

Storage of logs may be done under one of the three conditions, namely, in water in log pond or on land under an intermittent spray of water or dry after a prophylactic treatment. In water storage, the logs should be submerged by having a series of girders above the logs and keeping the water level above the girders. In floating logs, the wood above the water level remains in an ideal air-moisture ratio for decay to set in. However, in water storage, many logs may sink and this adds to recovery cost. Also, it involves high handling cost in comparison to land storage which also facilitates sorting of logs for selective conversion. The protection afforded in ponding may be obtained overland if logs are under water spray (Roff and Dobie 1968), either from top of the pile or as sprinklers from the ground as adopted by Western India Match Company in their storage yards at Calcutta and Bareilly where, in addition, ponding is also practised. For veneer logs, storage under water or under a continuous water spray keeps them in essentially original condition. Spray need not be continuous but can be intermittent as long as it does not permit the log surfaces to get noticeably dry. Except in locations where rainy days are continuous, spraying has to be done even in wet weather. Spraying increases moisture content in wood near the surface and thereby affords protection by restricting diffusion of oxygen into logs. It also exerts a cooling effect. In sprayed logs, numerous innocuous microorganisms colonise the surface at the ends and may intercept for their own needs a significant amount of oxygen that might otherwise enter through the ends (Scheffer 1969). Spraying also controls beetle attack. Water spraying for log protection is highly effective and economical and should be increasingly adopted if water is plentiful for economical spraying. It will be particularly useful for higher grades of hardwoods such as plywood and where the quality timber is scarcer and costly.

Protective chemical treatments should be applied as sprays to the whole log as soon as possible after felling. It is inadvisable to apply the treatments until the logs have moved to the storage site because hauling operations may expose patches of untreated wood to infection. In warm and humid weather, the treatment has to be given within about 48 hours the logs are cut and before decay and particularly stain fungi establish because such treatments are only surface protectants. A mixture of pentachlorophenol in oil (24 pounds in 100 gallons) and benzene hexachloride in oil (3.5 pounds in 100 gallons, concentration reduced to 1.75 pounds in 100 gallons if protection from pinhole borer is not needed) are the principal fungicide and insecticide respectively (Scheffer 1969). Sodium salt of pentachlorophenol in water may be used though oil solutions are said to be more effective than water solutions. This may be followed by an end coating with thick, viscous, bituminous preparation of asphalt or coal tar pitch to prevent checks which may penetrate deep into the wood exposing untreated surface to infection. Mud plaster may be indigenously used to coat ends after felling in the forest. The chemical treatment can be most effective if applied to the log pile with a mist sprayer to allow the chemical to settle down gradually even to the core of the pile (Scheffer 1969). Chemical treatment may not be necessary if the logs are converted within the safe period of storage, which will depend on the species of wood. Beyond this period, chemical protection is necessary for perishable woods, while resistant woods are expected to last for longer duration without chemical protection.

Biological protection of logs where a relatively harmless fungus inoculated into the logs will protect them from attack by decay fungi appears attractive, simple and effective because the growth of inoculant into the wood provides deeper protection than can be obtained through chemical treatment (Scheffer 1969). Spraying freshly cut logs with a 4 per cent solution of sodium fluoride or 5 per cent ammonium bifluoride to induce growth of *Trichoderma viride* retards decay (Lindgren 1953).

The storage yard should be on a well-drained ground, hygienically clean and free from weeds. The pile should not be laid directly on the soil particularly in a heavy rainfall area. The floor may be made of concrete or brick soling, slightly sloping towards a drain to remove the dripping from rain or the pile may be raised off the ground on creosoted timbers on concrete pillars. The logs should be stored systematically and not in a haphazard manner. For rough bolts large bolt diameter, long log length and tight piling retard drying of wood and hinder diffusion of oxygen into the wood necessary for development of decay fungi (Lindgren 1951). In peeled logs, conditions that favour rapid drying of bolts, like removal of bark, open piling, small bolt diameter and splitting large bolts are suggested (Lindgren 1953). The lay-out of the yard should be well planned and each lot of logs must be kept separately and used in sequence according to the length of time a lot is held in storage. One lot of logs should not be piled on another that arrived earlier, since the top layers may be utilised leaving the bottom layers untouched. Decayed logs as evidenced from the presence of fruiting bodies should be converted as early as possible.

DECAY OF PULPWOOD

Not much data is available on the loss in pulpwood due to degradation by fungi and insects. The loss is estimated at 15–20 per cent after 3 years of storage in Sweden and 11–15 per cent in southern pines after 1 year storage and 10 per cent in jack pine after 28 months storage in U.S.A. (Zabel 1959). Losses are much higher in tropical countries with more favourable conditions of degradation.

Till late, bamboo has been and still remains the most important source of pulpwood in India, though with limited supply of bamboo and increasing demand of pulpwood for the growing paper industry, hardwoods are used. Bamboo culms are stored in large piles in paper mills which have to hold the stock in storage for several months resulting in stain and decay and also attack by pin-hole borers. In high rainfall areas, decay may be particularly severe in piles stacked directly on ground when about 0.5 m of the pile from the ground may become completely decayed. Gregarious flowering in bamboo occurs at long intervals of about 25–40 years. After flowering, the clumps and culms die. Over an extensive area, the dead bamboo cannot be utilised within a reasonable time nor it is possible to extract them out of the forest and store in the mill yard. Since starch in the culms is utilised during flowering, the dead bamboo is not attacked by borers but can become severely decayed depending on the length of time it remains standing in the forest or during storage in the mill. Dead, dry and decayed bamboo standing in the forest may become a source of fire hazard and fire will destroy the natural regeneration of bamboo that comes up profusely after flowering. Gregarious flowering in *Bambusa arundinacea* which started around Dandeli in 1958, continued till 1966 (Bakshi, Guha, Sujan Singh, Pant and Taneja 1968) and sometime thereafter and this bamboo constituted the chief source of pulp wood for the West Coast Paper Mills, Dandeli. However, the bamboo culms were in different stages of decay mostly by white rotters and to a small extent by brown rotters. In white rot in advanced stages of decay the fibro-vascular strands separate out like broom-sticks (Pl. **18**, 3) due to disintegration of the easily perishable ground parenchyma. As a result of decay, there is loss of wood substance which may be evaluated by determining loss in weight and determining loss in specific gravity, both methods yield equivalent results (Feist, Springer and Hajny 1972). Decay results in numerous bore holes in fibres (Pl. **18**, 6, 7). In a quantitative assessment, Bakshi, Guha, Sujan Singh, Pant and Taneja (1968) found that the area of a bore-hole averages 2.06 per cent of the total wall area of a fibre. The authors further observed that the surface area occupied by bore holes ranged from 0–10 per cent in fibres with 0–5 bore holes, 10–20 per cent with 6–10 bore holes, 20–30 per cent with 11–15 bore holes, and 30–35 per cent with 16–25 bore holes. Since bore holes represent wood substance lost, decay will lower the yield and quality of pulp. The results obtained by the authors are reproduced in Table 2.

TABLE 2

Sulphate digestion of healthy and decayed bamboo, showing pulp yield and degradation data

Specimens		Unbleached pulp yield			$KMnO_4$ No.	Bleach consumption as available chlorine on O.D. pulp %	Bleached pulp yield*		
		per 200 gm	per 1000 cc	% Loss on volume basis			per 200 gm	per 1000 cc	% Loss on volume basis
Healthy		84.0	83.2	—	15.2	6.80	68.8	68.1	—
Decayed	White rot								
	Initial	76.0	80.0	3.8	18.0	8.60	62.0	65.3	4.1
	Moderate	86.0	61.9	25.6	22.3	11.60	76.4	54.9	19.4
	Advanced	101.6	74.7	10.2	20.0	9.36	92.6	68.1	—
	Brown rot	34.0	29.0	65.1	31.10	—	Unbleachable		—

* On raw material (oven dry).

The results show that there is a considerable reduction in yield of unbleached pulp on volume basis using bamboo with moderate and advanced decay, reduction being more in the former than in the latter. In bleached pulp, however, the yield is considerably less in moderate decay, while in advanced decay, the yield is same as in healthy bamboo. This is explained by the fact that in advanced decay, the ground parenchyma which is normally washed out during chemical pulping is lost through decay and consequently the pulpwood substance for the same volume is more. However, during decay, the cell wall substance in fibre is lost and this may be as high as 50 per cent due to bore holes only as stated above. This necessitates the use of a greater volume of decayed bamboo and therefore, the use of more digesters to obtain a pulp yield comparable to healthy bamboo The prevalent practice of buying the pulpwood by weight and indicating the pulp yield on the same basis aids in the concealment of this important agency of pulpwood loss. When the pulpwood is sufficiently decayed and friable, much of it is lost during chipping and screening. Decay may persist in mechanical pulp where the temperature is not high to kill the fungus, unlike that in chemical pulp (Campbell 1952). The permanganate number of unbleached pulp is higher in decayed than in healthy bamboo indicating a greater quantity of residual lignin in the former which needs more bleaching chemicals. Brown rotted bamboo containing a high percentage of lignin is unbleachable and is, therefore, a pulping liability. The strength properties of standard sheets are considerably lower when decayed bamboo is used as compared to healthy (Table 3). If, however, bamboo is used soon after flowering and before decay sets in to be of any consequence, the yield and quality are not affected.

TABLE 3

Sulphate digestion of healthy and diseased bamboo showing strength properties

Specimens		Strength properties of standard sheets at 65% R.H. and 25°C		
		Breaking length (Schopper) metres	Tear factor (Marx Elmendorf)	Burst factor
Healthy		7890	53.5	47.5
Decayed wood	White rot			
	Initial	3630	51.6	23.3
	Moderate	3470	48.1	31.4
	Advanced	3720	33.0	23.0
	Brown rot	Unbleachable		

From the results of studies on the effect of one white rotter and one brown rotter on the chemical composition of the cell wall, Campbell (1952) concluded that decay resulted in serious loss in weight and bulking problem in digester, excessive loss of carbohydrates and acidity of decayed wood. Increased alkali consumption in case of brown rots is due to acidity of decayed wood caused by rapid depolymerisation of carbohydrates (Cowling 1961).

In stained pulpwood, the yield is not appreciably affected though strength properties of pulp may be slightly lower than those of healthy bamboo (Bakshi, Guha and Gupta 1960). However, pulp brightness is significantly decreased and there is increase in bleach consumption to the extent of 10 per cent to bring stained pulpwood to acceptable brightness. One advantage of mould and sap stain in pulpwood is that the infection may increase the penetration of chemicals used in the chemi-groundwood process (Zabel 1959).

CHIP DETERIORATION

Lower handling cost in storing chips as compared to roundwood is the main advantage of outside chip storage for pulpwood, which is increasingly used in Sweden, U.S.A., Canada and other countries. However, losses in wood substance in chip storage are high. Pulp yield on weight basis is not affected though lower brightness for unbleached sulphite pulp and higher consumption of bleaching chemicals following chip storage are reported (Bergman, Nilsson and Jerkeman 1970). Chip piles tend to heat up due to respiration of microorganisms that develop in the pile and may sometimes result in spontaneous fires (Feist, Springer and Hajny 1973). These necessitate protection of chips. Water spraying has not shown any advantage especially for long term storage (Scheffer 1969). Kraft green liquor affords protection from decay and also prevents heat release from chips (Springer, Eslyn, Zoch and Hajny 1969). A spray of sodium hydroxide at 40 g per litre of water protects chips from deterioration (Bergman, Nilsson and Jerkeman 1970). Propionic acid used for protection of stored grains prevents fungus attack in chips, self heating in storge piles at a concentration of 2.0 per cent in water and also prevents water pollution (Esyln 1973). After a review of major chemical treatments and the cost of application, Smith and Hatton (1971) concluded that chemical treatment of coniferous wood chip piles is not yet economically feasible.

DECAY IN SERVICE

Wood may be used outdoors or indoors. Decay does not occur in wood below about 20 per cent moisture and hence the necessity of using seasoned wood for interior use. If the decay is already established in logs, there is no reliable means short of kiln-drying that can guarantee against decay continuing in the wood following conversion. To keep the wood continuously dry, adequate measures may be taken like moisture-proofing, ventilation or use of treated and durable woods where timber is liable to wetting. Treatment may be done by dipping with oil-soluble wood preservative containing water-repellents and painting doors and window panels after treatment (Verrall 1953, Scheffer and Verrall 1973).

In exterior use, wood is subject to wetting during rains and therefore use of seasoned wood is not necessary. In outdoors, timber is exposed to high decay hazards particularly in ground contact as in poles and railway sleepers. The use of durable timbers and preservative treated wood is recommended. Chemical treatment of soil in critical locations is suggested (Zabel and George 1960), like soil around poles. For control of soft rots, resistant wood among conifers should be selected which may be further protected through chemical treatments.

PRESERVATIVE TREATMENT OF WOOD

Wood may be treated with a preservative chemical to increase its service life. This becomes particularly necessary when wood is used under conditions of high decay hazard. If the wood is treated with a sufficient chemical and its penetration is adequate, then the service life of the treated wood will be prolonged. Insufficient chemical in the treated wood and incomplete penetration of the chemical may not protect the wood from decay. Compared to preservative treated wood, untreated wood has a shorter service life necessitating frequent replacements and, therefore, the use of treated wood is more economical in the long run, particularly in species of low natural durability which after treatment may have a comparable service life to that of naturally durable timbers.

Wood preservatives

An ideal wood preservative should possess a high toxicity against fungi and insects attacking wood, low volatility and high resistance to leaching and ability to penetrate deeply into the wood, which depends on low viscosity at temperatures at which the preservative is applied. It should be non-corrosive to metals and reasonably safe to handle.

Three main groups of chemicals may be recognised, namely, preservative oils, chemicals soluble in water and those soluble in organic solvents. Among preservative oils, creosote obtained from coal tar by distillation has remained most effective preservative for timbers in outdoor use like railway sleepers, telegraph and transmission poles and for marine piling due to its high toxicity, resistance to leaching by water and low volatility. The water-soluble preservatives like zinc chloride, mercuric chloride copper sulphate are highly leachable. The addition of soluble chromate to solutions containing copper, arsenic or mercury enhances toxicity and brings about a high degree of fixation of chemicals in the wood and reduces corrosion of metals. Copper-chrome-arsenic preparations under the trade name "ascu" has been a very effective wood preservative. Among the chemicals dissolved in organic solvents, pentachlorophenol, zinc and copper naphthenates are widely used. The toxic substance being more or less insoluble in water is not readily leached out of the treated wood.

Methods of treatment

The preservatives may be applied to the timber either on the surface mostly as a prophylactic for short term protection or may be impregnated into the wood for long term protection. The impregnation may be done by open tank treatment at atmospheric pressure or may be carried out under pressure, either by the full cell process where the cells of the wood are filled with the preservative oils ; or by the empty cell process, where the free oil from the wood cells is mostly recovered and thus the cells become empty to some extent. In addition, the sap replacement or Boucherie process, involves sap replacement with a preservative solution.

LABORATORY EVALUATION OF WOOD PRESERVATIVES

The effectiveness of a wood preservative in protecting timber in service is best determined from field tests or service trials. However, it takes long time to obtain significant results from field trials. Therefore, laboratory tests are developed for rapid evaluation of a timber preservative as to its toxicity to fungi, which is one of the most important qualities of a wood preservative. However, a chemical may be toxic to wood destroying fungi and yet may not remain effective when impregnated into wood due to weathering, leaching and degradation due to moulds and decay due to attack by fungi tolerant to the preservative. Therefore, while toxicity of the chemical to wood destroying fungi is most important, the permanence of the chemical in the treated wood is almost equally important. Resistance to weathering is difficult to assess in laboratory tests where the methods used for artificial weathering do not simulate conditions obtainable in nature. It is also not possible to correlate laboratory tests with actual performance of a preservative in the field. All the same laboratory tests indicate whether the chemical should be further tested as a preservative in the field.

Preliminary screening of chemicals as wood preservatives is done on malt agar (Schmitz 1931), which determines the minimum concentrations inhibiting growth of test fungi in agar. It is most useful for determining the relative toxicity of substances of same general chemical nature and of similar solubility in water (Anon. 1961 a). The laboratory tests for determining the toxicity of a chemical wood preservative are conducted either by the "soil block method" (Anon. 1961) or by the "agar block method" (Anon. 1961 a). The relative advantages and disadvantages of the two methods are outlined by Duncan (1953). From these tests the minimum amount of a preservative that is effective in preventing decay under the conditions of the tests by a particular fungus, termed the threshold value, is determined. The amount of preservative in terms of kg per cubic metre of wood is referred to as the threshold retention. The threshold is determined by estimating the point at which the decay fungus does not cause any weight loss in the wood. The threshold value of a preservative in laboratory tests depends on three main factors, namely, the method employed, the test fungi used and the test timber. In general, the value obtained by the soil block method is slightly higher than that in agar block method. In laboratory evaluation of copper resinate as wood preservative, Sharma, Puri, Mukerji and Tewari (1964) observed threshold values using chir sapwood to be 0.21 and 0.16 lbs per cu ft by soil block and agar block methods respectively; while for sapwood of sal, these values were 0.16 and 0.145 lbs per cu ft. The threshold values of low temperature tar creosote as determined by the agar block method was found to be 1.18-1.57 lbs per cu ft for chir sapwood and this value was considerably higher (3.9 lbs/cu ft) when semul was used as the test timber (Bakshi, Puri, Mukerji and Sujan Singh 1961).

Fungi tolerant to chemical preservatives

Decay fungi differ in their relative resistance to different preservatives. *Lenzites trabea* is resistant to both sodium fluoride and zinc chloride;

Poria monticola least resistant to sodium fluoride and *Lentinus lepideus* to zinc chloride (Richards 1924, 1925). *Lenzites trabea* is tolerant to arsenic but susceptible to copper, the reverse being true for *Coniophora cerebella* (Baechler 1941). Cartwright and Findlay (1958) observed *Lenzites trabea* as highly resistant to arsenic, *Lentinus lepideus* resistant to creosote and tar oils, *Coniophora cerebella* resistant to zinc and cadmium compounds and *Poria vaporaria* resistant to copper. *Lenzites trabea* is less inhibited by pentachlorophenol than other test fungi (Duncan and Richards 1950, Zabel 1954). *Poria xantha* exhibits tolerance to copper (Hirt 1949). *Poria monticola* possesses a high tolerance to copper and zinc as resinates and also copper sulphate (Puri, Sharma and Bakshi 1959). At the lower concentration of the preservatives, the growth of the organism may be stimulated to some extent (Madho Singh 1961).

Tolerance of wood destroying fungi to chemicals is dependent upon the fungus and upon the amount of chemical to which the fungus is exposed. In copper-chromium-arsenate treated wood, the chemical is distributed in the secondary wall. *Poria monticola* and *Poria vaillantii*, both copper-tolerant, are able to detoxify copper-chromium-arsenate preservatives (Levi 1973). *Lentinus lepideus* is creosote tolerant and attacks wood impregnated with creosote upto 4 per cent, equivalent to 12 kg of creosote per cubic metre of wood (Cartwright and Findlay 1958).

Soft rot fungi as a group have generally higher tolerance than decay fungi and treated wood which may show complete freedom from attack from decay fungi may exhibit soft rot attack (Duncan 1960). The toxic limit for the soft rot fungus, *Chaetomium globosum*, in pentachlorophenol was considerably higher than for *Polyporus versicolor*, a highly phenol-tolerant decay fungus (Savory 1955). Many soft rot fungi were found more tolerant of sodium fluoride, sodium chromate, sodium arsenate and zinc chloride than were the Basidiomycetes tested (Duncan 1960). Non-decay organisms tolerant to creosote and copper-chromium-arsenate have been isolated from treated wood but their ability to detoxify the chemicals have not been demonstrated (Levi 1973).

Moulds resistant to treated woods

Wood-inhabiting fungi not necessarily responsible for major decay, may degrade a toxic compound into a less potent form thus rendering it less effective in protecting wood from decay by less tolerant decay fungi.

It is generally felt that in service, decay fungi may not attack freshly treated wood but do so when the wood is conditioned due to leaching or prior colonisation by moulds. In the United States, *Hormodendrum resinae* commonly occurs on treated wood with high concentration of creosote and coal tar (Christensen, Kaufert, Schmitz and Allison 1942) which may serve as a soul carbon or nitrogen source (Marsden 1954). Pine sapwood treated with different fungicides favours development of mould growth, a specific treatment supporting growth of specific moulds which are tolerant to respective fungicides (Verrall 1949). In the decay of fence posts treated with a water-soluble preservative mixture containing arsenic, it is suggested (Madho Singh 1961) that *Fusarium* associated with decayed wood may reduce the toxicity of the wood preservative thus facilitating subsequent attack by the decay fungus, *Coprinus micaceous*, which is less tolerant to the preservative. Stored pulp impregnated with phenyl mercury acetate is colonised by *Penicillium requeforti* which degrades the preservative by absorbing the biologically active mercury and thus renders the wood liable to decay by *Stereum sanguinolentum* and *Peniophora pitya* (Russel 1955). Duncan and Deverall (1964) showed that wood inhabiting fungi may change preservatives e.g. ammonical copper arsenate, fluor-chrome-arsenate-dinitrophenol, creosote or pentachlorophenol, degrading them so as to render them less effective in protecting the wood from decay by less tolerant fungi. Under laboratory conditions, non-wood destroying fungi are able to detoxify pentachlorophenol Levi (1973).

Wood plays an important role in the economy of a country. With rapid rise in world population, the amount of wood per capita is declining rapidly, particularly in the less industrialised regions. Also, the supply of traditional timbers is dwindling and there is a paucity of industrial woods, particularly in the developing countries. Much wood is lost through biological degradation but the appraisal of such losses is lacking to justify the prevention and preservation cost in controlling biological degradation of wood. There is, therefore, an urgent need for developing adequate research bases in timber pathology on national, regional and international levels. Also, there is a need for creating an awareness amongst wood users about wood losses in order that research results are utilized towards protection of wood from biological degradation.

REFERENCES

Agarwal, G.P. & Ganguli, G. (1959). A leaf spot disease of *Anogeissus latifolia* Wall. due to *Pestalotiopsis versicolor* (Speg.) Steyaert. *Curr. Sci.* **28**, 295–296.

Agarwal, G.P. & Hasija, S.K. (1961). Fungi causing plant diseases at Jabalpur (M.P.). V. Proc. natn. Acad. Sci., India. Sec. B. **31**, 99–108.

Agarwal, G.P. & Hasija, S.K. (1961 a). Fungi causing plant diseases at Jabalpur (M.P.). VI. Some cercosporae. Proc. natn. Acad. Sci., India. Sec. B. **31**, 355–359.

Aggarwal, K.L. (1933). *Fomes annosus* on deodar. *Indian Forester* **59**, 239–242.

Aggarwal, K.L. (1936). Revised working plan for the Pabar forests ; 1935–53, Lahore.

Aggarwal, K.L. (1949). Fourth working plan for the Kulu and Seraj forests, 1949–50 to 1979–80, 308 pp.

Agnihothrudu, V. (1960). *Meliola albizziae* Hansford & Dieghton from Assam. *Curr. Sci.* **29**, 149.

Ainsworth, G.C. (1961). Ainsworth & Bisby's dictionary of the fungi, Fifth Ed. 547 pp. Commonwealth Mycological Institute, Kew, Surrey.

Akai, S. & Takeuchi, T. (1954). On damping-off disease of some coniferous seedlings caused by soil-inhabiting fungi. *Ann. Phytopath. Soc.* Japan **18**, 122–124.

Alexopoulos, C.J. (1962). Introductory Mycology, Second Ed. 613 pp. John Wiley & Sons, Inc., New York, London.

Altona, T. (1926). [Infection of *Tectona grandis* L.f. (Djati) by *Corticium salmonicolor* B. et Br. (Djamoer oepas)]. Tectona **19**, 31–53. R.A.M. **5**, 457, 1926.

Anahosur, K H. (1968). Studies in Indian Phyllachoraceae. XI. *Sydowia* **22**, 166–169.

Anahosur, K.H. (1969). *Kamatella*, a new genus of the Sphaeropsidales. *Bull. Torrey bot. Club* **96**, 207–208.

Anahosur, K.H. (1969 a). Ascomycetes of Coorg (India). VII. *Sydowia* **23**, 57–62.

Anahosur, K.H. (1970). Ascomycetes of Coorg (India). III. *Sydowia* **24**, 169–172.

Anahosur, K.H. (1970 a). Ascomycetes of Coorg (India). II. *Sydowia* **24**, 177–182.

Anon. (1933). Insect transmission of spike disease. *Nature* **132**, 3337, 592–593.

Anon. (1950). List of common names of Indian plant diseases. *Indian J. agric. Sci.* **20**, 107–142.

Anon. (1952). The National Forest Policy of India, Manager of Publications, Delhi.

Anon. (1955). Eucalypts for planting. FAO For. Forest Prod. Stud. No. 11 (Translated from French).

Anon. (1956). Mortality in forest species. Proc. VIII. silvic. conf. 1951. pt. I, 143–154 ; pt. II, 324–333.

Anon. (1956 a). Plant disease survey for the twelve months ending 30th June, 1956. 26th Annual report N.S.W. Dep. of Agric.. Biological branch. Div. of Sci. services, 36 pp. (mimeographed.) R.A.M. **36**, 576, 1957.

Anon. (1960). Agriculture Research. Rep. Indian Council Agricultural Research, 1957–58. p. 25.

Anon. (1961). Standard method of testing wood preservatives by laboratory soil-block cultures. Amer. Soc. for testing materials (ASTM : D1413–61), 1013–1026.

Anon. (1961 a). Methods of test for toxicity of wood preservatives to fungi. British Standards 838, 161, 28 pp.

Anon. (1961 b). 100 years of Indian forestry. **2**, 381 pp. Manager of Publications. Delhi.

Anon. (1962). Indian Standard: IS : 401-1961. Indian Standard Code of practice for preservation of timber (revised) 1962. 28 pp. Indian Standard Institution, New Delhi.

Anon. (1963). FAO World Forest Inventory, Rome. pp. 113.

Anon. (1964). FAO/IUFRO symposium on internationally dangerous forest diseases and insects, recommendations. *Unasylva* No. 78, 143-145.

Anon. (1964 a). Boron toxicity in rubber. *F.R.I. Plant. Bull.* **73**, 76-78.

Anon. (1965). Herbicide research unit annual report 1964. Regional Research Centre, Univ. of West Indies, Trinidad, 9 pp.

Anon. (1966). International rules for seed testing. Proc. int. Seed Test. Ass. **31**, 1-152.

Anon. (1966 a). Himachal Forests. Souvenir. X Central Board of Forestry meeting, Simla, 124 pp.

Anon. (1967). Important forest insects and diseases of mutual concern to Canada, the United States and Mexico. North American Forestry Commission, FAO. 248 pp.

Anon. (1970). *Fomes annosus*, a fungus causing butt-rot, root-rot and death of conifers. Leafl. For. Commn. No. 5, 10 pp. H M.S.O., London.

Anon. (1971). Report of Forest Research Institute, Rotorua for the period 1 Jan. to 31 Dec., 1970. 102 pp. N.Z. Forest Serv. Wellington.

Anon. (1972). Report of the Director General of Forests for the year ended 31 March, 1972. 72 pp. N.Z. Forest Serv. Wellington.

Anon. (1972 a). A new nursery leaf spot disease of *Hevea* caused by *Cylindrocladium quinqueseptatum.* Planters' Bull. of the Rubber Research Institute of Malaya. No. 119, 55-56. R.P.P. **52**, no. 221, 1973.

Anon. (1974). Monthly statistics of the foreign trade in India 1971-72. March 1974. Manager of Publications, Govt. of India, New Delhi.

Anon. (1975). Rubber and its cultivation. Rubber Board, India, 110 pp.

Arthur, J.C. & Cummins, G.B. (1933). Rusts of the North-west Himalayas. *Mycologia* **25**, 397-406.

Ashton, D.H. & Macaulay, B.J. (1972). Winter leaf spot disease of seedlings of *Eucalyptus regnans* and its relation to forest litter. *Trans. Br. mycol. Soc.* **58**, 377–386. R.P.P. **52**, no. 258, 1973.

Baechler, R.H. (1941). Resistance to leaching and decay protection of various precipitates formed in wood by double diffusion. Proc. Am. Wood Preserv. Ass. **37**, 23–31.

Bagchee, K. (1929). Investigation on the infestations of *Peridermium complanatum* Barclay on the needles and of *Peridermium himalayense* n. sp. on the stem of *Pinus longifolia* Roxb. Pt. I. Distribution, pathological study of the infection and morphology of the parasite. *Indian Forest Rec.* (Botany series) **14**, 24 pp.

Bagchee, K. (1933). Investigations on the infestations of *Peridermium himalayense* Bagchee on *Pinus longifolia*. Pt. II. *Cronartium himalayense* n. sp. on *Swertia* spp. Distribution, Morphology of the parasite, pathological study of the infection, biological relationship with the pine rust and control. *Indian Forest Rec.* (Botany series) **18**, 66 pp.

Bagchee, K. (1941). Contributions to our knowledge of the morphology, cytology and biology of Indian coniferous rusts. Pt. I. *Cronartium himalayense* Bagchee and *Peridermium orientale* Cooke on *Pinus longifolia* Roxb. *Indian Forest Rec.* (N.S.) Botany No. **1**, 7, 247–266.

Bagchee, K. (1944). Blister rust menace of Chir. *Indian Forester* **70**, 323–325.

Bagchee, K. (1947). Pathological notes: No. 3. An unrecorded parasite of teak (*Tectona grandis* L.f.) reported from Dehra Dun, U.P. *Indian Forester* **73**, 332–334.

Bagchee, K. (1950). Contributions to our knowledge of the morphology, cytology, and biology of Indian coniferous rusts. Pt. II. Observations on the occurrence of *Cronartium ribicola* Fischer and *Peridermium indicum* Colley and Taylor on *Pinus excelsa* Wall. in India with reference to their distribution, pathology, inoculation experiments and comparative morphology. *Indian Forest Rec.* (N.S.) Botany IV, No. 1, 41 pp.

Bagchee, K. (1950 a). Contribution to our knowledge of morphology, cytology, and biology of Indian coniferous rusts. Pt. III. The biology of *Peridermium brevius* (Barc.) Sacc. (*Coleosporium barclayense* sp. nov. on *Senecio rufinervis* DC.), and *Melampsora oblonga* sp. nov. on the needles of *Pinus excelsa* Wall. *Indian Forest Rec.* (N.S.) Botany IV, No. 2, 43–64.

Bagchee, K. (1950 b). Progress of Forest Pathology in India during the quinquennium 1944–49. *Indian Forester* **76**, 219.

Bagchee, K. (1952). A review of work on Indian tree diseases and decay of timber and methods of control. *Indian Forester* **78**, 540–546.

Bagchee, K. (1953). The fungal diseases of sal (*Shorea robusta* Gaertn.). I. Leaf spot (*Cercospora* sp.), stem canker (*Macrophoma shoreae* sp. nov.), sooty moulds (*Capnodium* sp. and *Meliola* sp.) and root and stem rot (*Xylaria* spp.) of sal. *Indian Forest Rec.* (N.S.) Mycol. **1**, 11–23.

Bagchee, K. (1953 a). A new and noteworthy disease of gamhar (*Gmelina arborea* Linn.) due to *Poria rhizomorpha* sp. nov. *Indian Forester* **79**, 17–24.

Bagchee, K. (1954). New and noteworthy diseases of trees in India. Pit canker disease of siris (*Albizia procera* Benth.) due to *Fusarium solani* (Mart.) App. et Wr. sensu Snyder and Hansen. *Indian Forest Rec.* **1**, 41–52.

Bagchee, K. (1954 a). The fungal diseases of sal (*Shorea robusta* Gaertn.). II. Secondary parasites of sal. *Indian Forest Rec.* (N.S.) Mycol. 1, 97–184.

Bagchee, K. (1959). The fungal diseases of sal (*Shorea robusta* Gaertn.). V. The heart-rot of sal caused by *Trametes incerta* (Curry) Cooke. *Indian Forest Rec.* (N.S.) Mycol. 2, 61–69.

Bagchee, K. (1961). The fungal diseases of sal (*Shorea robusta* Gaertn.). IV. *Fomes caryophylli* (Rac.) Bres., a destructive heart rot of sal. *Indian Forest Rec.* (N.S.) Mycol. 2, 25–59.

Bagchee, K. & Bakshi, B.K. (1951). *Poria monticola* Murr. on chir (*Pinus longifolia* Roxb.) in India. *Nature* **167**, 824.

Bagchee K. & Bakshi, B.K. (1954). Studies on Indian Thelephoraceae. I. Some species of *Stereum*, *Peniophora* and *Corticium*. *Indian Forest Bull.* (N.S.) Mycol. No. 166, 1–11.

Bagchee, K. & Ujagar Singh (1960). List of common names of fungi attacking Indian forest trees, timber and the herbaceous and shrubby undergrowths and list of cultures of forest fungi. *Indian Forest Rec.* (N.S.) Mycol. I, 1954, 199–348.

Bailey, I.W. & Vestal, M.R. (1937). The significance of certain wood destroying fungi in the study of enzymatic hydrolysis of cellulose. *J. Arnold Arbor.* **18**, 196–205.

Bakshi, B.K. (1951). Mortality of *Casuarina equisetifolia* Forst. *Indian Forester* **77**, 269-276.

Bakshi, B.K. (1952). *Oedocephalum lineatum* is the conidial stage of *Fomes annosus*. *Trans. Br. mycol. Soc.* **35**, 195.

Bakshi. B.K. (1953). Decay and sap stain of timber, their causes and prevention. *Indian Forester* **79**, 108–114.

Bakshi, B.K. (1954). Wilt disease of shisham (*Dalbergia sissoo* Roxb.) due to *Fusarium solani* Sensu Snyder and Hansen. *Nature* **174**, 278.

Bakshi, B.K. (1955). Wilt disease of shisham (*Dalbergia sissoo* Roxb.). Behaviour of *Fusarium solani*, the wilt organism, in soil. *Indian Forester* **81**, 276-281.

Bakshi, B.K. (1955 a). Diseases and decays of conifers in the Himalayas. *Indian Forester* **81**, 779–797.

Bakshi, B.K. (1957 a). Diseases of khair (*Acacia catechu* Willd.) and their prevention. *Indian Forester* **83**, 41–46.

Bakshi, B.K. (1957 b). Wilt disease of shisham (*Dalbergia sissoo* Roxb.). The effect of soil moisture on the growth and survival of *Fusarium solani* in the laboratory. *Indian Forester* **83**, 505–511.

Bakshi, B.K. (1957 c). Heart-rots in relation to management of sal. *Indian Forester* **83**, 651–661.

Bakshi, B.K. (1957 d). Forest Pathology notes from India. Red spruce and mycorrhiza. *Commonw. phytopath. News* **3**, 43.

Bakshi, B.K. (1958). Farm forestry—pathological considerations. Proc. Farm For. symp. Dehra Dun. 55–57 pp.

Bakshi, B.K. (1963). Dying of sal. II. Role of *Hypoxylon*. *Indian Forester* **89**, 265–268.

Bakshi, B.K. (1963 a). Selected dangerous forest diseases in India. Internationally dangerous forest diseases. U.S D.A. For. Serv. Misc. Publ. 939, 55–61.

Bakshi, B.K. (1964). Trees deserve better treatment. *Indian Fmg.* **14**, **6–7**.

Bakshi, B.K. (1964 a). Known and potential hazards from stem diseases–heart-rots. FAO/IUFRO symposium on internationally dangerous forest diseases and insects. FAO/FORPEST–64, 1–8.

Bakshi, B.K. (1964 b). Present status of chir pine blister rust. *Indian Forester* **90**, 695–697.

Bakshi, B.K. (1964 c). Diseases of *Tectona grandis* (Teak). In Diseases of widely planted forest trees. FAO/FORPEST–64, U.S.D.A., 234–240.

Bakshi, B.K. (1967). Diseases of man-made forests. World symposium on man-made forests and their industrial importance. FAO. FO/MMF 67–15a/I, 639–644.

Bakshi, B.K. (1967 a). Accelerated laboratory investigations on durability of wood. PL 480 Final Technical Report (1962–67), 57 pp.+ appendix.

Bakshi, B.K. (1971). Indian Polyporaceae on Trees and Timber. 246 pp. Indian Coun. Agric. Res., New Delhi.

Bakshi, B.K. (1972). Relative blister rust resistance of native and introduced white pines in Asia. Biology of rust resistance in forest trees : Proceedings of a NATO-IUFRO Advanced Study Institute, 1969. U.S. Dept. Agr. Forest Serv. Misc. Publ. No. 221, 251–255.

Bakshi, B.K. (1974). Mycorrhiza and its role in forestry. PL 480 Project Report. 89 pp. F.R.I. & C., Dehra Dun.

Bakshi, B.K., Arora, K.K. & Sujan Singh (1957). Root diseases of shisham (*Dalbergia sissoo* Roxb). V. Incidence of diseases in relation to soil pH and soil texture. *Indian Forester* **83**, 555–558.

Bakshi, B.K. & Boyce, J.S. (1959). *Polyporus shoreae* root rot in sal. *Indian Forester* **85**, 656–658.

Bakshi, B.K. & Boyce, J.S. (1959 a). Water blister in teak. *Indian Forester* **85**, 589–591.

Bakshi, B.K. & Dobriyal, N.D. (1970). Effect of fungicides to control damping-off on development of mycorrhiza. *Indian Forester* **96**, 701–703.

Bakshi, B.K., Guha, S.R.D. & Gupta, S. (1960). Effects of fungal damage to bamboo on the yield and quality of pulp. *Res. and Ind.* **5**, 38–39.

Bakshi, B.K., Guha, S.R.D., Sujan Singh, Pant, R. & Taneja, K. (1968). Decay in flowered bamboo and its effect on pulp. *Indian Pulp Pap.* **22**, 1–4.

Bakshi, B.K. & Puri, Y.N. (1971). Dwarf mistletoe on blue pine in the western Himalayas and its control. *Indian Forester* **97**, 547–552.

Bakshi, B.K., Puri, Y.N., Mukerji, H.K. & Sujan Singh (1961). Evaluation of wood preservatives in the laboratory. Low temperature tar creosote. *J. Scient. ind. Res.* **20** D, 273-276.

Bakshi, B.K., Puri, Y.N., Sharma, R.P., Sujan Singh & Balwant Singh (1956). New and noteworthy diseases of trees in India. Diseases due to *Polyporus tulipiferae, Hypoxylon ustulatum* and *Polyporus rubidus. Indian Forester* **82**, 449–454.

Bakshi, B.K., Puri, Y.N., & Sujan Singh (1967). Natural decay resistance of Indian timbers. I. Introduction and method. II. Decay resistance of sal (*Shorea robusta* Gaertn.) & Teak (*Tectona grandis* L.). *Indian Forester* **93**, 305–328.

Bakshi, B.K., Reddy, M.A.R. & Balwant Singh (1970). An unrecorded decay in living deodar. *Indian Forester* **96**, 72–74.

Bakshi, B K., Reddy, M.A.R., Puri, Y.N. & Sujan Singh (1972). Forest disease survey (Final Technical Report). Forest Pathology Branch, F.R.I., Dehra Dun, 117 pp.

Bakshi, B.K., Reddy, M.A.R. & Sujan Singh (1975). Ganoderma root rot mortality in khair (*Acacia catechu* Willd.) in reforested stands. *Eur. J. of Forest Pathology* (in press*).

Bakshi, B.K., Reddy, M.A.R., Sujan Singh & Pandey, P.C. (1970). Disease situation in Indian forests. 1. Stem diseases due to *Corticium salmonicolor and Monochaetia unicornis. Indian Forester* **96**, 826–829.

Bakshi, B.K., Reddy, M.A.R., Thapar, H.S. & Khan, S.N. (1972). Studies on silver fir regeneration. *Indian Forester* **98**, 135–144.

Bakshi, B.K., Rehill, P.S. & Choudhury, T.G. (1963). Field studies on heart-rot in sal (*Shorea robusta* Gaertn.). *Indian Forester* **89**, 1–17.

Bakshi, B.K., Sehgal, H.S. & Balwant Singh (1969). Cultural diagnosis of Indian Polyporaceae. I. Genus *Polyporus. Indian Forest Rec.* (N.S.) Forest Pathology **2**, 205–244.

*Since published in the journal **6**, No. 2, 30-38, 1976.

Bakshi, B.K., Sen, M. & Balwant Singh (1970). Cultural diagnosis of Indian Polyporaceae. II. Genera *Fomes and Trametes*. *Indian Forest Rec.* (N.S.) Forest Pathology **2**, 245–276.

Bakshi, B.K. & Sujan Singh (1954). Wilt disease of shisham (*Dalbergia sissoo* Roxb.). I. Introduction and host-parasite relatonship. *Indian Forester* **80**, 316–322.

Bakshi, B.K. & Sujan Singh (1959). Root diseases of shisham (*Dalbergia sissoo* Roxb.). VIII. Inoculation studies on wilt. *Indian Forester* **85**, 415–421.

Bakshi, B.K. & Sujan Singh (1960). A new genus in plant rusts. *Can. J. Bot.* **38**, 259–262.

Bakshi, B.K. & Sujan Singh (1961). New and noteworthy records of some mildews and rusts on Indian trees. *Indian Forester* **87**, 542–545.

Bakshi, B.K. & Sujan Singh (1964). Mortality of *Eucalyptus citriodora* Hook. *Indian Forester* **90**, 15–18.

Bakshi, B.K. & Sujan Singh (1967). Rusts on Indian Forest Trees. *Indian Forest Rec.* (N.S.). Forest Pathology **2**, 139–198.

Bakshi, B.K. & Sujan Singh (1968). Dothistroma blight—A potential threat to *Pinus radiata* plantations in India. *Indian Forester* **94**, 824–825.

Bakshi, B.K. & Sujan Singh (1970). Heart rots in trees. *Int. Rev. For. Res.* **3**, 197–251. Academic Press, N.Y.

Bakshi, B.K. & Sujan Singh (1972). Susceptibility of exotic pines to *Cronartium himalayense*. *Indian Forester* **98**, 239–240.

Bakshi, B.K., Sujan Singh & Balwant Singh (1963). A re-examination of *Fomes lignosus* and *Polyporus zonalis*. *Trans. Br. mycol. Soc.* **46**, 426–430.

Bakshi, B.K., Sujan Singh & Ujagar Singh (1966). A new root rot disease complex in teak. *Indian Forester* **92**, 566–569.

Balwant Singh & Bakshi, B.K. (1961). New records of *Fomes* from India. *Indian Forester* **87**, 302–303.

Banerjee, S. & Ghosh, T. (1942). Preliminary report on the occurrence of higher fungi on bamboos in and about Calcutta. *Sci. & Cult.* **8**, 194.

Banerjee, S. & Mukerjee, N. (1954). A disease of *Lagerstroemia speciosa* (L.) Pers. caused by *Marasmius campanella* Holterm. *Indian Phytopath.* **7,** 118–139

Barber, C.A. (1903). Report of spike disease in sandal wood trees in Coorg. *Indian Forester* **29,** 21–31.

Barber, C.A. (1906). Studies in root parasitism. The haustorium of *Santalum album.* Memoirs Dept. Agric. India.

Barclay, A. (1886). On a Uredine affecting the Himalayan spruce-fir (*Abies smithiana* Forbes). *J. Asiatic Soc. Bengal* **55,** 1–11.

Barclay, A. (1890). Descriptive list of the Uredineae occurring in the neighbourhood of Simla (western Himalayas). III. *J. Asiatic. Soc. Bengal* **59,** 75–112.

Basham, J.T. (1958). Decay of trembling aspen. *Can. J. Bot.* **36,** 491–505.

Bavendamm, W. (1928). Uber das Vorkommen und den Nachweis von oxydasen bei holzzerstorenden Pilzen. Z. *Pflanzenkrank. Pflanzenschutz* **38,** 257–276.

Baxter, D.V. (1943). Pathology in forestry practice. 618 pp. John Wiley & Sons, Inc., New York.

Bazan de Segura, C. (1969). [Root rot of *Eucalyptus globulus*]. *Rev. for. Peru* **3,** 58–60. R.P.P. **51,** no. 1972, 1972.

Bazan de Segura, C. (1970). [Leaf spots caused by *Cylindrocladium scoparium* on *Eucalyptus* spp. in Turrialba, Costa Rica]. *Turrialba* **20,** 365–366. R.P.P. **51,** no. 2890, 1972.

Benito Martinez, J. (1952). Conservacion de Maderas en sus Aspectos Teorico, Industrial y Economicos. 550 pp. Instituto Forestal de Investigaciones y Experiencias, Madrid.

Berbee, J.G., Berbee, Flora & Brener, W.H. (1953). The prevention of damping-off of coniferous seedlings by pelleting seed. (Abs.) *Phytopathology* **43,** 466.

Bega, R.V. (1963). Symposium on root diseases of forest trees. *Fomes annosus.* Fifty-fourth. Ann. meeting Amer. Phytopath. Soc., Oregon. *Phytopathology* **53,** 1120–1123.

Benzian, B. (1965). Experiments on nutrition problems in forest nurseries. Vol. I. Bull. No. 37. For. Commn, Lond. H.M.S.O. 251 pp.

Bergman, O., Nilsson, T. & Jerkeman, P. (1970). Reduction of microbial deterioration in outside chip storage by alkali treatment. *Svensk Papperstidning* **73**, 653–666.

Bessey, E.A. (1939). A Text Book of Mycology. 495 pp. Blakiston's Son & Co., Inc., Philadelphia.

Bhatt, V.V., Patel, M.K. & Thirumalachar, M.J. (1955). Two new *Xanthomonas* species on legumes. *Indian Phytopath.* **8**, 136–142.

Bier, J. E. (1959 a). The relation of bark moisture to the development of canker diseases caused by native, facultative parasites. I. Cryptodiaporthe canker on willow. *Can J. Bot.* **37**, 229-238.

Bier, J. E. (1959 b). The relation of bark moisture to the development of canker diseases caused by native, facultative parasites. II. Fusarium canker on black cottonwood. *Can. J. Bot.* **37**, 781-788.

Bier, J.E. (1959 c). The relation of bark moisture to the development of canker diseases caused by native, facultative parasites. III. Cephalosporium canker on western hemlock. *Can. J. Bot.* **37**, 1140-1142.

Bier, J.E., Foster, R.E. & Salisbury, P.J. (1946). Studies in Forest Pathology. IV. Decay of sitka spruce on the Queen Charlotte Islands. *Dom. Canada Dep. Agric. Tech. Bull.* **56**, 35 pp.

Bier, J.E., Salisbury, P J. & Waldie, R.A. (1948). Studies in Forest Pathology. V. Decay in fir, *Abies lasiocarpa* and *A. amabilis*, in the upper Fraser region of British Columbia. *Dom. Canada. Dep. Agric. Tech. Bull.* **66**, 28 pp.

Bilgrami, K.S. (1963). Leaf spot diseases of some common ornamental plants. Proc. natn. Acad. Sci. India, Sec. B. **33**, 429-452.

Birch, T.T.C. (1936). *Diplodia pinea* in New Zealand. New Zeal. State For. Ser. Bull. No. 8, 32 pp.

Björkman, E. (1942). [Conditions favouring the formation of mycorrhizae in pine and spruce.] (German, English summary) Symb. bot. upsal. **6**, 1-191. R.A.M. **24**, 465, 1945.

Björkman, E. (1949). The ecological significance of the ectotrophic mycorrhizal association in forest trees. *Svensk bot. Tidskr.* **43**, 223-262.

Blair, I.D. (1943). Behaviour of the fungus *Rhizoctonia solani* Kuhn in the soil. *Ann. appl. Biol* **30**, 118-127.

Boedijn, K.B. & Steinmann, A. (1930). [On the *Helicobasidium* and *Septobasidium* species occurring on tea and other cultivated plants in the Dutch East Indies.] Arch. voor Theecult. Nederl.-Indie. **1**, 3-59. R.A.M. **9**, 562, 1930.

Bond, G. (1957). The development and significance of the root nodules of *Casuarina*. *Ann. Bot.* **21**, 373-380.

Bondartzev, A. (1953). [Bracket Fungi of the European part of the U.S.S.R. and of the Caucasus]. 1106 pp. U.S.S.R. Academy of Sciences Moscow, Leningrad.

Bose, S.R. & Sarkar, S.N. (1937). Enzymes of some wood-rotting Polypores. Proc. R. Soc., B. **123**, 193-213.

Boyce, J.S. (1954). Forest plantation protection against diseases and insect pests. F.A.O. For. Dev. Pap. No. 3, 41 pp.

Boyce, J.S. (1959). Observations on Forest Pathology in India. A report to the F.R.I., Dehra Dun, India and to the U.S. Technical Cooperation Mission to India. 28 pp.

Boyce, J.S. (1961). Forest Pathology. Third Ed. 572 pp. Mc Graw-Hill Book Co., Inc., New York.

Boyce, J.S. & Bakshi, B.K. (1959). Dying of sal. *Indian Forester* **85**, 585-588.

Boyce, J.S. & Wagg, J.W.B. (1951). Conk rot of old-growth Douglas-fir in western Oregon. *U.S. For. Serv. For. Prod. Lab. Bull.* **4**, 96 pp.

Braun, H.J. (1958). Untersuchungen über den Wurzelschwamm *Fomes annosus* (Fr.) Cooke. *Forstwiss. Zent Bl.* **77**, 65-88.

Briscoe, C.B. (1959). Early results of mycorrhizal inoculation of pine in Puerto Rico. *Caribb. Forester* **20**, 73-77.

Brown, W. (1915). Studies in the physiology of parasitism. I. *Ann. Bot.* **29**, 313-348.

Brown, W. (1916). Studies in the physiology of parasitism. III. *Ann. Bot.* **30**, 399-406.

Brown, W. (1917). Studies in the physiology of parasitism. IV. *Ann. Bot.* **31**, 489-498.

Brown, W. (1922). Studies in the physiology of parasitism. VIII, IX. *Ann. Bot.* **36**, 101-119, 285-300.

Browne, F.G. (1968). Pests and Diseases of Forest Plantation Trees. 1330 pp. Clarendon Press, Oxford.

Buchanan, T.S. (1964). Diseases of white (5-needle) pines. In Diseses of widely planted forest trees. FAO/FORPEST–64. U.S. Dep. Agric. (Forest Service), 100-120.

Buchanan, T.S. (1967). Diplodia twig blight of pine. In Important forest insects and diseases of mutual concern to Canada, the United States and Mexico. Dept. of Forestry and Rural Development, Canada, 189–191.

Buckland, D.C. & Marples, E.G. (1952). Management of western hemlock infested with dwarf mistletoe. *B.C. Lumberman* **36**, 50.

Buller, A.H.R. (1931). Researches on fungi. Vol. 4, 329 pp. Longmans, Green and Co., London, New York.

Butler, E.J. (1903). Report on 'spike' disease among sandal wood trees. *Indian Forester* **29**, Appendix Series No. 1, 1–11.

Butler, E.J. (1905). Some Indian forest fungi. *Indian Forester* **31**, 487–496.

Butler, E.J. (1918). Fungi and Diseases in Plants. 547 pp. Thacker, Spink & Co., Calcutta.

Butler, E.J. (1928). Report on some diseases of tea and tobacco in Nyasaland. Dept. of Agriculture, Nyasaland, 30 pp. R.A.M. **8**, 202, 1929.

Butler, E.J. & Bisby, G.R. (1960). The Fungi of India, 552 pp. Revised by R.S. Vasudeva. Indian Coun. Agric. Res., New Delhi.

Butler, F.C. (1953). Saprophytic behaviour of some cereal root rot fungi. I. Saprophytic colonization of wheat straw. *Ann. appl. Biol.* **40**, 284–297. II. Factors influencing saprophytic colonization of wheat straw. *Ann. appl. Biol.* **40**, 298–304.

Campbell, W.G. (1952). The biological decomposition of wood. In Wood Chemistry 2, 1061–1116. Edited by Wise, L.E. and Jahn, E.C. Rheinhold Pub. Corp., New York.

Cartwright, K. St. G. & Findlay, W.P.K. (1958). Decay of Timber and its Prevention. 294 pp. H.M. Stationery Office, London.

Cervinkova, H. & Temmlova, B. (1966). [Diagnosis of internal decay in stems of standing trees]. *Lesn. Cas.* **12**, 1033–1046.

Champion, H.G. (1922). Notes on death of chir (*Pinus longifolia*) poles in the Almora plantations of Kumaon. *Indian Forester* **48**, 168-174, 232-246.

Champion, H.G. (1933). Regeneration and management of sal (*Shorea robusta* Gaertn. f.). *Indian Forest Rec.* **19**, 1-159.

Champion, H.G. & Seth, S.K. (1968). A revised survey of the forest types of India. 404 pp. Manager of Publications, Delhi.

Chanturiya, N.N. & Gvinepadze', M.S. (1972). [Trials of antibiotics in the control of leaf spot of mulberry]. *Mikologiya i Fitopatologiya* **6**, 533-535. R.P.P. **52**, no. 2992, 1973.

Chapman, A.D. & Scheffer, T.C. (1940). Effect of blue stain on specific gravity and strength of Southern pine. *J. agric. Res.* **61**, 125-133.

Chatterji, A.L. (1912). A new species of mildew. *Indian Forester* **38**, 28-30.

Chee, K.H. (1969). Variability of *Phytophthora* species from *Hevea brasiliensis*. *Trans. Br. mycol. Soc.* **52**, 425-436.

Chee, K.H. (1971). A new disease of *Hevea brasiliensis* caused by *Fusarium solani* and *Botryodiplodia theobromae*. *Pl. Dis. Reptr.* **55**, 152-153.

Chesters, C.G.C. (1948). A contribution to the study of fungi in the soil. *Trans. Br. mycol. Soc.* **30**, 100-117.

Chiddarwar, P.P. (1959). Contributions to our knowledge of Cercosporae of Bombay State. I. *Sydowia* **13**, 152-163.

Chiddarwar, P.P. (1959 a). Contributions to our knowledge of the Cercosporae of Bombay State. II. *Indian Phytopath.* **12**, 111-121.

Chiplonkar, Alaka (1969). New phytopathogenic Fungi Imperfecti from Maharashtra (India). *Sydowia* **23**, 106-109.

Chona, B.L., Kapoor, J.N. & Gill, H.S. (1960). Studies on powdery mildews from India. I. *Indian Phytopath.* **13**, 72-75.

Chowdhury, K.A. (1942). How to distinguish between the sapwood and heartwood of sal (*Shorea robusta*). *Indian Forest Bull.* **115**, 1-3.

Christensen, C.M. (1973). Loss of viability in storage : microflora. *Seed Sci. and Technol.* **1**, 547-562.

Christensen, C.M., Kaufert, F.H., Schmitz, H. & Allison, J.L. (1942). *Hormodendrum resinae* Lindau, an inhabitant of wood impregnated with creosote and coal tar. *Amer. J. Bot.* **29**, 552-558.

Cleland, J.B. (1935). Toadstools and Mushrooms and other Larger Fungi of South Australia. Part II, pp. 179-362. Published by favour of the Honourable the Premier (Hon. R.L. Butler, M.P.).

Cobb, F.W., Jr. & Schmidt, R.A. (1964). Duration of susceptibility of eastern white pine stumps to *Fomes annosus*. *Phytopathology* **54**, 1216-1218.

Cobb, F.W., Jr. & Barbar, H.W., Jr. (1968). Susceptibility of freshly cut stumps of redwood, Douglas fir and Ponderosa pine to *Fomes annosus*. *Phytopathology* **58**, 1551-1557.

Cochrane, V.W. (1958). Physiology of Fungi. 524 pp. John Wiley and Sons, Inc., New York.

Coleman, L.C. (1917). Spike disease of sandal. *Dept. Agr., Mysore State, Myc. Series Bull.* **3**, 52 pp.

Coleman, L.C. (1932). The transmission of sandal spike. *Indian Forester* **49**, 6-9.

Colley, R.H. & Taylor, M.W. (1927). *Peridermium kurilense* Diet. on *Pinus pumila* Pall. and *Peridermium indicum* n. sp. on *Pinus excelsa* Wall. *Jour. Agric. Res.* **34**, 327-330.

Comic, B. (1957). Neka zapazanja o pojavi *Dasyscypha willkommii* (Hart.) Rehm. na. arisu i *Trametes radiciperda* Hartig na smrci u sumama Mojstrane. *Sumarstvo* **10**, 629-631.

Corbett, N.H. (1963). Anatomical, ecological and physiological studies on microfungi associated with decaying wood. Ph. D. Thesis. University of London. (Quoted from Levy & Stevens 1966).

Corbett, N.H. (1965). Micro-morphological studies on the degradation of lignified cell walls by Ascomycetes and Fungi Imperfecti. *J. Inst. Wood Sci.* No. 14, 18-29.

Cowling, E.B. (1961). Comparative biochemistry of the decay of sweet-gum sapwood by white-rot and brown-rot fungi. U.S. Dep. Agric. Tech. Bull No. 1258, 79 pp.

Creamer, G.B. (1950). Wood decay. *Pulp Pap. Mag. Canada* **51,** 86-90.

Critchfield, W.B. & Little, E.L., Jr. (1966). Geographic distribution of the pines of the world. U.S. Dept. Agr. Forest Service. Misc. Publ. 991, 97 pp.

Cromer, D.A.N. (1935). The significance of mycorrhiza of *Pinus radiata. Bull. For. Bur. Aust.* **16,** 1-19.

Cunningham, G.H. (1965). Polyporaceae of New Zealand. 304 pp. N.Z. Dep. Sci. Ind. Res. Bull. **164**.

Da Costa, G.C. & Mundkur, B.B. (1948). A revision of the Genus *Phyllosticta* in India. Proc. natn. Inst. Sci. India **14**, 55-63.

Daft, M.J. & Nicholson, T.H. (1966). Effect of *Endogone* mycorrhiza on plant growth. *New Phytol.* **65,** 343–350.

Damle, Kamala (1960). *Uncinula tectonae* Salmon on *Tectona grandis* L. *J.Indian bot. Soc.* **39,** 243–258.

Das, S. (1958). A Cercospora leaf-spot of *Aegle marmelos* Corr. *Sci. Cult.* **24,** 140-141.

Davis, D. (1933). *Polyporus shoreae* (sal root fungus) in Bahraich Division. *Indian Forester* **59**, 507–512.

Day, W.R. (1928). Damage by late frost on Douglas fir, Sitka spruce and other conifers. *Forestry* **2,** 19–30.

Day, W.R. (1959). Observations on eucalypts in Cyprus. I. The character of gum veins and anatomical indications for their origin. *Emp. For. Rev.* **38,** 35–44. II. Root development in relation to soil conditions. *Ibid.* **38,** 186-197.

Davidson, D.A.G. (1945). Loranthus attack in sal plantations. *Indian Forester* **71,** 181–182.

Deogun, P.N. (1943). Relation between roots and methods of irrigation in dry tracts of Punjab. *Indian Forester* **69,** 97–106.

Desai, S.G., Gandhi, A.B., Patel, M.K. & Kotasthanae, W.V. (1966). A new bacterial leaf-spot and blight of *Azadirachta indica* A. Juss. *Indian Phytopath.* **19,** 322–323.

DeVay, J.E., Sinden, S.L., Lukezic, F.L., Werenfels, L.F. & Backman, P.A. (1968). Poria root and crown rot of cherry trees. *Phytopathology* **58,** 1239–1241.

Dijkstra, J. (1968). The occurrence of inclusion bodies in leaf epidermis cells of sandal affected with spike disease. *Neth. J. Pl. Path.* **74,** 101–105. R.A.M. **47,** no. 2855, 1968.

Dimitri, L. (1963). Untersuchungen über die Ausbreitung von *Fomes annosus* (Fr.) Cooke. *Phytopath. Z.* **48,** 349–369 (English summary).

Dimitri, L. (1966). Ausbreitung des Wurzelschwammes (*Fomes annosus* (Fr.) Cooke) und seine Beteiligung an der Rotfaule der Fichte. *Forst.-u. Holzw.* **21,** 191–194. R.A.M. **45,** no. 2266, 1966.

Dimond, A.E. (1955). Pathogenesis in wilt diseases. *A. Rev. Pl. Physiol.* **6**, 329–350.

Dinesh Kumar, Sujan Singh, Thapar, H.S. & Misra, B.M. (1968). A preliminary note on mycorrhiza formation in chir pine on sand at different N, P, K levels. *Indian Forester* **94,** 247–250.

Doo, S.C. (1968). Bacterial wilt of teak seedlings. *Union of Burma, J. of Life Sciences* **1**, p. 43.

Duncan, C.G. (1953). Soil-block and agar block techniques for evaluation of oil-type wood preservatives : creosote, copper naphthenate and pentachlorophenol. For. Path. spec. release. No. 37, U.S.D.A. Pub.

Duncan, C.G. (1960). Soft-rot in wood and toxicity studies on causal fungi. Proc. Am. Wood Preserv. Ass. **56,** 27–35.

Duncan, C.G. & Deverall, F.J. (1964). Degradation of wood preservatives by fungi. *Appl. Microbiol.* **12,** 57–62.

Duncan, C.G. & Richard, C.A. (1950). Evaluating wood preservatives by soil block tests. 1. Effect of the carrier on pentachlorophenol solution. 2. Comparison of a coal tar creosote, a petroleum containing pentachlorophenol or copper naphthenate and mixture of them. Proc. Wood Preserv. Ass. **46,** 131–145.

Ehrlich, J. (1941). Etiological terminology. *Chronica bot.* **6,** 248–249.

Eldin, H.L. (1964). Forestry practice. Bull. For. Commn. No. 14, Lond.

Elmer, G.L., Gosnell, J.M. & Stark, J.M. (1963). The role of boron and rainfall on the incidence of a wattle die-back in East Africa. *E. Afr. agric. for. J.* **29,** 31–38. R.A.M. **43,** no. 841, 1964.

Eslyn, W.E. (1959). Radiographical determination of decay in living trees by means of the thulium x-ray unit. *Forest Sci.* **5,** 37–47.

Eslyn, W.E. (1973). Propionic acid—a potential control for deterioration of wood chips stored in outside piles. *Tappi* **56,** 152–153.

Etheridge, D.E. (1956). Decay in subalpine spruce on the Rocky Mountain Forest Reserve in Alberta. *Can. J. Bot.* **34,** 805–816.

Etheridge, D.E. (1958). Decay losses in subalpine spruce (*Picea glauca, P. mariana* and *P. engelmannii*) on the Rocky Mountain Forest Reserve in Alberta. *For. Chron.* **34,** 116–131.

Etheridge, D.E. (1965). Report to the Govt. of Tanzania on forest tree diseases. FAO Expand. tech. Assist. Prog. Rep. 2056, 23 pp. (cyclostyled). R.A.M. **46,** no. 1745, 1967.

Evans, Elfed (1968). Plant Diseases and their Chemical Control. 286 pp. Blackwell Scientific Publications, Oxford & Edinburgh.

Faan, H.C., Lai, W.J., Chan, D.S. & Tom, W.F. (1964). A preliminary study of the incidence and transmission of mulberry dwarf disease in Kwangtung province. *Acta phytopath. sin.* **7,** 151–156.

Fassi, Bruno (1964). Evolution du pourridie blanc, du au *Fomes lignosus* (Kl.) Bres., dans une plantation de *Hevea brasiliensis* amenagee immediatement apres l'abattage de la foret. Publications de L'institut national pour L'etude agronomique due Congo. Serie scientifique No. 105, 1–54.

Feist, W.C., Springer, E.L. & Hajny, G.J. (1972). Determining loss of wood substance in outside chip storage: A comparison of two methods. U.S.D.A. For. Serv. Res. Pap. FPL 189, 9 pp.

Feist, W.C., Springer, E.L. & Hajny, G.J. (1973). Spontaneous heating in piled wood chips—contribution by bacteria. *Tappi* **56,** 148–151.

Figueiredo, M.B. & Namekata, T. (1967). [Record of *Calonectria quinqueseptata* n. sp., perfect state of *Cylindrocladium quinqueseptatum* on *Annona squamosa* and *Eucalyptus* sp.]. *Arq. Inst. biol. S. Paulo,* **34,** 91–96. R.A.M. **47,** no. 765, 1968.

Fischer, C.E.C. (1918). Cause of spike disease of sandal (*Santalum album* L.). *Indian Forester* **44,** 570–575.

Foott, J.H., Hendrickson, A.H. & Wilson, E.E. (1955). Walnut branch wilt. *Calif. Agric.* **9,** 11. R.A.M. **35,** 245. 1956.

Foster, R.E. (1954). Decay of alpine fir in the upper Fraser region. *Bi-mon. Prog. Rep. Div. Forest Biol., Ottawa* **10,** 93.

Foster, R.E. (1964). Known and potential hazards from root rot. FAO/IUFRO symposium on internationally dangerous forest diseases and insects, Oxford (cyclostyled), 11 pp.

Foster, R.E., Craig, H.M. & Willis, G.W. (1954). Studies in forest pathology. XII. Decay of western hemlock in the upper Columbia region, British Columbia. *Can. J. Bot.* **32,** 145–171.

Foster, R.E. & Johnson, A.L.S. (1963). The significance of root rot and frost damage in some Douglas fir plantations. *For. Chron.* **39,** 266–272.

Fox, R.A. (1964). A report on a visit to Nigeria undertaken to make preliminary study of root diseases of rubber. Docum. Res. Arch. Rubb. Res. Inst. Malaya **27,** 34 pp. (cyclostyled). R.A.M. **43,** no. 3003, 1964.

Fresa, R. (1958). [The presence of *Sphaceloma meliae* in the Beadtree (*Melia azedarach*)]. *Rev. Invest. agric., B. Aires,* **12,** 409–412. R.A.M. **38,** 548, 1959.

Froelich, R.C., Dell., T.R. & Walkinshaw, C.H. (1965). Factors associated with occurrence of *Fomes annosus* in pine plantations. *Phytopathology* **56,** 1058–1059.

Gadd, O. (1957). Wood decay resulting from rot fungi. *Paper and Timber* **39**, 363–374.

Garrett, S.D. (1944). Root Disease Fungi. 177 pp. Chronica Botanica Waltham, Mass., U.S.A.

Garrett, S.D. (1960). Biology of Root-infecting Fungi. 293 pp. Cambridge Univ. Press.

Garud, A.B. (1970). Some follicolus Fungi Imperfecti from India. *Sydowia* **24**, 69–72.

Gaumann, E. (1954). Toxins and plant disease. *Endeavour* **13**, 198–204.

Gaumann, E.A. & Dodge, C.W. (1928). Comparative Morphology of Fungi. 701 pp. McGraw-Hill Book Co., New York.

George, K. (1966). Selective control of Loranthus on teak. *Curr. Sci.* **35**, 444.

Gerdemann, J.W. & Trappe, J.M. (1974). The Endogonaceae in the Pacific Northwest. Mycol. mem. 5, 76 pp. N.Y. bot. Gdns.

Gerdemann, J.W. & Nicolson, T.H. (1963). Spores of mycorrhizal *Endogone* species extracted from soil by wet sieving and decanting. *Trans. Br. mycol. Soc.* **46**, 235–244.

Gibson, I.A.S. (1958). Dead top in Kenya pine plantations. *E. Afr. agric. J.* **24**, 47–52.

Gibson, I.A.S. (1964). The impact of disease on forest production in Africa. FAO/IUFRO symposium on internationally dangerous forest diseases and insects, Oxford. (cyclostyled).

Gibson, I.A.S. (1972). Dothistroma blight of radiata pine. *A. Rev. Phytopath.* **10**, 51–72.

Gibson, I.A.S., Christensen, P.S. & Munga, F.M. (1964). First observations in Kenya of a foliage disease of pines caused by *Dothistroma pini* Hulbary. *Commonw. For. Rev.* **43**, 31–48.

Gibson, I.A.S. & Corbett, D.C.M. (1964). Variation in isolates from *Armillaria* root diseases in Nyasaland. *Phytopathology* **54**, 122–123.

Gill, H.S., Munjal, R.L. & Chona, B.L. (1961). Studies on powdery mildews from India. II. *Indian Phytopath.* **14**,174–178.

Gill, L.S. & Andrews, S.R. (1942). Behaviour of dwarf mistletoe of western conifers in relation to control. *Phytopath.* **32**, 21–22.

Gill, L.S. & Hawksworth, F.G. (1961). The mistletoes—a literature review. U.S.D.A. tech. bull. No. 1242, 87 pp.

Gilmour, J.W. (1958). Chlorosis of Douglas fir (*Pseudotsuga taxifolia*). *N.Z. Jl.* **71**, 94-106.

Glasson, A.K. (1922). Mortality of sal in Buxa Division, Bengal. *Indian Forester* **48**, 22–31.

Golato, C. (1970). A serious disease of *Anacardium occidentale* in Tanzania. *Rev. Agric. subtrop. trop.* **44**, 334–340. R.P.P. **50**, no. 3059, 1971.

Goncalves, J.R.C. (1970). Canker of grafted rubber. *Instituto de Pesquisas e Experimentacao Agropecurias do Norte, Fitotecnia.* **1, 49–52.** R.P.P. **52**, no. 220, 1973.

Goncharenko, L.K. (1968). [The incidence of various causal agents of mulberry root collar infection and their relation to different conditions of nutrition.] [*Shovkivnitstvo. Resp. mizhvid Tem. nauk. Zb.*]1968 (4), 2631. R.A.M. **48**, no. 1242, 1969.

Gorrie, R.M. (1929). A destructive parasite of the Himalayan blue pine. *Indian Forester* **55**, 613–617.

Gosselin, R. (1944). Studies on *Polystictus circinatus* and its relation to butt rot of spruce. *Farlowia* **1**, 525–568.

Greaves, H. & Levy, J.F. (1965). Comparative degradation of the sap wood of Scots pine, beech and birch by *Lenzites trabea, Polystictus versicolor, Chaetomium globosum* and *Bacillus polymyxa. J. Inst. Wood Sci.* **15**, 55–63.

Greig, B.J.W. & Burdekin, D.A. (1970). Control and eradication of *Fomes annosus* in Great Britain. Proc. Third Int. Conf. on *Fomes annosus*, Aarhus, Denmark. 21–23.

Gremmen, J. (1963). De biologische Bestrijding van de Wortelzwan, *Fomes annosus* (Fr.) Cke. door Middel van *Peniophora gigantea* (Fr.) Massee (Eng. summary). *Tevens Verschenen in Let. Ned. Boschb Tijdschr.* **35**, 356–367.

Griffin, D.M. (1958). Influence of pH on the incidence of damping-off. *Trans. Br. mycol. Soc.* **41**, 483–490.

Gruschow, G. & Trousdell, K.B. (1958). Incidence of heart rot in mature loblolly pine in coastal North Carolina. *J. For.* **56,** 220–221.

Gwynne-Vaughan, H.C.I. & Barnes, B. (1930). The Structure and Development of the Fungi. 384 pp. Cambridge University Press, London, New York.

Hacskaylo, E. (1958). Mycorrhiza : what they are, what they do. *Natn. hort. Mag.* **37,** 119–122.

Hacskaylo, E. & Palmer, J.G. (1957). Effects of several biocides on growth of seedling pines and incidence of mycorrhizae in field plots. *Pl. Dis. Reptr.* **41,** 354–358.

Hacskaylo, E. & Snow, A.G. (1959). Relation of soil nutrients and light to prevalence of mycorrhizae. Northeast. For. Serv. Sta., Rept. 125, 1–13.

Hafiz Khan, A. (1910). Root infection of *Trametes pini* (Brot.) Fr. *Indian Forester* **36,** 559–562.

Hafiz Khan, A. (1928). Inoculation of chir (*Pinus longifolia*) with *Coleosporium campanulae* (Pers.) Lev. on *Campanula canescens* Wall. & *Coleosporium inulae* (Kunze) Ed. Fisch. on *Inula cappa* DC. *Indian Forester* **54,** 176–178.

Hahn, G.G. (1949). Further evidence that immune *Ribes* do not indicate physiologic races of *Cronartium ribicola* in N. America. *U.S. Dept. Agric. Pl. Dis. Reptr.* **33,** 291–292.

Haines, H.H. (1917). The unsoundness of sal. *Indian Forester* **43,** 306–312.

Handley, W.R.C. (1963). Mycorrhizal associations and *Calluna* heathland afforestation. *Bull. For. Commn. Lond.* **36,** 1–70.

Hansen, T.S., Kenety, W.H., Wiggin, G.H. & Stakman, E.C. (1923). A study of the damping-off disease of coniferous seedlings. *Minnesota Agric. Exp. Stn. Tech. Bull.* **15,** 28 pp.

Hansing, Earl D. (1970). Control of seed-borne fungi with systemic fungicides. Proc. int. Seed Test. Ass. **35,** 815–820.

Harley, J.L. (1940). A study of the root system of the beech in woodland soils with special reference to mycorrhizal infection. *J. Ecol,* **28,** 107–117.

Harley, J.L. (1972). The Biology of Mycorrhiza. 334 pp. Leonard Hill Ltd., London.

Harley, J.L. & Loughman, B.C. (1963). The uptake of phosphate by excised mycorrhizal roots of the beech. IX. The nature of phosphate compounds passing into the host. *New Phytol.* **62,** 350–359.

Harley, J.L. & Waid, J.S. (1955). The effect of light upon the roots of beech and its surface population. *Pl. Soil* **7,** 96–112.

Hartley, C. (1921). Damping-off in forest nurseries. *U.S. Dep. Agri. Bur. Plant Ind., Bull.* **934,** 99 pp.

Hartley, C. (1939). The clonal variety for tree planting : Asset or liabilities ? *Phytopathology* **29,** 9.

Hartley, C. & Boyce, J.S. (1933). The progress of forest pathology. 73rd Congr. 1st Sess., Senate Doc. **12,** 695–722.

Harris, H.A. (1934). Frost ring formation in some winter-injured deciduous trees and shrubs. *Am. J. Bot.* **21,** 485–498.

Hashimoto, H. (1968). Studies on the control of the anthracnose of *Acacia dealbata* caused by *Glomerella cingulata* Stonem. *Bull. Fukuoka. Ken. For. Exp. Stn.* **20,** 29 pp.

Hawker, L.E. (1950). Physiology of Fungi. 360 pp. University of London Press, Ltd.

Hawksworth, F.G. & Wiens, D. (1972). Biology and Classification of Dwarf Mistletoes (*Arceuthobium*). U.S.D.A. Agric. Handbook No. 401, 234 pp.

Heath, R.G. (1956). Annual Report of the Dept. of Agriculture, Malaya, for the year 1955, 100 pp. R.A.M. **37,** 8, 1958.

Heimburger, C. (1972). Breeding white pines for resistance to blister rust at the interspecies level. Biology of rust resistance in forest trees. Proc. NATO–IUFRO advanced study Inst., 1969 ; U.S.D.A. Misc. Pub. **1221,** 541–549.

Hemenway, A.F. (1926). Late frost injury to some trees in central Kentucky. *Am. J. Bot.* **13,** 364–366.

Hepting, G.H. (1971). Diseases of forest and shade trees of the United States. U.S.D.A. Handbook No. 386, 243–245.

Hepting, G.H. & Jemison, G.M. (1958). Forest protection. Timber resources for America's future. Forest Resour. Rept., U.S. Dept. Agric. For. Serv. **14,** 185–220.

Hirt, R.R. (1949). An isolate of *Poria xantha* on media containing copper. *Phytopathology* **39,** 31–36.

Hocking, D. (1966). Root rot of teak. 2. Further field observations, with new records of *Armillaria mellea.* Misc. Rep. Trop. Pestic. Res. Inst. 574 Mycol. Note. (For. Path. Div. E.A.A.F.R.O. 44) 4 pp. (Cyclostyled). R.A.M. **46,** no. 1348, 1967.

Hocking, D. (1968). Stem canker and pink stain of teak in Tanzania associated with *Fusarium solani.* *Pl. Dis. Reptr.* **52,** 628–629.

Hocking, D. & Jaffer, A.A. (1966). Field observations on root-rot of Teak and nursery disorders. Misc. Rep. (East Africa) Trop. Pestic. Res. Inst. Arusha. No. 567, 12 pp.

Hocking, D. & Jaffer, A.A. (1966 a). Field observations on root rot of Teak and nursery diseases. Mycol. Note. E. Afr. Agric. For. Res. Orgn. No. 41, 6 pp.

Hocking, D. & Jaffer, A.A. (1967). Field observations on root rot of Teak in Tanzania. *Pl. Prot. Bull. F.A.O.* **15,** 10–14.

Hodges, C.S. Jr. (1964). Seed and seedling diseases of forest trees of the world. FAO/IUFRO symposium on internationally dangerous forest diseases and insects, 1–8, Oxford.

Hodges, C.S. (1969). Modes of infection and spread of *Fomes annosus.* *A. Rev. Phytopath.* **7,** 247-266.

Hodges, C.S. & May, L.C. (1972). A root disease of pine, *Araucaria* and *Eucalyptus* in Brazil caused by a new species of *Cylindrocladium.* *Phytopathology* **62,** 898–901.

Hodges, C.S. & Reis, M.S. (1975). A canker disease of *Eucalyptus* in Brazil caused by *Diaporthe cubensis* Bruner. FAO-FORPEST, DI/ 75-16-1.

Hodgson, C.M. (1918). Spiked sandal wood. *Indian Forester* **44,** 66-71.

Hole, R.S. (1917). Cause of the spike disease of sandal (*Santalum album*). *Indian Forester* **43,** 429-442.

Hole, R.S. (1933). Plant Pathology in the forests of India. Part III. *Indian Forester* **59**, 638-649.

Holdeman, Q.L. (1970). Varietal and geographic distribution of walnut phloem canker in California. *Pl. Dis. Reptr.* **54**, 373-376.

Horsfall, J.G. (1956). The fight with the fungi or "The rusts and rots that rob us, the blasts and the blights that beset us" *Am. J. Bot.* **43**, 532-536.

Horsfall, J.G. and Dimond, A.E. (1959). The diseased plant. 1-17 pp. In Plant Pathology–an Advanced Treatise. Vol. I. Edited by J.G. Horsfall and A.E. Dimond, Academic Press, New York and London.

Hughes, S.J. (1952). Studies on microfungi. XIV. *Stigmella*, *Stigmina*, *Camptomeris*, *Polythrincium* and *Fuscicladiella*. *Mycol. Pap. Commonw. mycol. Inst.* **49**, 25 pp. R.A.M. **33**, 323, 1954.

Hull, R. (1971). Mycoplasma-like organisms in plants. *Rev. Pl. Path.* **50**, 121-130.

Hull, R., Plaskitt, A., Nayar, R.M. & Padmanabha, H.S.A. (1970). Electron microscopy of alternate host of sandal spike pathogen and of tetracycline–treated spike infected sandal trees. *J. Indian Acad. wood sci.* **1**, 62-64.

Hull, R., Horne, R.W. & Nayar, R.M. (1969). Mycoplasma-like bodies associated with sandal spike. *Nature* **224**, 1121-1122.

Huque, H. (1964). Plant quarantine situation in Pakistan. FAO/IUFRO symposium on internationally dangerous forest diseases and insects, Oxford. (cyclostyled).

Iizuka, N. (1956). Study on the judgement method of inner rot of the living tree (todo fir) by γ-ray. *Bull. Tokyo Univ. Forests* **52**, 143-163.

Imazeki, R. (1960). Introduced diseases with particular attention to the threat they pose forest protection in Asia. Proc. V World For. Congr. 879-882.

Imazeki, R. & Hongo, T. (1957). Coloured Illustrations of Fungi of Japan. 181 pp. Osaka Hoikusha, Japan.

Ingold, C.T. (1959). Notes on exudation and exudation pressure in birch. *New Phytol.* **34**, 437-441.

Ishiie, T. (1970). Mycoplasma-like organisms, causal agent of mulberry dwarf disease. *Jap. agric. Res. Q.* **5**, 48-53. R.P.P. **51**, no. 492, 1972.

Ishiie, T., Doi, H., Yora, K. & Asuyama, H. (1967). Suppressive effects of antibiotics of tetracycline group on symptom development of mulberry dwarf disease. *Ann. Phytopath. Soc. Japan* **33**, 267-275. R.A.M. **47**, no. 656, 1968.

Ito, K. (1949). Studies on 'murasaki-monpa' disease caused by *Helicobasidium mompa* Tanaka. Bull. For. Exp. Sta. Tokyo **43**, 126 pp. R.A.M. **30**, 337, 1951.

Ito, K. (1972). Cercospora needle blight of pines in Japan. Bull. Govt. Forest Exp. Stn. **246**, 21-33.

Ito, K. & Shibukawa, K. (1956). Studies on some anthracnose of woody plants. IV. A new anthracnose of Acacia with special reference to the life history of the causal fungus. Bull. For. Exp. Sta. Meguro, **92**, 51-64. R.A.M. **38**, 339, 1959.

Ito, K., Zinno, Y. & Suto, Y. (1975). Dothistroma needle blight of pines in Japan. Bull. Govt. For. Exp. Sta., Tokyo, Japan. No. 272, 123-140.

Ivory, M.H. (1967). A new variety of *Dothistroma pini* in Kenya. *Trans. Br. mycol. Soc.* **50**, 289-297.

Ivory, M.H. (1975). Heart-rot of *Eucalyptus* in Zambia. FAO-FORPEST DI/75/16-2.

Iyengar, A.V.V. (1960). The relation of soil nutrients to the incidence of the spike disease in sandalwood (*Santalum album* Linn.). *Indian Forester* **86**, 220-230.

Iyengar, A.V.V. (1960 a). Spike disease of plants. *Indian Forester* **86**, 489-496.

Jackson, L.W.R. (1940). Effects of H-ion and Al-ion concentrations on damping-off of conifers and certain causative fungi. *Phytopathology* **30**, 563–579.

James, W.O. & Baker, H. (1933). Sap pressure and the movement of sap. *New Phytol.* **32**, 317–343.

Jeevaratnam, A.J. (1967). A note on boron toxicity in young replantings. *Bull. Rubber Research Inst., Ceylon*, N.S. **2**, 22–23.

Jewell, F.F. (1959). Disease resistant studies in tree improvement research. V. Southern forest tree improvement conference, Raleigh, N.C. 1959, 3 pp.

Jones, D.R. (1933). Studies on a canker disease of cypresses in East Africa, caused by *Monochaetia unicornis* (Cooke & Ellis) Sacc. I. Observations on the pathology, spread and possible origins of the disease. *Ann. appl. Biol.* **40,** 323–343.

Jones, D.R. (1954). Studies on a canker disease of cypresses in East Africa, caused by *Monochaetia unicornis* (Cooke & Ellis) Sacc. II. Variation in the morphology and physiology of the pathogen. *Trans. Br. mycol. Soc.* **37,** 286-305.

Jones, D.R. (1954 a). Studies on a canker disease of cypresses in East Africa, caused by *Monochaetia unicornis* (Cooke & Ellis) Sacc. III. Resistance and susceptibility of species of *Cupressus* and allied genera. *Ann. appl. Biol.* **41,** 325–335.

Jorgensen, E. (1962). Observations on the formation of protection wood. *For. Chron.* **38,** 292–294.

Kadambi, K. (1954). On Loranthus control. *Indian Forester* **80,** 493–494.

Kalani, I.K. (1969). *Selenophoma kamatii* sp. nov. from India. *Sydowia* **23**, 203–206.

Kapoor, J.N. (1968). New microfungi from India. *Trans. Br. mycol. Soc.* **51,** 328–333.

Kar, A.K. & Maity, M.K. (1970). New *Asterina* spp. from West Bengal. *Trans. Br. mycol. Soc.* **54,** 435–444.

Kar, A.K. & Maity, M.K. (1971). Pyrenomycetes of West Bengal. III. *Trans. Br. mycol. Soc.* **56,** 189–193.

Kar, A.K. & Mandal, M. (1969). New *Cercospora* species from West Bengal. I. *Trans. Br. mycol. Soc.* **53,** 337–360.

Keilich, G., Bailey, P. & Liese, W. (1970). Enzymatic degradation of cellulose, cellulose derivatives and hemicelluloses in relation to the fungal decay of wood. *Wood Science and Technology* **4,** 273–283.

Kelman, A. (1953). The bacterial wilt caused by *Pseudomonas solanacearum*. Tech. Bull. N.C. Agric. Exp. Stn., No. 99.

Kerr, A. (1965). Report of Adviser in Plant Pathology. Rep. Tea Res. Inst. Ceylon. pt. II, 73-78.

Khan, A.H. (1951). Some diseases observed in the teak plantations of East Bengal. *Pakist. J. For.* **I,** 209–212.

Khan, A.H. (1955). A new disease of mulberry in the forest plantations of the Punjab. *Pakist. J. Sci. Res.* **7,** 92–99.

Kimmey, J.W. (1950). Cull factors for forest tree species in northwestern California. For. Surv. Release, U.S. For. Surv., Calif. For. Range Exp. Stn. **7,** 30 pp.

Kimmey, J.W. (1957). Dwarf mistletoes of California and their control. Calif. Forest & Range Expt. Sta. Tech. Paper 19, 12 pp.

Kimmey, J.W. & Hornibrook, E.M. (1952). Cull and breakage factors and other tree measurement tables for redwood. For. Serv. Release, U.S. For. Serv. Calif. For. Range Exp. Stn. **13,** 28 pp.

Kirk, T. Kent & Moore, Wayne E. (1972). Removing lignin from wood with white-rot fungi and digestibility of resulting wood. *Wood and fibre* **4,** 72–79.

Klyushnik, P.I. & Morgun, K.D. (1961). [Study of the biology of *Fomes annosus*]. [*Nauk. Pratzi. Ukr. nauk.-dosl. In-t lisoz Gospod & Agrolisomelior.*] **21,** 87-89. R.A.M. **43,** no. 1440, 1964.

Koch, R. (1882). Über die Milzbrandimpfung, eine Entgengung auf den von Pasteur in Genf gehaltenen Vortrag. Berlin, Fischer.

Korstian, C.F. & Long, W.H. (1922). The Western yellow pine mistletoe; effect on growth and suggestions for control. *U.S. Dept. Agr. Bull.* **1112,** 1–35.

Krasikova, A.V. (1966). Dimensions of heartrot of *Abies sibirica* in the Krasnoyarsk region. *Lesn. Z., Arhangel'sk* **9,** 15–17.

Kristensen, H.R. (1960). The sandal spike disease. Report to the Govt. of India. Expanded Tech. Assistance Progm. No. 1229. Food & Agriculture Organisation of United Nations, Rome.

Krstic, M.M. (1960). The significance of introduced diseases from the European point of view. Proc. V. World For. congr. 873-879.

Kuhlman, E.G. & Hendrix, F.F., Jr. (1964). Infection, growth rate, and competitive ability of *Fomes annosus* in inoculated *Pinus echinata* stumps. *Phytopathology* **54**, 556–561.

Kuijt, J. (1955). Dwarf mistletoes. *Bot. Rev.* **21**, 569–627.

Kumararaj, K. & Bhide, V.P. (1961). Damping-off of cashew nuts (*Anacardium occidentale* L.) seedlings caused by *Phytophthora palmivora* Butler in Maharashtra State. *Curr. Sci.* **31**, 23.

Lall, G. & Gupta, J.N. (1965). Notes on powdery mildews from India. III. *Indian Phytopath.* **18**, 210–213.

Lange, S. (1959). Das Erkennen von Stammfäulen mit Hilfe zerstörungsfreier Prüfverfahren. *Forstwiss. ZentBl.* **78**, 174–180.

Leach, C.M. (1967). The light factor in the detection and identification of seed-borne fungi. Proc. int. Seed Test. Ass. **32**, 565–589.

Leach, R. (1937). Observations on the parasitism and control of *Armillaria mellea*. Proc. R. Soc., B. **121**, 561–573.

Leach, R. (1939). Biological control and ecology of *Armillaria mellea* (Vahl) Fr. *Trans. Br. mycol. Soc.* **23**, 320–329.

Levi, M.P. (1964). The fungal degradation of wood. *J. Inst. Wood Sci.* No. 12, 56–66.

Levi, M.P. (1973). Decay of chemically treated wood. In Symposium on Decay of Resistant Wood. 2nd Int. Pl. Path. conf. Minnesota, USA.

Levisohn, I. (1956). Growth stimulation of forest tree seedlings by the activity of free-living mycorrhizal mycelia. *Forestry* **29**, 53, 59.

Levy, J.F. (1965). The soft rot fungi : Their mode of action and significance in the degradation of wood. In *Advances of Botanical Research* **2**, 323–357. (quoted from Levy and Stevens 1966).

Levy, J F. & Stevens, M.G. (1966). The initiation of attack by soft rot fungi in wood. *J. Inst. Wood. Sci.* No. 16, 49–55.

Liese, W. (1970). The action of fungi and bacteria during wood deterioration. Rec. a. Conv. Br. Wood Preserv. Ass. 1970, 81-94. R.P.P. **50**, no. 3236, 1971.

Liese, W. (1970 a). Ultrastructural aspects of woody tissue disintegration. *A. Rev. Phytopath.* **8**, 231-258.

Lindgren, R.M. (1951). Deterioration of southern pine pulpwood during storage. Forest Prod. Res. Soc. Proc. **5**, 169-180.

Lindgren, R.M. (1953). Deterioration losses in stored southern pine pulpwood. *Tappi* **36**, 260-263.

Link, G.K.K. (1933). Etiological Phytopathology. *Phytopathology* **23**, 843-862.

Lloyd, C.G. (1915). Synopsis of the Genus *Fomes*. pp. 211-288.

Low, J.D. & Gladman, R.J. (1960). *Fomes annosus* in Great Britain. An assessment of the situation in 1959. For. Rec. For. Commn., Lond. No. 41, 22 pp.

Lucas, R.L. (1955). A comparative study of *Ophiobolus graminis* and *Fusarium culmorum* in saprophytic colonization of wheat straw. *Ann. appl. Biol.* **43**, 134-143.

Lucy Channamma, K.A. & Rangaswami, G. (1968). On two new hosts of *Alternaria tenuis* in Mysore State. *Mysore J. agric. Sci.* **2**, 342-344.

Ludwig, R.A. (1952). Studies on the physiology of hydromycotic wilting in the tomato plant. Tech. Bull. MacDonald agric. Coll. 20, 39 pp. R.A.M. **33**, 56, 1954.

Ludwig, R.A. (1960). Toxins. In Plant Pathology—an Advanced Treatise. **2**, 315-357. Ed. J.G. Horsfall and A.E. Dimond. Academic Press, New York, London.

Lundquist, E. (1964). Report to the Govt. of India on expansion of plantation forestry. FAO Report 1874, 1-56.

Lushington, P.M. (1896). Report and working schemes of the Nilambur teak plantations.

Lyr, H. (1962). Detoxification of heartwood toxins and chlorophenols by higher fungi. *Nature* **4836**, 289–290.

Lyr, H. (1963). Enzymatische detoxification chlorerter phenols. *Phytopath. Z.* **47**, 73-83.

Madho Singh, C. (1961). Tolerance of some fungi to a water-soluble preservative and its components. *Forest Prod. J.* **11**, 20-22.

Magnani, G. (1964). Diseases of *Eucalyptus*. In Diseases of widely planted forest trees. FAO/FORPEST-64, Oxford, 159-167.

Mahakul, S.K. & Mohanty, N.N. (1969). Notes on some fungi hitherto not recorded from India. *Indian Phytopath.* **22,** 132-135.

Malencon, G. & Marion, J. (1951). Un parasite des suberaies nord–africaines: L' *Hypoxylon mediterraneum* (D. Ntrs.) Ces et D. Ntrs. *Rev. for. franc.* **11,** 681-686. R.A.M. **31,** 409, 1952.

Malencon, G. & Marion, J. (1952). L' *Hypoxylon mediterraneum* (D. Ntrs.) Ces. et D. Ntrs. et son comportement epiphytique dans les chenaies nord-africaines. *Rev. Mycol.* **17,** 49-73. R.A.M. **32,** 458-459, 1953.

Maramorosch, K., Shikata, E. & Granados, R.R. (1968). Structure resembling mycoplasma in diseased plants and in insect vectors. *Trans. N.Y. Acad. Sci.* **30,** 841-855.

Marsden, D.H. (1954). Studies on the creosote fungus, *Hormodendrum resinae*. *Mycologia* **46,** 161-183.

Marudarajan, D., Ramakrishnan, T.S. & Soumini, C.K. (1950). Wilt of Casuarina. *Curr. Sci.* **19,** 63-64.

Matuo, T. & Sakuri, Y. (1954). Hot water disinfection of the sapling of the mulberry tree infected by *Rosellinia necatrix* (Hart.) Berl. and the situation of the isolation ditch preventing the development of the disease in the mulberry farm. *J. seric. Sci. Tokyo* **23,** 271-277.

Marx, D.H. (1971). Ectomycorrhizae as biological deterrents to pathogenic root infections. In Mycorrhizae. Ed. E. Hacskaylo. U.S.D.A., For. Serv. Misc. Pub. 1189, 81-96.

McCarthy, C. (1899). Progress report of Forest administration in Coorg for 1898-1899 (unpublished report).

McLuckie, J. (1923). Studies in symbiosis. IV. The root nodules of root of *Casuarina cunninghamiana* and their significance. Proc. Linn. Soc. N.S.W. **48,** 194-204.

Meier, H. (1955). Über den Zellwandabbau durch Holzvermorschung-spilze und die submikroskopische struktur von Fichtentracheiden und Birkenholzfasern. *Holz Roh-u. werkstoff* **13,** 323–338.

Meier, H. (1964). General chemistry of cell walls and distribution of the chemical constituents across the wall. In The Formation of Wood in Forest Trees. Ed. M.G. Zimmermann, Academic Press, New York, pp. 137–151.

Melin, E. (1954). Growth factor requirements of mycorrhizal fungi of forest trees. *Svensk bot. Tidskr.* **48,** 86–94.

Melin, E. (1963). Some effects of forest tree roots on mycorrhizal basidiomycetes. In Symbiotic Associations. Ed. P.S. Nutman and B. Mosse. pp. 125–145. Cambridge Univ. Press, Lond.

Melin, E. & Nilsson, H. (1950). Transfer of radioactive phosphorus to pine seedlings by means of mycorrhizal hyphae. *Physiologia Pl.* **3,** 88–92.

Melin, E. & Nilsson, H. (1952). Transport of labelled nitrogen from an ammonium source to pine seedlings through mycorrhizal mycelium. *Svensk bot. Tidskr.* **46,** 281–285.

Melin, E. & Nilsson, H. (1953). Transfer of labelled nitrogen from glutamic acid to pine seedlings through the mycelium of *Boletus variegatus* (Sw.) Fr. *Nature, Lond.* **171,** 134.

Melin, E. & Nilsson, H. (1953 a). Transport of labelled phosphorus to pine seeedlings through the mycelium of *Cortinarius glaucopus* (Schaeff. ex Fr.) Fr. *Svensk bot.Tidskr.* **49,** 555–558.

Melin, E. & Nilsson, H. (1955). Ca^{45} used as an indicator of transport of cations to pine seedlings by means of mycorrhizal mycelium. *Svensk bot. Tidskr.* **49,** 119–122.

Melin, E. & Nilsson, H. (1957). Transport of C^{14} labelled photosynthate to the fungal associate to pine mycorrhiza. *Svensk bot. Tidskr* **51,** 166–186.

Melin, E. & Nilsson, H. (1958). Translocation of nutritive elements through mycorrhizal mycelia to pine seedlings. *Bot. Notiser* **111,** 251–256.

Melin, E., Nilsson, H. & Hacskaylo, E. (1958). Translocation of cations to seedlings of *Pinus virginiana* through mycorrhizal mycelium. *Bot. Gaz.* **119,** 243–246.

Meredith, D.S. (1959). The infection of pine stumps by *Fomes annosus* and other fungi. *Ann. Bot., Lond.* [*N.S.*] **23,** 455–476.

Meredith, D.S. (1960). Further observations on fungi inhabiting pine stumps. *Ann. Bot., Lond.* [*N.S.*] **24,** 63 78.

Merrill, W. (1970). Spore germination and host penetration by heart-rotting Hymenomycetes. *A. Rev. Phytopath.* **8,** 281–300.

Mhaiskar, V G. (1969). Some new Fungi Imperfecti on *Lagerstroemia flos-reginae* Retz. from India. *J. Univ. Poona* **36,** 103–105. R.P.P. **50,** no. 3845, 1971.

Mikola, P. (1948). On the black mycorrhizae of birch. *Arch. Soc. Zool.-Bot. Fenn. Vanamo* **1,** 81–85.

Mikola, P. (1965). Studies on ectendotrophic mycorrhiza of pine. *Acta for. fenn.* **79,** 1–56.

Mikola, P. (1969). Afforestation of treeless areas : Importance and technique of mycorrhizal inoculation. *Unasylva* **23,** No 92, 35–48.

Mikola, P. (1973). Application of mycorrhizal symbiosis in forestry practice. In Ectomycorrhizae, their Ecology and Physiology. Ed. G.C. Marks and T.T Kozlowski. pp. 383–411. Academic Press, New York.

Miller, E.C. (1938). Plant Physiology. McGraw Hill Book Co., Inc., New York.

Mishra, R.R. (1968). A new report of follicolous Ascomycetes from India. *Sci. Cult.* **34,** 355.

Mitchell, B.A. (1962). Bacterial wilt in teak, *Tectona grandis* Linn. *Malay. Forester* **25,** 164–166.

Mohanty, N.N. & Beraha, B.C. (1962). Occurrence of *Colletotrichum lindemuthianum* (Sacc. & Magn.) Briosi & Cav. On *Cassia fistula* L. Proc. 49th Indian Sci. Congr., pt. III, 248.

Mohanty, N.N. & Patnaik, S. (1970). Certain studies of *Rhizopus* rot of jack fruit. Proc. of 57 Indian Sci. congr. Kharagpur, 1970, pt. III, 547.

Moniz, L. & Raj, H.H. (1967). *Xanthomonas azadirachtii* n. sp. causing leaf spot disease in *Azadirachta indica* A. Juss. *Indian Jnl. Microbiol.* **7,** 159–160. R.P.P. 50, no. 289, 1971.

Morawski, Z.J.R., Basham, J.T. & Turner, K.B. (1958). A survey of pathological condition in the forests of Ontario. Rep. (For. Res. Cull Studies, Invent.) Div. Timb. Ont. Dep. Lands. For. **25,** 96 pp.

Mosse, B. (1963). Vesicular-arbuscular mycorrhiza: an extreme form of fungal adaptation. In Symbiotic Associations. Ed. P.S. Nutman and B. Mosse. 146–170 pp. Cambridge Univ. Press, Lond.

Mowry, H. (1933). Symbiotic nitrogen fixation in the genus *Casuarina. Soil Sci.* **36,** 409–425.

Müller, E. & Sanwal, B.D. (1959). On a new parasitic fungus on *Castanopsis. Indian Phytopath.* **12,** 29–32.

Mundkur, B.B. & Kheswalla, K.F. (1943). *Dasturella*—a new genus of the Uredinales. *Mycologia* **35,** 201–206.

Munjal, R.L., Chona, B.L. & Kapoor, J.N. (1959). Notes on miscellaneous Indian fungi. VI. *Indian Phytopath.* **12,** 176–181.

Munjal, R.L., Lall, G. & Chona, B.L. (1961). Some *Cercospora* species from India. VI. *Indian Phytopath.* **14,** 179–190.

Naphade, S.R. (1970). Some Ascomycetes from Maharashtra (India). *Sydowia* **24,** 241–244.

Napper, R.P.N. (1932). Observations on the root disease of rubber trees caused by *Fomes lignosus. J. Rubb. Res. Inst. Malaya* **4,** 5–33.

Narasimhan, M.J. (1918). A preliminary study of the root nodules of Casuarina. *Indian Forester* **44,** 265–268.

Narasimhan, M.J. & Thirumalachar, M.J. (1961). *Ravenelia esculenta,* an edible rust fungus. *Phytopath. Z.* **41,** 97–102.

Narayanamurti, D., George, J. & Verma, G.M. (1952). Studien Zur Biochemie holzzerstorender Pilze—Erste Mitteilung : Saure-produktion der Pilze. *Holz Roh-u. Werkstoff* **10,** 227–229.

Narayanamurti, D., George, J. & Verma, G.M. (1952 a). Studien Zur Biochemie holzzerstorender Pilze-Zweite Mitteilung : Pilzamylase. *Holz Roh-u. Werkstoff* **10,** 428-432.

Narayanamurti, D. & Verma, G.M. (1953). Studien Zur biochemie holzzerstorender Pilze—Dritte Mitteilung : cellulase und invertase von *Polystictus sanguineus* und über die mogliche wirkungsmechanik von Holzschutzmitteln. *Holz Roh-u. Werkstoff* **11,** 7–9.

Nattrass, R.M. (1950). Annual report of the Senior Plant Pathologist. 1948. Rep. Dep. agric. Kenya 1948, 95–98.

Nattrass, R.M., Booth, C. & Sutton, B.C. (1963). *Rhynchosphaeria cupressi* sp. nov., the causal organism of *Cupressus* canker in Kenya. *Trans. Br. mycol. Soc* **46,** 102–106.

Nayar, R. (1967). Sap transmission of sandal spike. *Indian J. expl. Biol.* **5,** 195–196.

Nayar, R. (1968). Chromatographic purification of a single viral protein in sandal spike. *Phytopath. Z.* **61,** 167–170.

Nayar, R. & Padmanabha, H.S.A. (1967). Serological diagnosis of sandal spike disease. Quoted from Nayar, R. & Pramila, M. (1968) : *Van Vigyan,* Sept.–Dec. Issue, pp. 56–61.

Nayar, R. & Padmanabha, H.S.A. (1970). Isolation, cultivation and pathogenicity trials with mycoplasma-like bodies associated with sandal spike disease. *J. Indian Acad. Wood Sci.* **1,** 59–61.

Nayar, R. & Srimathi, R.A. (1968). A symptomless carrier of sandal spike disease. *Curr. Sci.* **37,** 567–568.

Neergaard, P. (1973). Detection of seed-borne pathogens by culture test. *Seed Sci. & Technol.* 1, 217–254.

Neish, G A. (1972). When is a parasite not a parasite? *Mycologia* **64,** 1179–1180.

Newhook, F.J. & Podger, F.D. (1972). The role of *Phytophthora cinnamomi* in Australian and New Zealand forests. *A. Rev. Phytopath.* **10,** 229–326.

Nienstaedt, H. (1953). Tannin as a factor in the resistance of chestnut, *Castanea* spp., to the chestnut blight fungus, *Endothia parasitica* (Murr.) A. & A. *Phytopathology* **43,** 32–38.

Nisbet, J. (1895). The diseases of trees. *Indian Forester* **21,** 126–133.

Noble, M. & Richardson, M.J. (1968). An annotated list of seed-borne diseases. Proc. int. Seed Test. Ass. **33,** 1–191.

Nobles, M.K. (1948). Studies in forest pathology. VI. Identification of cultures of wood-rotting fungi. *Can. J. Res.* **26C,** 281–431,

Nobles, M.K. (1958). A rapid test for extracellular oxidase in cultures of wood inhabiting Hymenomycetes. *Can. J. Bot.* **36,** 91–99.

Nobles, M.K. (1965). Identification of cultures of wood-inhabiting Hymenomycetes. *Can. J. Bot.* **43,** 1097–1139.

Olson, R.V. (1944). The use of hydroponics in the practice of forestry. *J. For.* **42,** 264–268.

Ono, K. (1965). *Armillaria* root rot in plantations of Hokkaido : Effects of topography and soil conditions on its occurrence. Bull. Forest Exp. Stn. Meguro, Tokyo **179,** 62 pp. (English summary).

Overholts, L.O. (1963). The Polyporaceae of the United States, Alaska and Canada. 466 pp. Univ of Michigan Press, Ann. Arbor, Michigan.

Pai, H. Surendranath (1968). Two species of *Phyllosticta*, a common ascigerous stage on *Sapindus* L. *Mycopath. Mycol. appl.* **35,** 121–128.

Pande, A.K. & Patwardhan, P.G. (1967). *Maublancia indica* sp. nov. (Microthyriaceae) a new genus record to India. *Sydowia* **21,** 313–315.

Pandotra, V.R., Kachroo, J.L. & Sastry, K.S.M. (1968). Six powdery mildews from Jammu and Kashmir State. Proc. Indian Acad. Sci., Sec. B. **67,** 119–124.

Parthasarathi, K., Babu, D.R.C. & Rao, P.S. (1970). Studies on sandal spike. Pt. VIII. Polyphenolase activity and metabolism of sandal (*Santalum album* L.) in health and disease. Proc. Indian Acad. Sci. Sec. B. **72,** 277–284.

Parthasarathi, K., Gupta, S.K. & Rao, P.S. (1966). Studies on sandal spike. Pt. VII. Some useful stain tests for the diagnosis of the spike disease. Proc. Indian Acad. Sci. Sec. B. **64,** 152–156.

Parthasarathi, K., Gupta, S.K. & Rao, P.S. (1971). Studies on sandal spike. Pt. IX. Cation exchange capacity of sandal (*Santalum album* Linn.) in the health and disease. *Curr. Sci.* **40,** 640–641.

Parthasarathi, K., Ramaiah, P.K., Manjanatha, T.R. & Rao, P.S. (1962). Studies on sandal spike. Pt. III. The nitrate reductase activity in the normal and pathochemical states of sandal (*Santalum album* Linn.). Proc. Indian Acad. Sci. Sec. B. **55,** 285–289.

Parthasarathi, K., Ramaiah, P.K. & Rao, P.S. (1963). Studies on sandal spike. Pt. V. Ascorbic acid in the metabolism of sandal (*Santalum album* L.) in health and disease. Proc. Indian Aca. Sci. Sec. B. **57**, 68–71.

Parthasarathi, K. & Rao, P.S. (1962). Studies on sandal spike. Pt. II. Physiological significance of the disturbed iron balance in the spike disease of sandal (*Santalum album* L.). Proc. Indian Acad. Sci. Sec. B. **55**, 99–106.

Pascoe, T.A. & Scheffer, T.C. (1950). Jack pine pulpwood deterioration in yard storage. *Paper Trade J.* **131**, 16–21.

Patel, M.K., Kulkarni, Y.S. & Dhande, G.W. (1952). Some new bacterial diseases of plants. *Curr. Sci.* **21**, 345–346.

Patel, M.K., Kulkarni, Y.S. & Muniz, L. (1948). *Pseudomonas mangiferae-indicae* on mango. *Indian Phytopath.* **1**, 147–152.

Patel, M.K., Payak, M.M. & Kulkarni, N.B. (1951). Additions to fungi of Bombay. II. A new *Dasturella* on *Boswellia serrata* Roxb. *Indian Phytopath.* **4**, 71–73.

Patil, B.V. (1961). Powdery mildew on *Gmelina arborea*. *Curr. Sci.* **30**, 155–156.

Patwardhan, P.G. & Pande, A.K. (1969). Some Fungi Imperfecti from Maharashtra. *Sydowia* **23**, 95–101.

Pavgi, M.S. & Singh, R.A. (1971). Parasitic fungi from North India. VIII. *Mycopath. Mycol. appl.* **43**, 117–125.

Pavgi, M.S. & Singh, U.P. (1970). Parasitic fungi from North India. IX. *Sydowia* **24**, 113–119.

Pawsey, R.G. (1970). Forest diseases in Trinidad and Tobago with some observations in Jamaica. *Commonw. For. Rev.* **49**, 64–77.

Peace, T.R. (1938). Butt rot of conifers in Great Britain. *Q. Jl. For.* **32**, 81–104.

Peace, T.R. (1962). Pathology of Trees and Shrubs with special reference to Britain. 753 pp. Clarendon Press, Oxford.

Pearson, R.S. & Brown, H.P. (1932). Commercial Timbers of India. 2 Vols. 1150 pp. Govt. of India, Central Publication Branch, Calcutta.

Penfold, A.R. & Willis, J.L. (1961). "The Eucalypts, Botany, Chemistry, Cultivation and Utilization." 551 pp. Leonard Hill (Books) Ltd., London.

Perias, O.S. & Irugalbandara, Z.E. (1973). Histology of Hevea roots infected by *Fomes lignosus*. *Ann. appl. Biol.* **73,** 1–7.

Petch, T. (1921). Diseases and Pests of the Rubber Trees. 278 pp. Mac Millan & Co. Ltd., London.

Peterson, R.S. (1967). The *Peridermium* species on pine stems. *Bull. Torrey bot. Club* **94,** 511–542.

Pettifor, C.B. & Findlay, W.P.K. (1946). Effect of sap-stain on the tensile strength of corsican pine sapwood. *Forestry* **20,** 57–61.

Pinto-Lopes, J. (1952). Polyporaceae. Contribuicas para sua biotaxonomia. Mem. Soc. Broteriana **8,** 1–195.

Pirozynski, K.A. (1965). *Ovulariopsis* state of *Pleochaeta*. *Mycologia* **57,** 826–828.

Pole Evans, J.B. (1932) Arable farming and pasture problems. *Farming in South Africa* **7,** 341–352. R.A.M. **12,** 142, 1933.

Prain, D. (1890). Notes added to Dr. Barclay's paper (of Barbes 1890 c). Tratological effects. *J. Bombay nat. Hist. Soc.* **5,** 165–167.

Prasada, R., Desai, B.G., Bhatnagar, L.G., Goyal, J.P. & Pathak, V.N. (1971). Some new host records of *Leveilla* (*Oidiopsis*) *taurica*. *Indian Phytopath.* **24,** 405–406.

Preston, A. & Mclennan, E.I. (1948). The use of dyes in culture media for distinguishing brown and white wood-rotting fungi. *Ann. Bot. Lond.* (N.S.) **12,** 53–64.

Proctor, P. (1941). Penetration of the walls of wood cells by the hyphae of wood-destroying fungi. Yale Univ. School Forestry Bull. No. 47, 31 pp.

Pryor, L.D. (1956). Ectotrophic mycorrhiza in renantherous species of *Eucalyptus*. *Nature* **177,** 587–588.

Puri, Y.N., Sharma, R.P. & Bakshi, B.K. (1959). Toxic resistance of *Poria monticola* to copper and zinc in nutrient medium. *J. Scient. ind. Res.* **18** C, 149–152.

Puri, Y.N. & Taneja, K. (1967). Problem of decay in spruce and fir logs. *Indian Forester* **93,** 658–660.

Purnell, H. (1969). Collar rot or 'dead eye' of *Eucalyptus* spp. *For. Tech. Pap. Melbourne* **2,** 12–15. R.A.M. **40,** 492, 1961.

Quraishi, M.A. (1967). *Populus euphratica* Oliver and its fungal enemies. *Pakist. J. For.* **17,** 119–129.

Quraishi, M.A. & Jamal, S.M. (1969). Fungi on *Populus ciliata* Wall. in West Pakistan. *Pakist. J. For.* **19,** 429–441.

Qureshi, I.M. (1956). Mortality of *Casuarina equisetifolia* in plantations in Bombay State. Proc. 8th Silvic. Conf., Dehra Dun, 1951, Pt. 2, 329–330.

Qureshi, I.M., Kaul, O.N. & Srivastava, P.B.L. (1967). Nutrient requirements of some *Casuarina equisetifolia* plantations in Maharashtra and Gujarat. Proc. 11th Silvic. Conf., Dehra Dun, 1967.

Raciborski, M. (1900). Parasitischen Algen und Pilze Java's teil **1,** 28.

Raghavan, M.S. (1948). A brief note on the regeneration of sandal to supplement existing stock in forest or to replace sandal that may be killed by spike. *Indian Forester* **74,** 42–49.

Rajderkar, N.R. (1966). A new Alternaria leaf spot of *Sapindus emarginatus* Vahl from India. *Mycopath. Mycol. appl.* **29,** 62–64. R.A.M. **45,** no. 3191, 1966.

Rajkhowa, S. (1958). A working plan for the reserved forests of Goalpara West Division. Parts I and II, 1958–59. to 1972–73, 265 pp. (cyclostyled).

Ramaiah, P.K., Babu, D.R.C. & Rao, P.S. (1964). Studies on sandal spike. Pt. VI. A study of the amino-acid make-up and some organic acids of sandal (*Santalum album* L.) in health and disease. Proc. Indian Acad. Sci. **60** B, 281–286.

Ramaiah, P.K., Parthasarathi, K. & Rao, P.S. (1962). Studies on sandal spike. Pt. IV. A histochemical study of the sandal (*Santalum album* Linn.). Root haustoria in relation to mineral nutrition. Proc. Indian Acad. sci. **56** B, 111–113.

Ramakrishnan, T.S. (1955). Decline in cashew nut. *Indian Phytopath.* **8,** 58–63.

Ramakrishnan, T.S. (1957). Notes on some fungi. Proc. Indian Acad. Sci. **45**, 176–180.

Ramakrishnan, T.S. (1960). Notes on some fungi from South India. VIII. Proc. Indian Acad. Sci. **51**, 164–168.

Ramakrishnan, T.S. & Pillai, P.N.R. Collar rot, dry rot, or charcoal rot. *Rubber Board India Bull.* **5**, 2, 4 pp.

Ramakrishnan, T.S. & Pillai, P.N.R. (1961). Abnormal leaf fall of rubber (*Hevea brasiliensis*) caused by *Phytophthora palmivora* Butl. I. *Rubber Board Bull.* **5**, 11–20. II. *Rubber Board Bull.* **5**, 9 pp.

Ramakrishnan, T.S. & Pillai, P.N.R. (1963). Black stripe, black thread or bark rot caused by *Phytophthora palmivora* Butl. (*Phytophthora faberi* Maubl., *P. meadii* McRae), *P. heveae* Thompson. *Bull. Rubber Board India* **6**, (3 & 4), 1–3.

Ramakrishnan, T.S. & Pillai, P.N.R. (1963 a). Mouldy rot caused by *Ceratocystis fimbriata* Ell. & Halst. [*Ceratostomella fimbriata* (E. & H.) Elliot], *Sphaeronema fimbriatum* (E. & H.) Sacc. *Bull. Rubber Board India* **6** (3 & 4), 1–4.

Ramakrishnan, T.S. & Ramakrishnan, K. (1949). *Chaconia tectonae* Ramakrishnan. T.S. & K. sp. nov. on teak. *Indian Phytopath.* **2**, 17-19.

Ramakrishnan, T.S. & Ramakrishnan, K. (1950). Additions to fungi of Madras. VIII. Proc. Indian Acad. Sci. Sec. B. **32**, 97-111.

Ramakrishnan, T.S. & Sundaram, N.V. (1952). Notes on some fungi from South India. I. *Indian Phytopath.* **5**, 110-113.

Ramakrishnan, T.S. & Sundaram, N.V. (1953). Notes on some fungi from South India. II. *Indian Phytopath.* **6**, 27-38.

Ramakrishnan, T.S. & Sundaram, N.V. (1954). Notes on some fungi from South India. IV. *Indian Phytopath.* **7**, 140-151.

Ramakrishnan, T.S. & Sundaram, N.V. (1955). Additions to fungi of Madras. XVIII. Proc. Indian Acad. Sci. Sec. B. **42**, 58-64.

Rama Rao, M. (1911). Host plants of sandal trees. *Indian Forest Rec.* **2**, pt. IV, 159-207.

Rangaswami, S. & Griffith, A.L. (1940). Entomological investigations on the spike disease of sandal. Further studies of spike disease of sandal. *Indian Forest Rec., Ent.* **6,** 85-196.

Rao, B.S. (1971). Avoiding secondary leaf fall disease of rubber by chemical defoliation. *PANS* **17,** 461-463. R.P.P. **51,** no. 1851, 1972.

Rao, L.N. (1942). Parasitism in the Santalaceae. *Ann. Bot. Lond.* [*N.S.*] **6,** 131-150.

Rao, P.S. (1965). The scientific study of sandal : physiological and pathological aspects. In Advancing Frontiers in the Chemistry of Natural Products, 237-250. Hindustan Pub. Corp., India.

Rao, P.S., Srimathi, R.A. & Nag, K.C. (1972). Response of spike disease of sandal to benlate treatment. *Curr. Sci.* **41,** 221-222.

Rao, V.G. (1962). A new Phyllosticta leaf spot of *Sapindus emarginatus* Vahl from India. *Curr. Sci.* **34,** 475-476.

Raychaudhuri, S.P., Chatterjee, S.N. & Dhar, H.K. (1961). Preliminary note on occurrence of yellow net vein disease of mulberry. *Indian Phytopath.* **14,** 94-95.

Raychaudhuri, S.P., Chatterjee, S.N. & Dhar, H.K. (1962). A mosaic disease of mulberry. *Indian Phytopath.* **15,** 187-189.

Raychaudhuri, S.P., Chenulu, V.V., Ghosh, S.K., Varma, A., Rao, P.S., Srimathi, R.A. & Nag, K.C. (1972). Chemical control of spike disease of sandal. *Curr. Sci.* **41,** 72-73.

Rayner, M.C. (1945). Trees and Toadstools. 71 pp. Faber & Faber Ltd., London.

Reddy, M.A.R. (1969). Damping-off in conifer nurseries in India. *Indian Forester* **95,** 475-479.

Reddy, M.A.R. & Misra, B.M. (1970). Fungicidal soil treatments to control damping-off diseases in pines. *Indian Forester* **96,** 270-275.

Reddy, M.A.R. & Pandey, P.C. (1973). Cercospora needle blight of radiata pine in India. *Indian Forester* **99,** 308-309.

Reddy, M.A.R., Tewari, R.K. & Bakshi, B.K. (1971). Target leaf spot disease of *Rauvolfia serpentina* and its chemical control. *Indian Forester* **97,** 487-491.

Reddy, S.M. (1969). Some new pathogenic fungi from India. *Indian Phytopath.* **22,** 353-358.

Reis, M.S. & Hodges, C.S. (1975). Status of Forest diseases and insects in Latin America. FAO/FORPEST, DI/75.

Ranjhen, P.L., Venketesh, M.V. & Joshi, M.C. (1962). Plant Quarantine in India. *Sci. Cult.* **28,** 215-218.

Richards, C.A. (1924). The comparative resistance of 17 species of wood destroying fungi to sodium fluoride. Proc. (20th) Ann. Meet. Wood Preserv. Ass. 37-44.

Richards, C.A. (1925). The comparative resistance of 18 species of wood destroying fungi to zinc chloride. Proc. Am. Wood Preserv. Ass. **21,** 18-22.

Richardson, S. (1960). The role of physiology in forest tree improvement. Proc. 5th World For. Congr. Seattle, U.S.A. **2,** 733-741.

Rishbeth, J. (1950). Observations on the biology of *Fomes annosus*, with particular reference to East Anglian pine plantations. I. The outbreaks of disease and ecological status of the fungus. *Ann. Bot. Lond.* (N.S.) **14,** 365-383.

Rishbeth, J. (1951). Observations on the biology of *Fomes annosus*, with particular reference to East Anglian Pine Plantations. II. Spore production, stump infection, and saprophytic activity in stumps. *Ann. Bot. Lond.* (N.S.) **15,** 1–21.

Rishbeth, J. (1952). Control of *Fomes annosus* Fr. *Forestry* **25,** 41-50.

Rishbeth, J. (1959). Stump protection against *Fomes annosus*. I. Treatment with creosote. *Ann. appl. Biol.* **47,** 519–528.

Rishbeth, J. (1959 a). Stump protection against *Fomes annosus*. Treatment with substances other than creosote. *Ann. appl. Biol.* **47,** 529–541.

Rishbeth, J. (1963). Stump protection against *Fomes annosus*. III. Inoculation with *Peniophora gigantea*. *Ann. appl. Biol.* **52,** 63–77.

Robinson, G.W.F. (1968). Prevention of wood chip discolouration at Bowaters Mersey. *Pulp and Paper Mag. Can.* **69,** T72–T76.

Robinson, R.A. (1971). Vertical resistance. *Rev. Pl. Pathology* **50,** 233–239.

Robinson, R.A. (1973). Horizontal resistance. *Rev. Pl. Pathology* **52,** 483–501.

Roff, J.W. & Dobie, J. (1968). Water sprinklers check biological deterioration in stored logs. *Br. Columbia Lumberm.* **52,** 6 pp.

Roldan, Emiliano F. & Andres, Primo P. (1953). Bacterial wilt of teak seedlings (*Tectona grandis* Linn.). *Phill. J. For.* **9,** 133–144.

Rosch, R., Liese, W. & Berndt, H. (1969). Untersuchungen über die Enzyme von Blauepilzen. I. Cellulose–, Polygalak turonase–, Pektinesterase–und Laccase–Aktivitiat. *Arch. Mikrobiol.* **67,** 28–50 (cited from Liese 1970 a).

Ross, E.W. (1968). Duration of stump susceptibility of loblolly pine to infection by *Fomes annosus*. *Forest Sci.* **14,** 206–212.

Ross, E.W. (1973). *Fomes annosus* in the southeastern United States : Relation of environmental and biotic factors to stump colonization and losses in the residual stand. *U.S.D.A. Tech. Bull.* **1459,** 1–26.

Roth, E.R. & Hepting, G.H, (1943). Origin and development of oak stump sprouts as affecting their likelihood to decay. *J. For.* **41,** 27–36.

Roth, L.F. & Riker, A.J. (1943). Life history and distribution of *Pythium* and *Rhizoctonia* in relation to damping-off of red pine seedlings. *J. agric. Res.* **67,** 129–148.

Roth, L.F. & Riker, A.J. (1943 a). Influence of temperature, moisture and soil reaction on the damping-off of red pine seedlings by *Pythium* and *Rhizoctonia*. *J. agric. Res.* **67,** 273–293.

Roth, L.F. & Riker, A.J. (1943 b). Seasonal development in the nursery of damping-off of red pine seedlings caused by *Pythium* and *Rhizoctonia*. *J. agric. Res.* **67,** 417–431.

Roth, E.R. & Sleath, B. (1939). Butt rot in unburned sprout oak stands. Tech. Bull. U.S. Dep. Agric. **684,** 42 pp.

Russell, P. (1955). Inactivation of phenyl mercuric acetate in groundwood pulp by mercury-resistant strain of *Penicillium roqueforti* Thom. *Nature* **176,** 1123-1124.

Saccardo, P.A. (1883). Sylloge fungorum **2,** 594.

Saccardo, P.A. (1926). Sylloge fungorum **24,** 284, 560.

Saccardo, P.A. (1931). Sylloge fungorum **25,** 47.

Sadasivan, T.S. (1961). Physiology of wilt disease. *Rev. Pl. Physio.* **12,** 449–468.

Sadasivan, T.S. & Subramanian, C.V. (1960). Interaction of pathogen, soil, other microorganisms in the soil and host. In Plant Pathology-An Advanced Treatise **2,** 273–313. Ed. J.G. Horsfall and A.E. Dimond. Academic Press, New York, London.

Saharan, G.S. (1971). A new record of powdery mildew pimples on *Cedrela toona* for Himachal Pradesh, India. *Indian J. Mycol. and Pl. Path.* **1,** 88–89.

Sahni, V.P. (1966). Deuteromycetes from Jabalpur. II *Mycopath. Mycol. appl.* **29,** 226–244.

Sakuri, Y. (1952). Pathologico-anatomical observations on the white root-rot of mulberry trees caused by *Rosellinia necatrix* (Hart.) Berl. Res. Rep. Fac. Text Sihinshu Univ. **2,** 15-26. R.A.M. **34,** 796, 1955.

Salam, M.A. & Rao, P.N. (1958). Fungi from Hyderabad (India). II. *Oidium* spp. *Indian Phytopath.* **11,** 126–129.

Sarbhoy, A.K., Girdhari Lal & Varshney, J.L. (1975) Fungi of India 1967–71. 148 pp. Navyug Traders, New Delhi.

Sarma, Y.N. & Nayudu, M.V. (1970). New hosts of *Corynespora cassiicola* (Berk. & Curt.) Wei. *Curr. Sci.* **39,** 495–496.

Sastry, K.S.M., Thakur, R.N., Gupta J.H. & Pandotra, V.R. (1971). Three virus diseases of *Eucalyptus citriodora. Indian Phytopath.* **24,** 123–126.

Satchuthananthavale, V. (1971). Review of the Plant Pathology. Dept. A. Rev. Rubb. Res. Inst. Ceylon 1970, 43–55. R.P.P. 51, no. 2803, 1972.

Savory, J G. (1954). Damage to wood caused by micro-organisms. *J. appl. Bact.* **17**, 213–218.

Savory, J.G. (1955). The role of the microfungi in the decomposition of wood. *Rec. Br. Wood Preserv. Ass.* **5,** 3–19.

Savory, J.G. & Pinion, L.C. (1958). Chemical aspects of decay of beechwood by *Chaetomium globosum*. *Holzforschung* **12,** 99–103.

Schaad, N.W. & Wilson, E.E. (1968). Pathological anatomy of the phloem canker disease of Persian walnut. *Phytopathology* **58,** 1066.

Scheffer, T.C. (1969). Protecting stored logs and pulpwood in North America. *Material und Organismen* **4,** 167–199.

Scheffer, T.C. & Cowling, E.B. (1966). Natural resistance of wood to microbial deterioration. *A. Rev. Phytopath.* **4,** 147–170.

Scheffer, T.C. & Hopp, H. (1949). Decay resistance of black locust heartwood. Tech. Bull. U.S. Dep. Agric. 984, 37 pp.

Scheffer, T.C. & Lindgren, R.M. (1940). Stains of sapwood and sapwood products and their control. U.S. Dep. Agric. Tech. Bull. No. 714, 124 pp.

Scheffer, T.C. & Verrall, A.F. (1973). Principles for protecting wood buildings from decay. U.S.D.A. For. Serv. Res. Pap. F.P.L. 190, 56 pp.

Schmitz, U. (1931). A suggested toximetric method for wood preservation. Proc. Am. Wood Preserv. Ass. **27,** 81–86.

Schreiner, E. J. (1960). Objectives of pest resistance improvement in forest trees and their possible attainment. Proc. V World For. Congr. **2,** 721–727, Seattle, U.S.A.

Schwarz, M.B. (1926). [Cobweb disease of mahogany and teak]. Korte Meded. Inst. Plziekt., Buitenz. **2,** 9 pp.

Sehgal, H.S., Sen, M. & Bakshi, B.K. (1966). Temperature relation of Indian species of Polypores. *Indian Forest Rec.* (N.S.) Forest Pathology **2,** 131–137.

Sen, M. (1973). Cultural diagnosis of Indian Polyporaceae. III. Genera *Daedalea, Favolus, Ganoderma, Hexagonia, Irpex, Lenzites, Merulius* and *Poria*. *Indian Forest Rec.* (N.S.) Forest Pathology **2,** 277–304.

Seshadri, V.S. (1968). Studies on Indian Phyllachoraceae X. *Sydowia* **22,** 284–287.

Seth, S.K., Khan, M.A.W. & Yadav, J.S.P. (1960). Sal mortality in Bihar. *Indian Forester* **86,** 645–679.

Shain, L. & Bolland, L. (1974). *Fomes annosus* in Australia. Proc. IV Int. Conf. on *Fomes annosus* (IUFRO) U.S D.A. For. Serv. 25–29.

Sharma, R.P., Puri, Y.N., Mukerji, H.K. & Tewari, M.C. (1964). Evaluation of the toxicity of copper resinate under laboratory conditions. *J. Timb. Dr. Preserv. Ass. India* **10,** 1–17.

Sharples, A. (1936). Diseases and Pests of the Rubber Trees. 480 pp. MacMillan & CO. Ltd., London.

Sher Singh (1924). The liability of deodar to the attack of *Trametes pini* (Brot.) Fr. in Lolab, Kashmir. *Indian Forester* **50,** 361–365.

Shields, J.K. (1969). Inhibition of fungi in a softwood chip pile. Can. Forest Serv., Bi-mon. Res. Notes **25,** 3–4.

Shigo, A.L. (1972). Succession of microorganisms and patterns of discolouration and decay after wounding in red oak and white oak. *Phytopathology* **62,** 256–259.

Shigo, A.L. (1973). Biology of decay and wood quality. II. Int. Congress of Plant Pathology, Minnesota, U.S.A.

Shishkina, A.K. (1969). [New causal agents of seedling diseases of *Cedrus deodara* Loud. in Georgia]. Proc. Inst. Georgian S.S.R. **21,** 95–100. R.P.P. **50,** no 2083 *l*, 1971.

Shishkina, A.K. & Tsanava, N.I. (1970). [On new diseases of ornamental plants in Georgia]. *Mikol. i. Fitopatol.* **4,** 76–79. R.P.P. **49,** no. 2203, 1970.

Shorrocks, V.M. (1964). Boron toxicity in *Hevea brasiliensis*. *Nature* **204,** 599–600.

Shreemali, J.L. (1970). Three new leaf spot diseases caused by phytopathogenic fungi. *J. appl. Sci.* **2,** 58–60. R.P.P. **51,** no. 4058, 1972.

Sinclair, W.A. & Risley, J.H. (1964–65). Annosus root rot, new hope for control. *The Conservationist* **19,** 20–21.

Singh, Bahadur (1962). Studies in angiospermic parasites. 1. *Dendrophthoe falcata* (L.f.) Ettingsh., its life history, life of hosts and control measures.Bull. natn. bot. Gdns. No. 69, 75 pp.

Singh, D.V. & Mathur, R.S. (1963). *Cercospora glandulosa* Ellis & Kellerman on *Ailanthus excelsa* Roxb.—a new record for India. *Sci. & Cult.* **29,** 612–613.

Slankis, V. (1957). The role of auxins and other exudates in the mycorrhizal symbiosis of forest trees. In Physiology of Forest Trees (Symposium). Ed. K.V. Thimann. pp. 427–443. Ronald Press Co., New York.

Slankis, V. (1961). On the factors determining the establishment of ectotrophic mycorrhiza of forest trees. Recent Advances in Botany, 1738–1742, Univ. Toronto Press.

Smith, R.S. & Hatton, J.V. (1971). Economic feasibility of chemical protection for outside chip storage. *Tappi* **54**, 1638–1640.

Smythies, E.A. & Howard, S.H. (1923). A Sal (*Shorea robusta*) yield table for the United Provinces, with an account of the types and distribution of sal forests in the United Provinces. *Indian Forest Rec.* **10**, 1–25.

Sobotka, A. (1968). [Influence of the use of some biocides in forest nurseries on the formation of mycorrhizas of Scots pine]—Pr. vyzk. Ust. lesn. Hospod. Mysliv. **36**, 63–76. R.A.M. **48**, no. 2143, 1969.

Sohi, H.S. & Nayar, S.K. (1969). Some new records of fungi in India. I. *Indian Phytopath.* **22**, 410–412.

Sohi, H.S. & Nayar, S.K. (1969 a). New records of fungi from Himachal Pradesh. II. *Indian Phytopath.* **22**, 462-465.

Sohi, H.S. & Nayar, S.K. (1971). New records of fungi from Himachal Pradesh. III. *Research Bull. Panjab Univ.* **22**, 243-245.

Sommer, N.F. (1955). Sun burn predisposes walnut trees to branch wilt. *Phytopathology* **45**, 607–613.

Spaulding, P. (1958). Diseases of foreign forest trees growing in the United States. Agric. Handbook **139**, U.S. Dept. Agric. 118 pp.

Spaulding, P. (1961). Foreign diseases of forest trees of the world. Agric. Handbook **197**, U.S. Dept. Agric. 361 pp.

Springer, E.L., Eslyn, W.E., Zoch, L.L. & Hajny, G.J. (1969). Control of pulp chip deterioration with kraft green liquor. U.S.D.A. For. Serv. Res. Pap. F.P.L. 110, 4 pp.

Spurr, S.H. (1945). A new definition of silviculture. *J. For.* **43**, 44.

Sreenivasaya, M. (1930). Contributions to the study of spike disease of sandal (*Santalum album* Linn.). Pt. XI. New methods of disease transmission and their significance. *J. Indian Inst. Sci.* **13,** 113–117.

Sreenivasaya, M. (1933). Quinquennial survey of investigation in spike disease of sandal. Pt. I. Laboratory results found applicable to silvicultural conditions. *Indian Forester* **59,** 473–478.

Sreenivasaya, M. & Naidu, G.G. (1928). Contributions to the study of spike disease of sandal (*Santalum album* Linn.). Pt. V. Transmission of spike by budding. *J. Indian Inst. Sci.* **11** A, 244–247.

Srivastava, S.K. & Patel, P.N. (1969). Epidemiology of bacterial leaf spot blight and shot hole disease of neem in Rajasthan. *Indian Phytopath.* **22,** 237–244.

Stakman, E.C. & Harrar, J.G. (1957). Principles of Plant Pathology. 581 pp The Ronald Press Co., N.Y.

Stevens, F.L. & Young, P.A. (1927). On the use of terms saprophyte and parasite. *Phytopathology* **17,** 409–411.

Streets, R.J. (1962). Exotic Forest Trees in the British Commonwealth. 750 pp. Clarendon Press, Oxford.

Subba Rao, M.K. (1936). Pink disease. *United Planters Assoc. South India, Tea Scient. Dep. Bull.* **10,** 25 pp.

Subramanian, C.V. & Ramakrishnan, K. (1956). List of fungi 1952–1956. *J. Madras Univ.*, B. **26,** 327–421.

Sujan Singh, Balwant Singh & Bakshi, B.K. (1961). Fungus flora of South Andamans. *Indian Forester* **87,** 248-250.

Sujan Singh & Pandey, P.C. (1971). *Tunicopsora,* a new rust genus on bamboo. *Trans. Br. mycol. Soc.* **56,** 301–303.

Sujan Singh, Puri, Y.N. & Bakshi, B.K. (1973). Decay in relation to management of dry coppice teak forests. *Indian Forester* **99,** 421–430.

Sujan Singh, Sehgal, H.S., Pandey, P.C. & Bakshi, B.K. (1967). Anthracnose disease of cashew (*Anacardium occidentale* Linn.) its cause, epidemiology and control. *Indian Forester* **93,** 374–376.

Sujan Singh & Tewari, R.K. (1970). Role of a precursor fungus in decay in standing teak. *Indian Forester* **96,** 874-876.

Sutton, J.C. & Barron, G.L.. (1972). Population dynamics of Endogone spores in soil. *Can. J. Bot.* **50,** 1909-1914.

Sydow, H. & Mitter, J.H. (1933). Fungi Indici. I. *Annls. mycol.* **31,** 84–97.

Sydow, H., Mitter, J.H. & Tandon, R.N. (1937). Fungi Indici. III. *Annls. mycol.* **35,** 222–243.

Tandon, R.N. & Bilgrami, K.S. (1957). Leaf spot disease of *Artocarpus heterophyllus* caused by *Phyllosticta artocarpina*. Proc. Acad. Sci. India, Sect. B. **27,** 204-209.

Tandon, R.N. & Bilgrami, K.S. (1961). Some leaf-spot diseases. *Curr. Sci.* **30,** 348-349.

Tandon, R.N. & Chandra, S. (1963-64). Supplement to the List of Indian Fungi (1957-1962). 246 pp. University of Allahabad Publication, Senate House, Allahabad.

Tandon, R.N. & Srivastava, M.P. (1964). Fruit rot of *Emblica officinalis* Gaertn. caused by *Pestalotia cruenta* Syd. in India. *Curr. Sci.* **33,** 86-87.

Terashita, T. (1962). Studies on the diseases of *Acacia dealbata*. I. Isolation of pathogens and some aspects of their overwintering. II. Control of the seedling diseases by chemicals. *Bull. For. Exp. Sta. Meguro* **147,** 119-127, 129-136. R.A.M. 42, 495, 1963.

Tewari, I. (1968). On the morphology and taxonomy of *Elsinoe kamatii* spec. nov. on *Syzygium cumini* from India. *Nova Hedwigia* **15,** 213-221. R.P.P. 49, no. 1102, 1970.

Thind, K.S. & Rattan, S.S. (1971). The Polyporaceae of India. V. *Indian Phytopath.* **24,** 50-57.

Thirumalachar, M.J. (1949). Telia of the leaf-rust on teak. *Curr. Sci.* **18,** 175-177.

Thirumalachar, M.J. (1950). Some new or interesting fungi. II. *Sydowia* **4,** 63-73.

Thirumalachar, M.J. (1951). Critical notes on some plant rusts. II. *Sydowia*, **5,** 23-29.

Thomas, G.P., Etheridge, D.E. & Paul, G. (1960). Fungi and decay in aspen and balsam poplar in the boreal forest region, Alberta. *Can. J. Bot.* **38,** 459-466.

Thomas, G.P. & Thomas, R.W. (1954). Studies in forest pathology. XIV. Decay of Douglas fir in the coastal region of British Columbia. *Can. J. Bot.* **32,** 630-653.

Tint, H. (1945). Studies in the *Fusarium* damping-off of conifers. I. The comparative virulence of certain Fusaria. *Phytopathology* **35,** 421-439.

Tint, H. (1945 a). Studies in the *Fusarium* damping-off of conifers. II. Relation of age of host, pH, and some nutritional factors to the pathogenicity of *Fusarium. Phytopathology* **35,** 440-457.

Toole, E.R. (1956). Hispidus canker. *Forest Fmr.* **16,** 7.

Toole, E.R. (1959). Canker-rots in southern hardwoods. Forest Pest Leafl. U.S. For. Serv. **33,** 4 pp.

Toole, E.R. (1959 a). Decay after fire injury to southern bottomland hardwoods. *Tech. Bull. U.S. Dep. Agric.* **1189,** 25 pp.

Troup, R.S. (1912). *Peridermium cedri* as a destructive fungus. *Indian Forester* **38,** 222-223.

Troup, R.S. (1914). *Peridermium cedri* as a destructive fungus. *Indian Forester* **40,** 469-472.

Troup, R.S. (1921). The Silviculture of Indian Trees. 3 vols., 1195 pp. Oxford Univ. Press (Clarendon), London and New York.

Tsuchizaki, T., Hibino, H. & Saito, Y. (1971). Mulberry ring spot virus isolated from mulberry showing ring spot symptoms. *Ann. Phytopath. Soc., Japan* **37,** 266-271. R.P.P. **51,** no. 2697, 1972.

Truman, R. (1974). Die-back of *Eucalyptus citriodora* caused by *Xanthomonas eucalypti* sp. n. *Phytopathology* **64,** 143-144.

Uemura, S. (1964). Isolation and properties of microorganisms from root nodules of non-leguminous plants. *Bull. Govt. For. Exp. Stn. Meguro* **167,** 59-91.

Ullasa, B.A. (1970). Taxonomy of some Indian Ascomycetes. *Sydowia* **24,** 161-163.

Vaartaja, O. (1952). Forest humus quality and light conditions as factors influencing damping-off. *Phytopathology* **42,** 501-506.

Vaartaja, O. (1964). Chemical treatment of seed beds to control nursery diseases. *Bot. Rev.* **30,** 91 pp.

Vail, J.W., Calton, W.E. & Strang, R.M. (1957). Die-back of wattle—a boron deficiency. *E. Afr. agric. J.* **23,** 100–103. R.A.M. **37,** 319, 1958.

Vamos, R. (1954). [Damping-off of pine seedlings]. Erdö **3,** 34–40. R.A.M. **34,** 760, 1955.

Van der Goot, P. (1936). [Diseases and pests of cultivated crops in the Dutch East Indies in 1935]. Meded. Inst. PlZiekt., Batavia **87,** 106 pp. R.A.M. **16,** 160, 1937.

Van der Plank, J.E. (1963). Plant Diseases : Epidemics and Control. 349 pp. Academic Press, New York & London.

Van der Westhuizen, G.C.A. (1965). *Cytospora eucalypticola* sp. nov. on *Eucalyptus saligna* from northern Transval—a disease of young *Eucalyptus saligna* in northern Transval. *S. Afr. For. J.* **54,** 8-11. R.A.M. **45,** no. 620, 1966.

Van Hall, C.J.J. (1924). [Diseases and pests of economic crops in the Dutch East Indies in 1923]. Meded. Inst. voor Plantenziekten **64,** 47 pp. R.A.M. **4,** 79, 1925.

Van Vloten, H. (1962). Present and potential significance of root rots in intensive forest management. Proc. V World For. Congr., Seattle, Wash. 1960. **2,** 887–890.

Varma, A., Chenulu, V.V., Raychaudhuri, S.P., Prakash, N. & Rao, P.S. (1969). Mycoplasma-like bodies in tissues infected with sandal spike and brinjal little leaf. *Indian Phytopath.* **22,** 289–291.

Vasudeva, R.S. (1963). Indian Cercosporae. 245 pp. Indian Coun. Agric. Res., New Delhi.

Venkata Ram, C.S. (1962). Report of the Plant Pathologist (1961–62). *Rep. Tea. Sci. Dep. un. Pl. Ass. S. India.* 1961–62, 31–41. R.A.M. 42, 402, 1963.

Venkata Ram, C.S. (1963). Report of the Plant Pathologist. *Rep. Tea Sci. Dep. un. Pl. Ass. S. India.* 1962–63, 22–39 a. R.A.M. **43,** no. 558, 1964.

Venkata Ram, C.S. (1967). Report of the Plant Pathologist. *Rep. Tea Sci. Dep. un Pl. Ass. S. India*, 1966–67, 23–41. R.A.M. **47,** no. 621 a, 1968.

Venkataramani, K.S. (1963). Boron deficiency in the silver oak. *Rep. Tea Sci. Dep. un. Pl. Ass. S. India*, 1962–63, 70–71. R.A.M. no. 584, 1964.

Venkata Rao, M.G. (1938). The influence of host plants on sandal and on spike disease. *Indian Forester* **64,** 656–669.

Venkata Rao, M.G. & Iyengar, K. (1934). Studies in spike disease of sandal. Methods of inoculation and variation of results under different methods. *Indian Forester* **60,** 689–701.

Verrall, A.F. (1949). Some molds on wood favoured by certain toxicants. *J. agric. Res.* **78,** 695–703.

Verrall, A.F. (1953). Decay prevention in wooden steps and porches through proper design and protective treatments. *J. Forest Prod. Res. Soc.* **3,** 54–60.

Vidal Hall, M.P. & Williams, G.H.D. (1956). Teak as a plantation tree in the Sudan. Forest Mem. Sudan 8, 14 pp. R.A.M. **39,** 354, 1960.

Wagener, W.W. & Davidson, R.W. (1954). Heart rots in living trees. *Bot. Rev.* **20,** 61–134.

Waid, J.S. & Woodman, M.J. (1957). A non-destructive method of detecting diseases in wood. *Nature* **180,** 47.

Wakefield, E.M. (1922). Fungi exotici. XXVI. *Kew Bull.* 1922, 161–165.

Waksman, S.A. (1917). Is there any fungus flora in soil? *Soil Sci.* **3,** 565–589.

Walker, J. & Bertus, A.L. (1971). Shoot blight of *Eucalyptus* spp. caused by an undescribed species of *Ramularia*. Proc. Linn. Soc. N.S.W. **96,** 108–115. R.P.P. **51,** no. 1971, 1972.

Wallis, G.W. (1960). Survey of *Fomes annosus* in East Anglian pine plantations. *Forestry* **32,** 203–214.

Wani, D.D. & Thirumalachar, M.J. (1969). Studies on *Elsinoe* and *Sphaceloma* diseases of plants in Maharashtra (India). IV. *Sydowia* **23,** 252–256.

Wani, D.D. & Thirumalachar, M.J. (1969 a). Studies on *Elsinoe* and *Sphaceloma* diseases of plants in Maharashtra (India). V. *Sydowia* **23,** 257–260.

Wani, D.D. & Thirumalachar, M.J. (1970). Studies on *Elsinoe* and *Sphaceloma* diseases of plants in Maharashtra State (India). VII. *Sydowia* **24**, 317–321.

Ward, J.C., Hann, R.A., Baltes, R.C. & Bulgrin, E.H. (1972). Honeycomb and ring failure in bacterially infected red oak lumber after kiln drying. U.S.D.A. Forest Serv. Res. Pap. F.P.L. 165, 36 pp.

Wardrop, A.B. & Dadswell, H.E. (1953). The development of the conifer tracheid. *Holzforschung* **7**, 33–39. (Quoted from Principles of Wood Science and Technology. I. Solid Wood, by F.F.P. Kollmann & W.A. Cote Jr. 1968, 592 pp. Spinger-Verlag, New York Inc.).

Wastie, R.L. (1973). Influence of weather on the incidence of Phytophthora leaf fall of *Hevea brasiliensis* in Malayasia. *J. Rubb. Res. Ins. Malaya* **23**, 381–390. R.P.P. **53**, no. 2297, 1974.

Waterman, A.M. (1943). *Diplodia pinea*, the cause of a disease of hard pines. *Phytopathology* **33**, 1018–1031.

Wellman, F.L. & Grant, T.J. (1951). An apparent virus disease of *Casuarina* in the American tropics. *Pl. Dis. Reptr.* **35**, 498–499.

White, L.T. (1953). Studies in forest pathology. X. Decay of white pine in the Timagami Lake and Ottawa Valley areas. *Can. J. Bot.* **31**, 175–200.

White, I.G. (1955). Toxin production by the oak wilt fungus, *Endoconidiophora fagacearum*. *Am. J. Bot.* **42**, 759–764.

Wilcox, W.W. (1968). Changes in wood microstructure through progressive stages of decay. U.S. Forest Serv. Res. Pap. F.P.L. 70, 46 pp.

Wilde, S.A. & Persidsky, D.J. (1954). Effect of biocides on the development of ectotrophic mycorrhizae in monterey pine seedlings. Agron. Abstr. p. 56, Nov. 8–12.

Wilson, C.L. (1975). Emerging tree disease problems in urban ecosystems. Second World Consultation of Forest Diseases and Insects. FAO/IUFRO/DI/75/17—April, 1975, New Delhi.

Wilson, E.E., Starr, M.P. & Berger, J.A. (1957). Bark canker, a bacterial disease of the Persian walnut tree. *Phytopathology* **47**, 669–673.

Wilson, E.E., Zeitoun, F.M. & Fredrickson, D.L. (1967). Bacterial phloem canker, a new disease of Persian walnut trees. *Phytopathology* **57**, 618–621.

Wilson, M. & Henderson, D.M. (1966). British Rust Fungi. 384 pp. Cambridge University Press.

Wolar, E.L. (1972). [Root rot of *Melia azedarach* caused by *Rosellinia bunodes*]. *Revista Forestal Argentina* 16, 185. R.P.P. **53,** no. 1578, 1974.

Worley, J F. & Hacskaylo, E. (1959). The effect of available soil moisture on the mycorrhizal association of Virginia pine. *Forest Sci.* **5,** 267–268.

Wright, E. (1957). Influence of temperature and moisture on damping-off of American and Siberian elm, Black locust and Desert willow. *Phytopathology* 47, 658–662.

Yde-Anderson, A. (1962). Seasonal incidence of stump infection in Norway spruce by air-borne *Fomes annosus* spores. *Forest Sci.* **8,** 98–103.

Yokota, S. (1957). Wood decay of *Abies sachalinensis* in the Tokyo University Forest, Hokkaido. III. On wood decay of oni-todo. *Bull. Tokyo Univ. Forests* **53,** 139–148 (English summary).

Young, H.E. (1954). Brown rot disease of *Hevea* caused by *Fomes lamaensis* (Murr.) Sacc. ex Trott. Adv. Circ. Rubber Res. Inst., Ceylon, 48, 2 pp. R.A.M. **35,** 392, 1956.

Zabel, R.A. (1954). Variations in preservative tolerance of wood-destroying fungi. *J. Forest Prod. Res. Soc.* **4,** 166–169.

Zabel, R.A. (1959). Fungus losses in the paper industry and related research needs. *Tappi* **42,** 5, 28A–34A.

Zabel, R.A. & George, R.A., St. (1960). Wood protection from fungi and insects during storage and use. Proc. V World For. Congr. **3,** 1530–1539.

Zeijlemaker, F.C.J. (1963). The artificial induction of gummosis in black wattle. Rep. Wattle Res. Inst., Univ. Natal **16,** 39–41.

Zeijlemaker, F C.J. (1969, 1970 & 1972). Plant physiology and pathology. Rep. Wattle Res. Inst. Univ. Natal, 1968–69 (22), 13–14; 1969–1970 (23), 28–29; 1971-72, 49 pp.

Zeijlemaker, F.C.J. & Margot, P. (1971). Wattle research, *Eucalyptus* research, plant physiology and pathology. Black-butt disease of wattle. Rep. Wattle Res. Inst. Univ. Natal, 1970–71; 23–24, 37, 49–50, 51–53.

Zycha, H. & Dimitri, L. (1962). Erfahrungen mit einem Gerät zur Fäuleermittlung an stehenden Stämmen. *Forstwiss. ZentBl.* **81,** 222–230.

INDEX